Surveillance and Diagnostics of Next Generation Nuclear Reactors

Other related titles:

You may also like

- PBPO052 | Wood | Nuclear Power | 2007

We also publish a wide range of books on the following topics:
Computing and Networks
Control, Robotics and Sensors
Electrical Regulations
Electromagnetics and Radar
Energy Engineering
Healthcare Technologies
History and Management of Technology
IET Codes and Guidance
Materials, Circuits and Devices
Model Forms
Nanomaterials and Nanotechnologies
Optics, Photonics and Lasers
Production, Design and Manufacturing
Security
Telecommunications
Transportation

All books are available in print via https://shop.theiet.org or as eBooks via our Digital Library https://digital-library.theiet.org.

IET Energy Engineering 233

Surveillance and Diagnostics of Next Generation Nuclear Reactors

Edited by
Imre Pázsit, Hoai-Nam Tran and Zsolt Elter

Institution of Engineering and Technology

About the IET

This book is published by the Institution of Engineering and Technology (The IET).

We inspire, inform and influence the global engineering community to engineer a better world. As a diverse home across engineering and technology, we share knowledge that helps make better sense of the world, to accelerate innovation and solve the global challenges that matter.

The IET is a not-for-profit organisation. The surplus we make from our books is used to support activities and products for the engineering community and promote the positive role of science, engineering and technology in the world. This includes education resources and outreach, scholarships and awards, events and courses, publications, professional development and mentoring, and advocacy to governments.

To discover more about the IET please visit https://www.theiet.org/.

About IET books

The IET publishes books across many engineering and technology disciplines. Our authors and editors offer fresh perspectives from universities and industry. Within our subject areas, we have several book series steered by editorial boards made up of leading subject experts.

We peer review each book at the proposal stage to ensure the quality and relevance of our publications.

Get involved

If you are interested in becoming an author, editor, series advisor, or peer reviewer please visit https://www.theiet.org/publishing/publishing-with-iet-books/ or contact author_support@theiet.org.

Discovering our electronic content

All of our books are available online via the IET's Digital Library. Our Digital Library is the home of technical documents, eBooks, conference publications, real-life case studies and journal articles. To find out more, please visit https://digital-library.theiet.org.

In collaboration with the United Nations and the International Publishers Association, the IET is a Signatory member of the SDG Publishers Compact. The Compact aims to accelerate progress to achieve the Sustainable Development Goals (SDGs) by 2030. Signatories aspire to develop sustainable practices and act as champions of the SDGs during the Decade of Action (2020-30), publishing books and journals that will help inform, develop, and inspire action in that direction.

In line with our sustainable goals, our UK printing partner has FSC accreditation, which is reducing our environmental impact to the planet. We use a print-on-demand model to further reduce our carbon footprint.

Published by The Institution of Engineering and Technology, London, United Kingdom

The Institution of Engineering and Technology (the "**Publisher**") is registered as a Charity in England & Wales (no. 211014) and Scotland (no. SC038698).

The Institution of Engineering and Technology
Futures Place,
Kings Way, Stevenage,
Herts, SG1 2UA,
United Kingdom

www.theiet.org

Whilst every reasonable effort has been undertaken by the Publisher and its licensors to acknowledge copyright on material reproduced, if there has been an oversight, please contact the Publisher and we will endeavour to correct this upon a reprint.

Trade mark notice: Product or corporate names referred to within this publication may be trade marks or registered trade marks and are used only for identification and explanation without intent to infringe.

Where an author and/or contributor is identified in this publication by name, such author and/or contributor asserts their moral right under the CPDA to be identified as the author and/or contributor of this work.

British Library Cataloguing in Publication Data

A catalogue record for this product is available from the British Library

ISBN 978-1-83953-708-0 (hardback)
ISBN 978-1-83953-709-7 (PDF)

Typeset in India by MPS Limited

Cover image credit: Cover Image: XH4D/E+ via Getty Images

"One man's noise is another man's signal"
(Oszvald Glöckler, 1984, paraphrasing Hector Urquhart, 1860)

Contents

List of abbreviations

1D	One-dimensional
2D	Two-dimensional
3D	Three-dimensional
ACNEM	Average current nodal expansion method
ADS	Accelerator-driven system
AMR	Advanced modular reactor
AMS	Analysis and measurement services
ANN	Artificial neural network
ANM	Analytical nodal method
APSD	Auto power spectral density
BWR	Boiling water reactor
BWXT	BWT technologies
CANDU	CANadian Deuterium Uranium reactor
CCF	Cross-correlation function
CMFD	Coarse mesh finite difference
CORTEX	Core Monitoring Techniques & Experimental Verification and Demonstration (EU project)
CPSD	Cross power spectral density
CRDM	Control rod drive mechanism
CTM	Coordinate transformation method
DDAA	Differential die-away analysis
DDSI	Differential die-away self-interrogation
DoE	Department of Energy
ECFM	Eddy current flow meter
EM	Electromagnetic (flow meter)
EPR	European power reactor
EPRI	Electric Power Research Institute
FDM	Finite difference method
FEM	Finite element method
FFT	Fast Fourier transform
FHR	Fluoride salt-cooled high-temperature reactor
FNPP	Floating nuclear power plant
Gen-IV	Generation IV
GCR	Gas-cooled reactor
GFEM	Galerkin finite element method
GFR	Gas-cooled fast reactor
GUI	Graphical user interface

HACNEM	High-order average current nodal expansion method
HALEU	High assay low enrichment uranium
HTGR	High-temperature gas-cooled reactor
HWR	Heavy water reactor
IAEA	International Atomic Energy Agency
ICFM	In-core fuel management
I&C	Instrumentation and control
IEC	International electrotechnical commission
IMSR	Integral molten salt reactor
iPWR	Integrated pressurised water reactor
IRF	Impulse response function
JEFF	Joint Evaluated Fission and Fusion Nuclear Data Library
KP	Kairos power
KWU	KraftWerk union AG
LAR	License amendment request
LCSR	Loop current step response
LFR	Lead-cooled fast reactor
LWR	Light water reactor
MCNP	Monte Carlo N-particle
MOC	Method of characteristics
MOX	Mixed oxide
MSDR	Molten salt demonstration reactor
MSFR	Molten salt fast reactor
MSR	Molten salt reactor
MSRE	Molten salt reactor experiment
MTC	Moderator temperature coefficient
NASA	National Aeronautics and Space Administration
NEA	Nuclear Energy Agency
NEM	Nodal expansion method
NRC	Nuclear Regulatory Commission
NRMS	Normalised root mean square
NRT	Negative reactivity transient
OECD	Organisation for Economic Co-operation and Development
OLM	Online monitoring
ORNL	Oak Ridge National Laboratory
PFBR	Prototype fast breeder reactor
PM	Power method
PSD	Power spectral density
PWR	Pressurised water reactor
RCS	Reactor coolant system
RCCA	Reactor control cluster assembly
R&D	Research and development
RTDs	Resistance temperature detectors
SCWR	Supercritical water-cooled reactor
SFR	Sodium-cooled fast reactor
SMR	Small modular reactor

TCs	Thermocouples
TH	Thermal–Hydraulic or Thermal Hydraulics
TIP	Traversing in-core probe (neutron detector)
TR	Topical report
TRISO	TRistructural ISOtropic
UDV	Ultrasound Doppler velocimetry
VHTR	Very-high temperature reactor
VVER	Water-cooled water-moderated (pressurised water) power reactor

Preface

Diagnostics of nuclear power reactors with noise analysis of process signals, primarily that of the neutron noise, have been developed over many decades. The research started as early as the 1960s, and the first pilot applications were made in the early 1970s. The main reactor types considered were the dominating light water reactors, pressurised water reactors (PWRs) and boiling water reactors (BWRs). Since its development in the 1970s, this technology has been crucial for light and heavy water reactors, aiding in early detection and identification of failures and operational parameter monitoring.

It was seen already at the appearance of accelerator-driven subcritical systems (ADSs), which a couple of decades ago held the promise to become the next generation of nuclear power reactors, that the existing diagnostic methods for source-driven subcritical systems had to be revised and suited to the new systems. This concerned primarily the zero-power noise methods, such as the Feynman-alpha and Rossi-alpha methods for the measuring and online monitoring of the subcritical reactivity of the ADS during operation. The need for updating the traditional methods arose because of the differences between the traditional systems (stationary source with simple Poisson statistics) and the ADS (pulsed source with compound Poisson statistics). Apart from a few singular papers, power reactor diagnostic methods specifically for ADS were not investigated during that period.

In the meantime, Generation-IV (Gen-IV) systems and small modular reactors (SMRs) took over the status of next generation systems. In many aspects, these systems also have differing characteristics from the traditional Gen-II or Gen-III reactors: fast energy spectrum, higher count rates, different range of the local component, different noise sources, different domains of the validity of the reactor kinetic approximations, etc. These differences concern in the first place the field of power reactor diagnostics. Yet, no similar studies were made regarding the tailoring or extending the diagnostic methods developed for the current fleet of Gen-II systems for the specific characteristics of Gen-IV reactors and SMRs, as was done for the development of zero-power noise methods for ADS. A major part of his book is devoted to an attempt to fill this gap. Naturally, the extension of zero-power noise methods for the future systems is also discussed from this perspective.

One possible reason for the lack of such studies is the large variety of the planned designs and the fact that a majority of the designs are not yet fixed in sufficient detail. Further, since there is no operational experience with such reactors, not all the possible scenarios for faults to be detected or parameters to monitor are known. The types of designs are also evolving dynamically, which represents a further problem.

To handle the situation, a few generic types of next generation reactors were selected, and the possible anomalies (noise sources) were extrapolated from the existing reactors. The corresponding diagnostic methods were developed and discussed from this perspective, and the properties of the various systems were analysed for power reactor noise diagnostics. No doubt, some of the problems discussed in this book might become obsolete already by the publication of this book. Nevertheless, efforts were made to capture the main characteristics of next generation reactors, which will persist despite the continuously changing landscape of new reactor designs.

A final goal of this book is to increase the awareness of designers about the need of considering the application of monitoring and surveillance systems at a very early stage. As the designs of Gen-IV reactors, as well as small and medium modular reactors, evolve, integrating diagnostics from the design phase is vital to avoid retrofitting issues. By analysing the possibilities and effectiveness of noise diagnostics for the main types of new designs, the material in this book might contribute to earlier and more effective integration of noise diagnostic methods into the new designs.

Acknowledgements

We are grateful to Christoph von Friedeburg, Senior Books Commissioning Editor at the IET, for his initiative and proposal to produce this book. Without his vision and continued insistence, this book would have never been written. Thanks are also due to the whole Editorial team at the publisher, including Christoph himself, as well as Assistant Editors Olivia Wilkins and Megan Mc Gill for their patience and flexibility regarding the uncountable number of shifting of deadlines and regrouping both the book editors and authors of the chapters.

We appreciate very much the help received from people who kindly provided us with their input files with detailed Monte Carlo continuous energy models of the various systems, from which we could generate two-group cross sections and other parameters for the analyses in Chapters 3 and 4. Our thanks go to Dr Valeria Raffuzzi of the University of Cambridge for the input data to the Allegro, the Advanced Light Water SMR and the MSFR inputs; to Dr Thanh Mai Vu and her students at the University of Sharjah for the LFR SMR and the MSDR input files as a part of their senior design projects; and to Dr Sandra Bogetic of the University of Tennessee for the input files to the fluoride salt-cooled high-temperature reactor (FHR). Their kind co-operation is highly appreciated. We are indebted to Dr Máté Szieberth, Gergely Klujber and Máté Boros for their contribution to the reactivity measurement methods in Chapter 6 with measured and simulated data.

About the editors

Imre Pázsit is a professor at the Division of Subatomic, High Energy and Plasma Physics, Department of Physics at Chalmers University of Technology, Sweden, and an adjunct professor at the Department of Nuclear Engineering & Radiological Sciences of the University of Michigan, Ann Arbor, USA. His research interests include fluctuations in neutron transport, reactor diagnostics and the theory of multiplicity in nuclear safeguards. He has published over 200 articles in international journals and, together with L. Pál, is the author of the book *Neutron Fluctuations – A Treatise on the Physics of Branching Processes.* He is a fellow of the American Nuclear Society, a member of the Royal Swedish Academy of Engineering Sciences and the Royal Society of the Arts and Sciences in Gothenburg, and an Honorary Doctor at the Budapest University of Technology and Economics. He is also the holder of the Leo Szilárd Medal of the Hungarian Nuclear Society, the Order of the Rising Sun from the Japanese government, and the E. P. Wigner Reactor Physicist Award and the Don Miller Award from the American Nuclear Society.

Hoai-Nam Tran is a professor at the Institute for Advanced Study, Phenikaa University, Vietnam. Professor Tran's research focuses on burning strategies and fuelling schemes for various reactor types, including high-temperature gas reactors and pebble bed modular reactors, their neutron characteristics and calibration, and diagnostics. Past positions include Chalmers University of Technology, Nagoya University and Tokyo Institute of Technology, Japan.

Zsolt Elter works as a reactor physicist in the Swedish nuclear industry and is part-time employed as a researcher at the Division of Applied Nuclear Physics at Uppsala University, Sweden, where he teaches reactor physics. He earned his PhD in nuclear engineering under the supervision of Prof. Imre Pázsit from Chalmers University of Technology. His professional interests include Monte Carlo-based neutronics and radiation shielding simulations, as well as deterministic lattice physics methods.

List of contributors

Yasunori Kitamura   Institute for Integrated Radiation and Nuclear Science, Kyoto University, Japan

Hash M. Hashemian   Analysis and Measurement Services, Knoxville, USA

Belle Upadhyaya   Department of Nuclear Engineering, The University of Tennessee, USA

Chapter 1

Introduction

Imre Pázsit[1] and Hoai-Nam Tran[2]

Reactor diagnostics, in particular diagnostics based on neutron noise analysis, has been developed in the early 1960s and 1970s, and its application has become increasingly widespread to traditional light water and heavy water reactors [1–8]. These are the most commonly used commercial plants, belonging to Generation II (Gen-II). Experience shows the significant benefits of reactor diagnostics in terms of determining operational parameters in the normal and abnormal state, and in the detection, identification and quantification of incipient failures at an early stage.

There are two relatively different areas of neutron noise diagnostics. At low power, such as in an educational or a research reactor, or in a power reactor during startup, the sources of neutron fluctuations of technological origin (vibration of core internals, boiling of the coolant, etc.) are absent, and the cause of the neutron fluctuations is solely the branching process, i.e. the neutron multiplication process in a constant medium. These types of neutron fluctuations are called zero-power reactor noise, and they are used mostly to determine the subcritical reactivity of the system while it is still driven by an external neutron source.

The other area, called power reactor noise diagnostics, utilises the neutron fluctuations caused by processes of technological origin, for example, movement of control rods, fuel assemblies, two-phase flow, etc., called perturbations or noise sources. These processes are inherently random, and their effect will be to induce spatial and temporal fluctuations of the neutron flux, which is called power reactor noise. If the neutronic transfer between the cross-section fluctuations, represented by the perturbation at a certain point of the core and the induced neutron noise at some distance (at the detector position), is understood and can be quantified, then there is a chance that by analysing the detector response and in possession of the transfer function, one can detect, identify and quantify the noise source. The noise source/perturbation can either be present already in normal operation (such as two-phase flow in a boiling water reactor (BWR)) or may be due to an incipient failure, such as vibrations due to a mechanical failure. Power reactor diagnostics can be used to determine parameters

[1] Division of Subatomic, High Energy and Plasma Physics, Department of Physics, Chalmers University of Technology, Sweden
[2] Phenikaa Institute for Advanced Study, Phenikaa University, Hanoi, Vietnam

of the processes already present (such as two-phase flow parameters), determine reactivity coefficients, detect changes in the value of these parameters and detect noise sources due to incipient failures at an early stage.

Both the zero-power and power reactor diagnostic methods were developed specifically for Gen-II reactors, mostly pressurised water reactors (PWRs) and BWRs, and to some extent Canadian deuterium uranium (CANDU) reactors as well. These reactors have a thermal spectrum with a relatively long prompt neutron generation time. For the theory of zero-power methods, a one-speed (energy-independent) neutronic framework was fully satisfactory. For the power reactor noise methods, several applications were based on simple one-group theory. Two-group theory was also used for a better description of real inhomogeneous cores, accounting for reflectors, and determining two-phase flow parameters, but the diagnostics were still based on using the thermal neutron detector signals.

A substantial part of the next generation of power plants, notably most of the Gen-IV reactors, will deviate significantly from the light and heavy water-moderated thermal reactors by having a fast spectrum and other type of coolant, in particular liquid metal, fluoride salts containing the fuel, or gas. The concept of small and medium large modular reactors is also emerging, some of which will be Gen-II/III and some of them Gen-IV type. The substantial difference in the reactor physics of the present and next generation nuclear systems raises the question as to what extent the methods used so far will be applicable in the new systems.

The different types of fuel and coolant will definitely affect the character of the neutronic transfer of the core, which is the prime tool to perform diagnostics. Characteristic scales, such as the range of the local component of the noise, the validity of point kinetics, etc., may be rather different from what we have experienced so far. Another significant difference is the fast spectrum of many of the new designs. Methods based on one-group theory and basing the diagnostics on measuring the noise in the thermal group will not be sufficient, or even applicable in those reactors. Regarding zero-power noise methods, partly the traditional methods based on one-group theory need to be modified, and partly in fast cores, one can count on much shorter prompt neutron generation times and higher count rates. And last, but not least, in some designs it may not be possible to use in-core detectors, which may raise the question of using gamma detectors instead, which can be used at larger distances from the core than neutron detectors.

These questions have not been addressed so far, or only to a very limited extent. One reason for this may be that the new systems are still essentially in the design phase. Although it might appear that it is only practical to focus attention on diagnostics and monitoring once the design is fixed, this approach has large potential drawbacks. Namely, as experience with the recent fleet of reactors showed, inclusion of new sensors into a design beyond those planned originally is practically impossible. Hence, questions of diagnostics and monitoring should be included already in the design phase.

It appears therefore useful to take a thorough inventory of the applicability of the presently used methods in the next generation systems, and explore possible ways for developing these further to take into account the different conditions. Concerning zero-power noise, the existing methods, such as the Feynman- and Rossi-alpha methods, are be extended to the use of two-group theory, to use gamma counts instead of neutron counts, and to use continuous (time-resolved) detector signals instead of pulse counting for extracting the prompt neutron decay constant. For the power reactor noise diagnostics, the space- and frequency-dependent transfer functions of several Gen-IV reactors and small modular reactors (SMRs) are quantitatively investigated using a simple one-dimensional (1D) two-group model. For the application of the two-group methods for fast reactors, the theory had to be developed to be suitable for practical applications. Such transfer properties can be modelled and studied qualitatively and quantitatively for the various construction types, even without an exact knowledge of the construction details. The rationale behind using such simple models is explained in the next chapter.

One might argue that to expedite the planning of effective diagnostics, one should have a good knowledge of what type of problems might occur and also what methods are suitable for detecting and quantifying them and with what efficiency. Regarding the type of problems, these are mostly revealed only during operation. However, as a first step, one could extrapolate the experience from the current fleet of Gen-II reactors, as well as make qualified guesses of the new type of problems that a specific new design might incur, or parameters of interest that should be monitored.

One of the main purposes of this book is to report on a thorough study of the dynamic transfer properties of the main Gen-IV and SMR types, and also to investigate the possibilities of the diagnostics of the main types of perturbations that are known to have occurred in the current systems. This should expedite the design of efficient diagnostic systems for the various Gen-IV and SMR types, and give some guidelines regarding the types of diagnostic methods that are usable, or not usable, in the various types.

The content of the book is organised as follows. In Chapter 2, the theory and formalism of both zero-power noise and power reactor noise, used in the book, are described with illustrations. The 1D two-group framework, in which either the development of the methods is performed, or in which the qualitative and quantitative investigations of the core transfer properties are made, is described in detail. Molten salt reactors (MSR) get their own section, due to their very different reactor physics properties, because of the movement of the delayed neutron precursors. For MSRs, in two-group theory, analytical solutions exist only for thermal systems in the infinite fuel recirculation velocity (prompt recirculation) approximation and only for thermal MSR. With analytical and quantitative analysis, it is demonstrated that the dynamic behaviour and spatial response of an MSR can be, to a very good approximation, simulated with the assumption of zero fuel velocity (as with traditional reactors), but reducing the effective delayed neutron fraction. This opens up the possibility to investigate even MSRs with a fast spectrum, since the traditional two-group theory was already extended to fast systems in the book. The analysis leading to the conclusion of how to simulate an MSR by a modified traditional core with reduced effective delayed neutron fraction is new and is hence included in detail.

Chapters 3–5 concern power reactor diagnostics. In Chapter 3, the simple analytical methods described in Chapter 2 are used to qualitatively and quantitatively analyse the neutronic transfer properties of several Gen-IV reactors (sodium-cooled, gas-cooled and MSR) as well as to investigate the possibilities of diagnosing three basic perturbation types known from the current reactors: oscillations of the strength of an absorber, a vibrating fuel rod or absorber rod, and propagating perturbations.

Chapter 4 presents a similar survey for a number of known SMR designs. These comprise both water-moderated and Gen-IV-type systems, including the lead-cooled fast reactor (LFR). The methods and tools used, as well as the scenarios studied, are the same as in the preceding chapter.

One significant recent development in power reactor diagnostics is the use of numerical methods for the calculation of the space- and frequency-dependent complex transfer function for real inhomogeneous cores, with the same spatial and energy resolution as the in-core fuel management codes. These codes are often called "noise simulators". The use of such tools can significantly improve the accuracy of diagnostic methods in concrete applications. So far these codes have also been developed mostly for thermal reactors, using two energy groups. However, work is going on to develop noise simulators for Gen-IV reactors, using a larger number of energy groups. Chapter 5 describes the principles of such noise simulators and gives a survey of the international status of development of such codes by several research groups, as well as the specific characteristics and potentials of the various approaches.

It will not go unnoticed that such tools were not used in the investigations in Chapters 3 and 4. This was deliberate, for simple reasons. One has to keep in mind that noise simulators are useful in giving accurate quantitative results for concrete inhomogeneous cores, but they are not optimal for investigating tendencies and qualitative behaviour. The accurate results come at the price of longer running times, and the advantage of being able to model inhomogeneous cores is largely lost by the fact that the concrete designs of the next generation systems are not yet available. Another point is that with the detailed spatial dependence of the noise when heterogeneities are taken into account, i.e. showing spatial variations at the assembly or even pin level, it is impossible to make a qualitative assessment on the level of deviation from point kinetics, or the relative contribution of the local component of the noise, which are important parts of judging the efficiency of the diagnostic step. And last, but not least, the existing noise simulators are not suited yet to handle MSRs, i.e. reactors with moving fuel, which are one of the important types of Gen-IV reactors, whereas the simple analytical models can be used to analyse them, with the added advantages listed earlier in this section. In order to have a fair comparison between the different systems, they need to be analysed by similar looks, which gives yet another argument of using the analytical tools mentioned.

One can also put the validity and applicability of the simple analytical methods in the power reactor field into another perspective. Because of the simplifications inherent in the simple models, the quantitative results do not supply accurate guidance on the efficiency of the various noise techniques in terms of absolute statements. However, in relative terms, the conclusions drawn hold with large likelihood even in practice. The analysis shows the differences between various designs, and since the same simplifications were applied to each system, the relative differences predicted are applicable in practice. When the new generation reactors start to deliver power, operational experience will be gained on the applicability of the noise methods. Then, when experience is gained about a certain core, the studies presented in this book give reliable guidance on what one can expect on the other designs, still under construction, by extrapolating the comparative results presented in these conceptual studies.

A side benefit of the extensive use of analytical solutions in 1D two-group theory is that all the exact formulae used for both the traditional thermal reactors, as well as their extension to fast systems (this latter not having been published elsewhere yet), are described in detail. Although the traditional theory has been used for decades, it was mostly used for calculating the thermal noise in thermal systems. Even there, most often by using approximations. The formalism for the noise in the fast group was developed much later, but no complete list of the full formulae was given. Several of the earlier publications also contain typos. This book contains a complete list of the full analytical formulae for the fast and thermal noise in both thermal and fast systems, including those for MSRs in the infinite fuel velocity approximation (renamed as the prompt recirculation approximation in this book). Therefore, this book contains the most complete collection of these formulae that are found scattered in the literature, contain approximations and are incomplete. This way, it also serves as a handbook of two-group power reactor noise theory, both for traditional and fast systems.

The formulae referred to here are used in Chapters 3 and 4 for the analysis of the dynamical response of Gen-IV systems and MSRs. The calculations, including the plotting, were made by the numerical option of the symbolic manipulation language Wolfram Mathematica [9] and are collected in a Wolfram Mathematica notebook. To amplify the usefulness of the collection of the formulae, as well as their potential use, the Mathematica notebook, as well as the database on which the calculation were made (the group constants and kinetic parameters of the various cores), are being made available to the readers via open sources. The notebook [10] is uploaded to the Wolfram Notebook Archive [11]. The nuclear and geometrical data of the individual reactors are uploaded to the public domain repository Mendeley Data [12]. More details and instructions to use the notebook are found in Appendix A of this book.

With the information provided in Appendix A and within the Notebook itself, it is fully possible for the readers to reconstruct all results and plots in the book, and also to use the notebook with new data, and it is even possible to modify the code for new functionality. One interesting and easily implemented possibility is to change the size of the default systems available in the notebook by default, since the size of the system is an input data. When changing the size, the notebook automatically recalculates the cross sections to keep the system critical. One could then investigate how the dynamic properties of a given system change if, for instance, a certain full-size Gen-IV system is downsized to smaller dimensions or when the dimensions of a certain small-size SMR are increased.

Data sets for new cores in the data read format of the notebook (described in Appendix A) can directly be run. It would be appreciated if readers testing to run the notebook with some new systems uploaded their .json files to the public GitHub site given in the notebook. After checking them for their format, they will be authorised and shared publicly. Alternatively, such files could be sent to the Editors to upload them to the public repository.

It also has to be noted that up to this point, the term reactor diagnostics was synonymous with neutron or gamma noise diagnostics, i.e. fluctuations in nuclear quantities. This is obvious for zero-power noise, which is solely based on the fluctuation analysis of neutron or gamma detector signals. However, in the case of power reactors, the diagnostics can be based also on other process signals, including those whose fluctuation leads to neutron noise. Such signals include temperature, pressure and possibly displacement sensors, the latter at least during pre-operational tests. As an example, with the correlation of in-core thermocouples, the flow velocity of the coolant can be determined from the cross-correlation of the thermocouple signals.

These methods are not the main focus of the present book, for two reasons. One is that unlike neutrons, which, during the fission chains, carry information about disturbances at some distance from the detection of the neutron noise, those signals have only local information and are characteristic only to the measurement point. Therefore, no transfer functions need to be calculated, and the signals can be directly processed to extract their information content. The second reason is that those signals, and the corresponding sensors, are most likely very similar to the ones used in the current systems, at least for their functioning. Probably they need to be developed for enduring higher temperatures and more aggressive chemical environments, but this requires technological development, which lies outside the scope of this book.

On the other hand, analysis of such signals is also actual for next generation systems, although the principles are largely the same as in the present systems. Therefore, to complement the mainly theoretical and conceptual character of the chapters so far, two more chapters were added. Chapter 7 reports on the status of online monitoring systems, mostly used for non-nuclear signals (temperature, flow and pressure). These are highly effective and are deployed at a large number of reactors in the US and worldwide. The company of the author of that chapter leads the development of current online monitoring methods, and is involved also in the preparations for instrumentation of SMRs. This work is supported by the US DoE.

Finally, Chapter 8 reports on the development of SMRs, with a focus on instrumentation and monitoring approaches being developed for efficient operation. It gives a survey of various SMR systems, design options and vendors. It also gives an account of the challenges in SMR instrumentation and the general approaches for SMR monitoring.

As is clear from this description, the character of Chapters 2–6, which are rather theoretical and method development oriented, deviates substantially from that of the last two, whose approach is rather pragmatic and application oriented. Because of this, the style of the book may feel somewhat uneven. Yet, it was felt that the two parts complement each other, and presenting both will give a more complete outlook on the way ahead for surveillance and diagnostics of next generation nuclear systems.

Chapter 2
Principles of traditional reactor diagnostics and noise analysis

Imre Pázsit[1]

2.1 Introduction

The principles of reactor diagnostics for the current generation of reactors are available in a few monographs and review articles [1–8]. For a wider perspective on the subject, the reader is recommended to explore these sources. However, with regards the methods used in this book, we tried to make the text self-contained, such that all the basics are explained that are needed to follow the derivations and the extensions introduced to make the methods suitable for applications to next generation reactors. This is the subject of this chapter.

As it was already mentioned in the Introduction to the book and is reflected in the aforementioned references, neutron noise diagnostics has two different branches, which differ in many aspects, such as the origin of the neutron fluctuations, the operational domain in which the two types dominate, the mathematical methods used for their description and the application area. For this reason, the applications in the future systems will also be treated in separate chapters. In this overview chapter for the methods used so far in traditional systems, they will both be included.

2.2 Zero-power reactor noise diagnostics

2.2.1 General principles

Zero-power noise is a terminology used for the fluctuations of the detector counts in a stationary system whose properties are constant in time. Actually, in the theoretical derivations, it is assumed that the properties of the system are constant also in space, which is not true in reality; it is only used to simplify the mathematical treatment. The important thing is that the reason for the temporal fluctuations of the number of neutrons in the system, or the number of detector counts during a time interval, is due solely to branching, i.e. the generation of two or more neutrons simultaneously in fission. Due to the simultaneous birth of several neutrons, the evolution of

[1]Division of Subatomic, High Energy and Plasma Physics, Department of Physics, Chalmers University of Technology, Sweden

the individual chains started by these neutrons will be time-correlated. Because of these correlations, the statistics of the number of neutrons or the detector counts will deviate from the Poisson statistics, this latter depending only on one single parameter, and hence the expectation (first moment) carries all information on the process. The higher moments of a Poisson process, such as the variance, do not contain any information that could not be obtained already from the first moment.

This is different for branching processes. Due to the aforementioned temporal correlations, the statistics of the number of the neutrons in the system will deviate from that of a Poisson process, and each individual moment carries information which is different from that of the other moments. For example, the variance will not be equal to the mean as it would be for a Poisson process; it will be larger than the mean, and the deviation from the mean carries new (excess) information. Likewise, the temporal autocovariance function,* which would be zero for independent detector counts, will be larger than zero for a branching process. To put it in a simplified way, using higher moments besides the expectation makes it possible to determine more unknown parameters of the process.

In zero-power reactor diagnostics, the main task is to determine the subcritical reactivity of the system from the measurement of individual detector counts. Since the system is in a subcritical state, a constant expectation is maintained by driving the system with an extraneous neutron source. Then either the dependence of the variance to mean of the detector counts on the detection gate length is used (Feynman-alpha method) or the normalised (with the mean) dependence of the autocovariance of two counts on the time leg τ between the two counts (Rossi-alpha method) is used to determine the subcritical reactivity ρ, which is contained in the prompt neutron decay constant α in the form

$$\alpha = \frac{\rho - \beta}{\Lambda} \tag{2.1}$$

where β is the delayed neutron fraction and Λ is the prompt neutron generation time. These are usually assumed to be known; therefore, determining α is sufficient to determine the reactivity.

However, in practice neither the source intensity nor the detector efficiency is known; hence, only measuring the expectations would not make it possible to determine the reactivity.† This problem can be solved by using higher-order moments of the detector counts. In principle, three different moments would be necessary. In practice, it is sufficient to use, in addition to the expectation, one higher moment (the variance or the autocovariance), but adding time as an extra parameter will make

---

*In agreement with Reference [6], the word "autocovariance" is used here to indicate that it is based on the joint expectation of two values of the same process at two different time instants, to distinguish it from the cross-covariance, which is based on the joint expectation of two random variables belonging to two different processes. Often the word "covariance" is used for either of these two, which might lead to some confusion; hence, it is important to define in what context the term covariance is used. The word covariance will occasionally also be used in this book, for instance in general statements that are valid both for the auto- and the cross-covariance.

†In pulsed neutron experiments, such as the area ratio or Sjöstrand method [13], it is sufficient to measure expectations. However, the pulsed neutron measurement is not a stationary noise method. Moreover, it is not applicable at power reactors during startup since it requires a neutron generator.

it possible to extract the reactivity. Namely, since both the variance and the auto-covariance, as well as the expectation, depend linearly on the noise strength, it is cancelled in ratios such as the variance to mean (Feynman-alpha) or the autocovariance to mean (Rossi-alpha method). In these "normalised" expressions, the detector efficiency appears as a multiplying factor, whereas their time-dependence will be a non-linear function, depending on the prompt neutron decay constant α, which thus can be determined by simple curve fitting.

The aforementioned will be illustrated by quoting the corresponding formulae of the Feynman- and Rossi-alpha methods. In the Feynman-alpha experiment, the relative variance $\sigma_Z^2(T)/\langle Z(T)\rangle$ of the number Z of the counts is determined as a function of the measurement time T. In a simplified notation, when using one averaged delayed neutron group with a decay constant λ, it has the analytical form

$$\frac{\sigma_Z^2(T)}{\langle Z(T)\rangle} \equiv 1 + Y(T) = 1 + \varepsilon A_1 \left(1 - \frac{1 - e^{-\alpha T}}{\alpha T}\right)$$

$$+ \varepsilon A_2 \left(1 - \frac{1 - e^{-\alpha_d T}}{\alpha_d T}\right) \tag{2.2}$$

Here, α is the prompt neutron decay constant of (2.1), whereas α_d is the delayed neutron decay constant, related to the reactivity ρ, and the delayed neutron fraction β and decay constant λ in the form (valid for $\rho<<\beta$) [6]

$$\alpha_d = -\frac{\lambda\rho}{\beta - \rho} \tag{2.3}$$

The coefficients A_1 and A_2 contain the delayed neutron fraction β, the subcritical reactivity, and the first and second factorial moments of the number of neutrons generated in induced fission. Explicit expressions for these coefficients are found in the literature, see e.g. [4,6,7]. As is seen in (2.2), the source intensity S does not appear in the expression. The sought parameter α can be obtained by a curve fitting of the experimental results to the analytical formulae, which does not require the knowledge of the detector efficiency ε.

An illustration of Feynman-alpha curves corresponding to different levels of subcritical reactivity is shown in Figure 2.1.

An alternative method to determine the prompt neutron decay constant from the statistics of neutron counts is the so-called Rossi-alpha method. In contrast to the variance to mean method, the Rossi-alpha method is based on the dependence of the relative temporal autocovariance of detections in two infinitesimal time intervals dt as a function of the time separation τ between the two detections. Based on the fact that for infinitesimal time intervals the expectation

$$\langle Z(t, dt)\rangle = Z\,dt \tag{2.4}$$

depends linearly on dt, with the proportionality parameter Z being the stationary detection intensity, one defines the Rossi-alpha function $R(\tau)$ as follows. Define the autocovariance of the detections $C_Z(\tau)$ as

$$C_Z(\tau)\,dt\,d\tau = \langle Z(t, dt)\,Z(t + \tau, d\tau)\rangle - \langle Z(t, dt)\rangle\langle Z(t + \tau, d\tau)\rangle \tag{2.5}$$

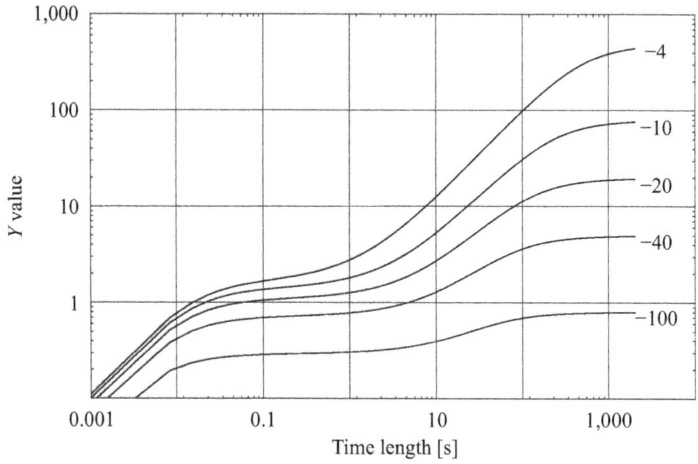

Figure 2.1   Typical Feynman-alpha $Y(t)$ values for a light water core as functions of the measurement time length for different negative reactivities, indicated in 10^{-4} units on the right of the curves

Then, the normalised autocovariance (autocovariance to mean or Rossi-alpha) function $R(\tau)$ is given as

$$R(\tau)\, d\tau = \frac{C_Z(\tau)\, dt\, d\tau}{\langle Z(t,\, dt)\rangle} = \frac{\langle Z(t, dt)\, Z(t+\tau, d\tau)\rangle - \langle Z(t,\, dt)\rangle\langle Z(t+\tau,\, d\tau)\rangle}{\langle Z(t,\, dt)\rangle} \quad (2.6)$$

In the same model and with the same notations as for the Feynman-alpha expression (2.2), the Rossi-alpha expression $R(\tau)$ is given as

$$R(\tau) = \frac{1}{2}\,\varepsilon\big(\alpha A_1\, e^{-\alpha\tau} + \alpha_d A_2\, e^{-\alpha_d\tau}\big) \quad (2.7)$$

Similarly to the Feynman-alpha method, the source intensity does not appear in the expression, and the prompt neutron decay constant α can be extracted from a measurement by fitting the measured data to the theoretical formula, without need of the knowledge of the detector efficiency.

There is a considerable conceptual resemblance between (2.2) and (2.7), since they contain the same parameters multiplying the time-dependent functions, i.e. that on the measurement time and the time lag in the autocovariance, respectively. The same exponential decays appear also in both formulae, although in different combinations. This similarity is not a coincidence; rather, it is due to the fact that there exist simple relationships between the two formulae, such that they can be mutually derived from one another. Thus, it can be shown [4,14] that the autocovariance function $C_Z(\tau)$ can be obtained from the variance of the counts collected during a period T as

$$C_Z(\tau) = \frac{1}{2}\,\frac{d^2\,\sigma_Z^2(T)}{d\,T^2}\bigg|_{T=\tau} \quad (2.8)$$

Accounting for the fact that in (2.2) the expectation of the number of counts during the measurement time T is given as

$$\langle Z(T) \rangle = Z \cdot T \tag{2.9}$$

with, as told earlier, Z being the mean detection intensity, taking $\sigma_Z^2(T)$ from (2.2) and performing the operation indicated in (2.8), using the $C_Z(\tau)$ so obtained in (2.6), with making use of (2.4) indeed yields the Rossi-alpha expression (2.7).

Conversely, from the general expression for the variance in terms of the auto-covariance function of a continuous parametric discrete process (here the number of counts as a function of the measurement time), one can reconstruct the Feynman-alpha formula from the Rossi-alpha. One has (see [15,16])

$$\sigma_Z^2(T) = \langle Z(T) \rangle + \int_0^T \mathrm{d}t_1 \int_0^T \mathrm{d}t_2\, C_Z(|t_1 - t_2|)$$

$$= \langle Z(T) \rangle + 2 \int_0^T \mathrm{d}t \int_0^t \mathrm{d}\tau\; C_Z(\tau) \tag{2.10}$$

Then, noting from (2.6) and (2.4) that

$$C_Z(\tau) = Z \cdot R(\tau) \tag{2.11}$$

making use of (2.7), performing the operations in (2.10) and dividing the variance obtained with the mean $Z \cdot T$ indeed leads to the Feynman-alpha formula (2.2).

The relationships between the Feynman- and Rossi-alpha formulae will be utilised later in Chapter 6.

2.2.2 Methodology: master equations

Derivation of the statistical properties, such as the variance and the autocovariance of the neutron counts, is made by use of the master equations (also called Kolmogorov–Chapman, or simply Kolmogorov equations). The master equation is a probability balance equation, which describes the evolution of the probability distribution of the process. It is used for the family of processes that are called Markovian processes. The Markovian property states that if the state of the system is known at a certain moment in time, then its subsequent evolution is fully determined by this state, and the previous history of the process does not have any influence on it. In a simplified way, this can be formulated that as the past of the process does not have any influence on the future evolution of the process, it does not possess a memory.

Because this tool is used in Chapter 6, we illustrate the principles of writing down the equations whose solution leads to the Feynman-alpha formula. Here, we only show how to write down the master equation and how the equations for the various moments are derived from it. The solution process will not be displayed here, since it is straightforward, and the interested reader will find it in the references. In the example shown here, we treat the probably most classical case: a homogeneous infinite material in one-speed (one energy group) theory, with accounting for one (averaged) group of delayed neutrons. The detector is thought to be homogeneously dispersed into the medium. The medium is described with its macroscopic cross sections Σ_i, with i standing either for f (fission), c (capture in the medium) or d (capture in the detector), respectively.

The parameters needed for setting up a probability balance equation are the reaction intensities λ_f of fission, λ_c of capture in the medium and λ_d of capture in the detector. These are given, respectively, as

$$\lambda_f = v\Sigma_f, \quad \lambda_c = v\Sigma_c \quad \text{and} \quad \lambda_d = v\Sigma_d \tag{2.12}$$

Here, v is the neutron velocity of thermal neutrons. These reaction intensities refer to a single neutron.

In addition, we need the decay constant λ of the delayed neutron precursors, and the number distribution $p(n, m)$ of generating n prompt neutrons and m delayed neutron precursors in a fission event. Finally, in the subcritical system, a stationary neutron distribution is maintained with an extraneous source with a simple Poisson statistics of constant intensity S; that is, in first order of dt, the probability of emitting one neutron during the infinitesimal time dt is equal to $S\,dt$.

To arrive at the Feynman-alpha formula, we need to set up an equation for the probability $P(N, C, Z, T)$ that at time T, there are N neutrons and C delayed neutron precursors in the system, and that a total of Z counts were detected between time $t = 0$ and $t = T$. It is assumed that the source was switched on at $t = -\infty$, at which time there were no neutrons and delayed neutron precursors in the system. This means that at time $t = 0$, when the measurement starts, the transient following the switching on of the source has already died out and the system is in a stationary state.

With these preliminaries, the procedure of deriving a forward-type master equation goes as follows. We set up a balance equation for predicting the distribution at $T + dT$, starting from its state at time T, based on its Markovian property that its history for $t < T$ does not influence its evolution for $t \geq T$. We sum up the probabilities of all transitions which during dT bring the system to the desired state. Due to dT being an infinitesimal quantity, in first order of dT only any one of the mutually exclusive events can take place: either any of the reaction types for *one* neutron, or a decay of only one of the delayed neutron precursors, or a source emission of one single neutron, or neither of these. For the reactions, the probability of any of the N neutrons undergoing a reaction, or the C precursors to decay, the corresponding intensity has to be multiplied with the number of the corresponding entity. This leads us to the following balance equation: we add up for the probabilities of the mutually exclusive possible events for the system between T and $T + dT$, which will lead to

$$
\begin{aligned}
P(N, C, Z, T + dT) = \ &\lambda_c P(N + 1, C, Z, T)(N + 1)\ dT + \\
&\lambda_d P(N + 1, C, Z - 1, t)(N + 1)\ dT + \lambda_f \sum_n \sum_m P(N + 1 - n, C - m, Z, T) \times \\
&(N + 1 - n)\, p_f(n, m)\ dT + S\, P(N - 1, C, Z, T)\ dT + \\
&\lambda P(N - 1, C + 1, Z, T)(C + 1)\ dT + \\
&P(N, C, Z, T)\Big[1 - \{(\lambda_f + \lambda_c + \lambda_d)\,N + \lambda C + S\}\ dT\Big]
\end{aligned}
\tag{2.13}
$$

The first four lines represent the cases when the system was in a different state at time $t = T$ but through one of the possible transitions arrived to the proper state, and the last line corresponds to the case when the system was already in the desired state, and none of the transition events took place.

After rearranging, dividing by dT and letting $dT \to 0$, one arrives at the differential equation

$$\frac{dP(N, C, Z, T)}{dT} = \lambda_c P(N+1, C, Z, T)(N+1) +$$

$$\lambda_d P(N+1, C, Z-1, t)(N+1) +$$

$$\lambda_f \sum_n \sum_m P(N+1-n, C-m, Z, T)(N+1-n) \, p_f(n, m) + \quad (2.14)$$

$$S \, P(N-1, C, Z, T) + \lambda P(N-1, C+1, Z, T)(C+1) +$$

$$P(N, C, Z, T) \left[N(\lambda_f + \lambda_c + \lambda_d) + \lambda C + S \right]$$

Equation (2.13) was included in this derivation only to give insight into the logics of the derivation. Later, in Chapter 6, when we make derivations for new cases, we will sometimes immediately start with the differential equations of the type (2.14).

In general, master equations of the above type, referring to the complete probability distribution, cannot be solved analytically. This is not necessary either, because we only need the first two moments (mean/expectation and variance) of this distribution. For these, equations can be derived from (2.14), which are much simpler to solve. To expedite obtaining these equations, it is practical (and is common practice) to switch to the generating functions of the appearing probability distributions. Thus, we introduce the generating functions

$$G(x, y, z, T) = \sum_N \sum_C \sum_Z x^N y^C z^Z P(N, C, Z, T) \quad (2.15)$$

and

$$g_f(x, y) = \sum_n \sum_m x^n y^m p_f(n, m) \quad (2.16)$$

With the help of these definitions, the following equation is obtained from (2.14):

$$\frac{\partial G(x, y, z, T)}{\partial T} = \left\{ \lambda_f [g_f(x, y) - x] - \lambda_c (x - 1) - \lambda_d (x - z) \right\} \frac{\partial G(x, y, z, t)}{\partial x}$$

$$+ \lambda(x - y) \frac{\partial G(x, y, z, t)}{\partial y} + (x - 1) S \, G(x, y, z, t) \quad (2.17)$$

with the initial condition

$$G(x, y, z, T = 0) = x^{N_0} y^{C_0} \quad (2.18)$$

Equation (2.17) is a single equation, from which equations for all sought quantities, such as $\langle N \rangle$, $\langle C \rangle$ ... $\langle Z(t) \rangle$, $\langle Z^2(t) \rangle$, etc., can be derived by taking derivatives with respect to the parameters x, y and z. From the definition of the generating function (2.15), it follows that

$$\langle N \rangle = \left. \frac{\partial G(x, y, z, T)}{\partial x} \right|_{x=y=z=1} ; \quad \langle (N(N-1) \rangle = \left. \frac{\partial^2 G(x, y, z, T)}{\partial x^2} \right|_{x=y=z=1} \quad (2.19)$$

and so on. The goal is to obtain equations for $\langle Z(T) \rangle$ and $\sigma_Z^2(t)$ which appear in (2.2). This is achieved by taking the corresponding derivatives of (2.17). Due to the already existing derivatives in the equation, there will be cross-derivatives, which will lead to a coupled system of differential equations which need to be solved simultaneously.

We show here only the equations for the first moments, which are obtained from (2.17) by taking the derivatives of both sides w.r.t. to x, y and z, respectively, and substitute $x = y = z = 1$. In this process, the derivatives of the fission neutron number generating function $g_f(x, y)$ will also appear, which are identified with the expectation of the number of prompt neutrons and delayed neutron precursors as

$$\left. \frac{\partial g_f(x, y)}{\partial x} \right|_{x=y=1} = \sum_n \sum_m n \, p_f(n, m) \equiv \langle \nu_p \rangle \equiv \langle \nu \rangle \, (1 - \beta) \tag{2.20}$$

$$\left. \frac{\partial g_f(x, y)}{\partial y} \right|_{x=y=1} = \sum_n \sum_m m \, p_f(n, m) \equiv \langle \nu_d \rangle \equiv \langle \nu \rangle \, \beta \tag{2.21}$$

where $\langle \nu \rangle$ is the average total number of neutrons per fission and β is the effective delayed-neutron fraction. We also introduce the standard notations

$$\rho = \frac{\langle \nu \rangle \lambda_f - (\lambda_f + \lambda_c + \lambda_d)}{\langle \nu \rangle \lambda_f} \tag{2.22}$$

$$\Lambda = \frac{1}{\langle \nu \rangle \lambda_f} \quad \text{and} \quad \varepsilon = \frac{\lambda_d}{\lambda_f} \tag{2.23}$$

where ρ, Λ and ε stand for reactivity, prompt neutron generation time and detector efficiency, respectively. By introducing simplified notations for the expectations such as

$$\langle N(t) \rangle \equiv N(t), \qquad \langle C(t) \rangle \equiv C(t) \quad \text{and} \quad \langle Z(t) \rangle \equiv Z(t) \tag{2.24}$$

the equations for the first moments will read as

$$\frac{dN(t)}{dt} = \frac{\rho - \beta}{\Lambda} N(t) + \lambda C(t) + S \tag{2.25}$$

$$\frac{dC(t)}{dt} = \frac{\beta}{\Lambda} N(t) - \lambda C(t) \tag{2.26}$$

and

$$\frac{dZ(t)}{dt} = \lambda_d N(t) = \varepsilon \lambda_f N(t), \qquad t \geq 0 \tag{2.27}$$

Derivation of the higher moments goes in a completely analogous manner. This is a standard procedure, which will not be pursued further here. The derivation of the traditional Feynman-alpha has a vast literature; for the new cases needed for the next generation of reactors, the derivations are given in Chapter 6.

As is also known from the literature, an alternative way of deriving the Feynman-alpha is through the backward-type master equations. These are derived by setting up a probability balance equation for considering the possible events that can happen with a starting neutron between $t = 0$ and $t = dt$. Then another equation is needed, connecting this probability to the case when neutrons are emitted into the system continuously by an extraneous source with exponential distribution between $t = 0$ and $t = T$. This procedure will lead to equations which are very different in character from the forward-type equations shown in the foregoing. Instead of getting coupled differential equations for the auto- and cross-moments of the various variables, any moment can be obtained separately from all the others from a generic quantity, the expectation of the number of neutrons in a system due to one single starting neutron, by multiple nested integrals. The backward equation has definite advantages in a number of situations, such as when treating space- and energy-dependent cases. However, for the simple one-speed point models used in this book, the use of the forward method is more straightforward, and this method will be used in Chapter 6. Hence, we shall not describe this method or use it for the derivation of the Feynman-alpha formula. The derivation can be found in e.g. [6], and a detailed discussion of the differences between the two methods, together with a discussion of the reasons of these differences is found in [8].

2.2.3 Overview and outlook

Reactivity measurements with the Feynman- and Rossi-alpha methods are a very well-established standard methodology. There exist even other, neutron noise-based methods, such as the Cf-232 method and the Bennett method, which will not be touched upon here. In Chapter 6, we primarily deal with the generalisation of the Feynman- and Rossi-alpha methods for next generation systems. The following aspects will be investigated or elaborated:

- the applicability and feasibility of the traditional Feynman- and Rossi-alpha methods with the material parameters of the next generation reactors;
- extending the traditional methods to accounting for two energy groups, since most next generation systems will have a fast spectrum;
- extending the neutron counting-based traditional methods to include, or use only, gamma photons. This may be necessary due to decreased access to in-core detectors or detectors immediately adjacent to the core;
- use of non-Poissonian neutron sources, such as Cf-232 and pulsed neutron sources, used in ADS (accelerator driven systems);
- to extend the traditional, pulse counting-based method to consider the continuous signal of fission chambers or scintillation detectors, in order to overcome the difficulties caused by the dead time at high count rates.

2.3 Power reactor noise diagnostics

2.3.1 General principles

Power reactor noise concerns the temporal and spatial fluctuations of the neutron flux, which are induced by some technological processes in the core that lead to temporal and spatial fluctuations of the reactor material. Examples of such processes are the vibrations of fuel assemblies, control rods, or the core barrel, density fluctuations of the coolant due to boiling in a boiling water reactor (BWR), and inlet temperature fluctuations in a pressurised water reactor (PWR). Some of these stochastic processes are part of normal operation, such as the presence of turbulent two-phase flow; some of them only appear during abnormal situations, such as excessive stochastic flow-induced vibrations of the core components that are amplified due to material degradation. Because these random material changes are the reason for the fluctuations of the neutron flux, they are commonly called perturbations, or noise sources, whereas the temporal fluctuations of the neutron flux around its stationary value, induced by the perturbations, are called the neutron noise.

Power reactor noise analysis based on neutron fluctuations has been used for several decades, and a large experience has been built up. A collection of PWR and BWR developments and applications in Swedish power plants is described in [17] and the references therein (see also [18,19]). The properties of the neutron noise induced by various perturbations in the different systems were studied in a large number of cases. For a long time, these calculations were made by simple models, allowing analytical solutions, and relatively simple unfolding procedures for extracting the parameters of the noise source. Simplified models may not always supply quantitative results of sufficient accuracy, although in some cases they were surprisingly efficient for practical applications (see [20,21]). However, they were instrumental in understanding the factors influencing the character of the induced noise, as well as investigating the possibilities of unfolding the parameters of the noise source. This is what will be described in the following.

Relatively recently, numerical methods were developed for the calculation of the space- and frequency-dependent complex transfer function of real inhomogeneous cores, as well as the induced noise for various perturbations. An overview of these methods developed worldwide by various groups is given in Chapter 5. These codes supply high-accuracy results, but lend less insight, not the least since the unfolding is often mad by "black box" methods, such as artificial neural networks (ANNs). The experience and understanding gained from the simple models is essential for the correct and efficient use of the numerical codes and ANN-based inversion methods.

As it was seen in the previous section, the variance of the zero-power neutron fluctuations (due to the branching process) is in the same order of magnitude as the mean. On the other hand, as it will be seen soon, the variance of the power reactor noise, due to cross-section fluctuations, is proportional to the square of the mean. Hence, at high power, power reactor noise will dominate, and the presence of the branching noise can be ignored. Thereby we say that in a power reactor, in the absence of perturbations, the system is in a steady state, and the neutron noise vanishes.

Since the neutron noise induced by a certain perturbation is characteristic to the given perturbation, power reactor noise can be used to detect and quantify the perturbation. Quantification may mean that one locates the position of a given (local) perturbation, such as the position of a vibrating fuel assembly, or determines some other parameter either in the normal or abnormal state, such as the void fraction of the two-phase flow in BWRs, or the value of some safety parameter, such as the moderator temperature coefficient (MTC) of PWRs. Although the information on the perturbation is rather implicit in the induced neutron noise, on the other hand neutron detectors collect information about perturbations within a certain, relatively wide field of view, due to the fact that the fission chains propagate the information. That is, unlike that of a temperature or displacement sensor, the information content of the neutron noise is not local. Besides, neutron detectors are one of the very few, and often the only in-core sensors which endure the high pressure, temperature and radiation environment, and which are therefore often the only in-core sensors.

The methodology of power reactor noise diagnostics is as follows. The starting point is the "direct task", to be able to predict (calculate) the neutron noise induced by the perturbation. Any possible perturbation needs therefore to be represented in terms of the corresponding fluctuations of the macroscopic cross sections, because it is these latter which appear in the diffusion of transport equation. The final goal is naturally the opposite: given the induced (measured) neutron noise, find out the perturbation which induced it. This is referred to as the inverse task, or the unfolding procedure.

Formally, one can put this as follows. Since the perturbations (i.e. the fluctuations of the macroscopic cross sections) are much smaller than their mean values (they are first-order small quantities), linear theory is applicable: the products of any two first-order quantities can be neglected. This means, that the induced response of the neutron flux, more exactly the neutron noise $\delta\phi(\boldsymbol{r},\omega)$, can be expressed (for simplicity described here in a one energy group setting) as a spatial convolution of the noise source $S(\boldsymbol{r},\omega)$ with the space- and frequency-dependent complex transfer function $G(\boldsymbol{r},\boldsymbol{r}',\omega)$ as

$$\delta\phi(\boldsymbol{r},\omega) = \int_{V_R} G(\boldsymbol{r},\boldsymbol{r}',\omega)\,S(\boldsymbol{r}',\omega)\,\mathrm{d}\boldsymbol{r}' \tag{2.28}$$

Concrete representations and derivations of these quantities, even in two-group theory, will be given in the later subsections. Here, we only use these for illustration of some principles. In (2.28), the transfer function $G(\boldsymbol{r},\boldsymbol{r}',\omega)$ belongs to the unperturbed core, hence it can be calculated irrespective of the perturbation, for all $\boldsymbol{r}$, $\boldsymbol{r}'$ and ω values.

The task is now that in possession of such an expression and a concrete realisation of its components, and measuring the neutron noise $\delta\phi(\boldsymbol{r},\omega)$ induced by the perturbation represented in (2.28), to determine the noise source $S(\boldsymbol{r},\omega)$, which contains the information on the perturbation. One might be tempted to think that this can be achieved with some kind of fitting the measured data to the theoretical expression, similarly to the case of the Feynman- and Rossi-alpha methods. Here, however, we note a fundamental difference between the two methods. The ease of the application of the unfolding of the sought parameter in the zero-power case depends partly on the availability of a simple analytical expression for the Feynman- and Rossi-alpha formulae. And partly and more important, there is no space-dependence involved, and in the measurement one can determine the corresponding quantities (mean and variance) as continuous functions of time (or, in practice, with a high temporal resolution).

For the case of power reactor noise, even if a simple analytical expression existed for the noise $\delta\phi(\boldsymbol{r},\omega)$, such a fitting would be possible only if the same quantity could be measured as a continuous function of space (or at least with a high spatial resolution). This is obviously not possible; one has access to the measured neutron noise only at a few, sparsely spaced detector positions. The number of available detector positions differs quite markedly between the different reactor types: a BWR usually contains a relatively large number of neutron detectors in fixed positions, whereas many PWRs contain only movable detectors, out of which a limited number (typically 5) can be inserted into the core at a time, and even that only during a limited time period. But even those cores that contain several neutron detectors, such as the Swedish BWRs and the water-cooled water-moderated (pressurised water) power reactor (VVER)-type reactors, do not have detectors dense enough such that inverting (2.28) would be possible for a general noise source. Not the least because the effect of a local perturbation is not felt everywhere in the core, only in the vicinity of the perturbation, where it is larger than the background noise and noise from other noise sources.

In view of the limited number of neutron detectors available, the only viable approach is to abandon the idea of unfolding for an arbitrary noise source. Rather, for each known perturbation, one constructs a simple analytical model, which only contains a few parameters, such as modelling the two-dimensional (2D) vibrations of a control rod by the random movement of a point absorber on the horizontal plane. In that case there is a chance to unfold those few parameters from the signals of a few neutron detectors.

With regard to the availability of simple analytical formulae for the expression of the type of (2.28), such can only be obtained in utterly simple models, such as a homogeneous bare core in one- or two-group theory. Such solutions have been used in the past, and they still play a role in investigating the possibility of solving the inverse task of unfolding the parameters of the perturbation from the measured noise. They give an overall estimate of the spatial relaxation of the space-dependent component of the noise, or on the dominance (or not) of its point kinetic component.[‡] These aspects are decisive what regards the possibilities of unfolding and are not strongly influenced by the spatial fine structure and inhomogeneities of the unperturbed core. Therefore, such basic studies will be presented for some of the planned Generation IV (Gen-IV) and small modular reactor (SMR) systems in Chapters 3 and 4.

On the other hand, an accurate quantitative unfolding of the parameters of the perturbation from the neutron noise requires an accurate calculation of the transfer function of the core, and a faithful representation of the perturbation in terms of the cross-section fluctuations. In this respect there has been a very spectacular development in the past decade or so. Both deterministic (diffusion or transport theory) and stochastic (Monte Carlo) methods were developed which are capable to calculate the space- and frequency-dependent complex transfer functions of real, inhomogeneous cores in two energy groups with the same spatial resolution as that of the static in-core fuel management codes [22,23]. These tools were primarily developed for the present light water reactors (LWRs), but can be adjusted to be able to calculate also Gen-IV and SMR cores, and the work is already going on in this direction. A survey

---

[‡] These concepts are explained in the next subsection.

of the methods and codes developed by various groups internationally will be given in Chapter 5.

In the realistic cases, as no analytical or semi-analytical solution is possible, only the numerical solution by the aforementioned noise simulators exists. However, this is not an obstacle any longer for performing the unfolding, i.e. extracting parameters of the noise source from the measured noise, in possession of the numerical values of the transfer function. With the emergence of machine learning, inverse tasks can be solved easily by training an ANN to identify input parameters that led to a certain solution. Their training requires only simulated measurement data, which can be produced by solving only the direct task (i.e. calculating the noise induced by a given perturbation for a wide range of parameters of the perturbation), see, e.g., [24].

The mathematical treatment of power reactor noise, which will be given shortly, is very different from the master equation method of zero-power noise. There are several independent reasons for this. One is that it would be rather difficult to extend the master equation approach to space-dependent cases. The methodology would become rather complicated, with no chance for simple solutions. The space-dependent treatment of stochastic transport in a static medium with backward equations in itself is not an insurmountable problem, as it is demonstrated in some recent publications [25,26]. The real stumbling block is the treatment of the perturbations, i.e. handling the stochastic nature of the fluctuations of the macroscopic cross sections. In the master equation approach, they appear as transition probabilities (reaction intensities). To assume that they themselves are random processes would mean that one has to deal with a so-called doubly stochastic process. This is doable, but the formalism would become hopelessly complicated. This is illustrated in [6], where an infinite homogeneous medium is treated as a binary random process (in time, without space dependence). To extend this formalism to describe a continuous random variation of the medium runs into fundamental mathematical difficulties, which will not be described here, let alone trying to make it space-dependent.

But there is no need for such a complicated formalism either. One can instead choose the much simpler path to start with the traditional deterministic diffusion or transport equations, which are written down for the expectation of the neutron flux. In these equations, the macroscopic cross sections appear as mere coefficients, which may depend on space. Then, a perturbation is represented by the random variation of the cross sections, which are affected by the perturbation. Hence, the original deterministic differential or integro-differential equation is turned to a stochastic differential or integro-differential equation, whose coefficients are random processes. The fluctuations of the cross sections are usually specified through spectral methods (auto- and cross-correlations, or auto-and cross-power spectra), and from these the auto- and cross-spectra of the neutron detector signals can be calculated by solving the corresponding equations. This strategy will now be described in the continuation.

Similarly to the static calculations for in-core fuel management (ICFM), for the present commercial reactors the use of two-group diffusion theory is sufficient. Transport theory treatment, or the use of Monte Carlo methods for power reactor noise problems, also occurs in the literature, but the majority of the work is performed in one- or two-group diffusion theory. In the beginning, analytical methods were applied to simple systems, starting with one-dimensional (1D) or 2D bare homogeneous systems in one-group theory. Such methods, while generally not applicable quantitatively to real problems, have the advantage that they lend a very good insight into the problem, the character of the dynamic response of the core, and for devising inverse methods for unfolding the parameters of the noise source. In fact, such methods were successfully applied to localise excessively vibrating control rods in a PWR [20,27], and to estimate the location of an unseated fuel element in a BWR, which led to a local channel instability [21].

Later, the interpretation of the in-core noise measurements in BWRs prompted the need to account for the local component, which could be made by turning to two-group theory. By the use of two-group theory, also reflected systems could be treated. The two-group theory used was suited to thermal systems, in which the slow group is the thermal group. Such an approach is fully suitable for noise problems in LWRs. In that approach both prompt fission neutrons and delayed neutrons appear in the fast group.

For reactors with a fast spectrum, such a description is not satisfactory. In most fast reactors the thermal flux is negligible, and a two-group representation with group 2 being the thermal neutrons is completely unsatisfactory. In order that a two-group approach be meaningful, the energy separating Groups 1 and 2 has to be put much higher. One recommendation in the literature [28] is to have the energy separation at the fission threshold for fast fission in the fertile component. In this case both prompt and delayed fission neutrons will appear in both groups, with different weights. This approach has not been used in the literature before. The formalism will be developed in this chapter and applied in the analysis made in Chapters 3 and 4.

The description so far concerns only reactors with solid fuel. One of the planned Gen-IV-type reactors is the molten salt reactor (MSR). The reactor physics and noise diagnostics of MSR differ from those with a solid fuel due to the movement of the delayed neutron precursors. It has already been seen that the movement of the precursors changes the neutronic coupling, leading to changed properties of the transfer function. One consequence is that the weight of the point kinetic or reactivity term is amplified. Therefore, to study the dynamic properties of such systems, the corresponding two-group theory needs to be used, either for thermal or for fast-spectrum MSRs. The way now the MSRs will be treated in Chapters 3 and 4 is described in Section 2.3.6.2 in this chapter.

The organisation of the rest of this chapter is as follows. First, the basic power reactor noise diagnostic concepts and methods will be explained in one-group diffusion theory. Then the traditional two-group theory for thermal systems will be described, and its use demonstrated on some classical examples. After that the two-group theory of neutron noise in fast reactors will be derived and illustrated. Finally, the theory of neutron fluctuations in an MSR will be derived. The tools developed in this section will then be used for the analysis of the Gen-VI and SMR systems in the subsequent two chapters.

2.3.2   *Noise equations in one-group theory*

For transparency, the basic principles will be illustrated in a one-group model, assuming one group of delayed neutrons.

We assume that the unperturbed reactor is in a critical state. With obvious notations, the static equation, which only contains the neutron flux, reads as

$$\nabla D(r)\nabla \phi_0(r) + [\nu \Sigma_f(r) - \Sigma_a(r)]\phi_0(r) = 0 \qquad (2.29)$$

with the boundary conditions of vanishing flux at the extrapolated boundaries, i.e.

$$\phi_0(r_B) = 0 \qquad (2.30)$$

where r_B is an arbitrary point on the extrapolated boundary. To simplify the description, we assume a homogeneous, or piecewise homogeneous system, in which case (2.29) is simplified to

$$D\nabla^2 \phi_0(r) + [\nu \Sigma_f - \Sigma_a]\phi_0(r) = 0 \qquad (2.31)$$

The perturbations will manifest themselves by the fact that the cross sections will become time-dependent, and in most cases the time-dependence will also have a spatial dependence. As a consequence, the neutron flux will also be space- and time-dependent. For this case then, one needs to use the space- and time-dependent diffusion equations, in which now the delayed neutron precursors will also appear, together with the corresponding equation for the delayed neutron precursors. One will hence have the two coupled equations, which read as

$$\frac{1}{v}\frac{\partial \phi(r,t)}{\partial t} = D\nabla^2\phi(r,t) + [\nu\Sigma_f(r,t)(1-\beta) - \Sigma_a(r,t)]\phi(r,t) + \lambda C(r,t) \qquad (2.32)$$

and

$$\frac{\partial C(r,t)}{\partial t} = \beta\nu\Sigma_f(r,t)\phi(r,t) - \lambda C(r,t) \qquad (2.33)$$

Here, for simplicity and according to general praxis, the time-dependence of the diffusion coefficient was disregarded. It can be shown that for small perturbations this is a good approximation, although there are cases when the time dependence of the diffusion coefficient needs to be taken into account [29].

In power reactor noise problems, it is assumed that the cross sections fluctuate around their stationary value in a way that the time-integrated reactivity effect of the perturbations is zero, so the reactor remains critical.[§] As a consequence, the neutron flux will also fluctuate around the critical flux, in a space-dependent manner, where the space-dependence will be determined by both the properties of the static (unperturbed) system as well as the space dependence (and frequency) of the perturbation. The transfer properties and the neutronic response to current reactors have been investigated in much detail. It is the same aspects that are explored in the subsequent chapters of this book.

Assuming that both the perturbations and the induced neutron noise are first-order small quantities whose product can be neglected, it is worth switching to the fluctuations and deriving an equation directly for the neutron noise driven by the

---

[§]A possible non-zero but small reactivity effect of the perturbation will be handled by the control system to keep the reactor critical.

perturbation. To this end, we split up the equations into static and time-dependent parts with zero expectation as follows:

$$\phi(r,t) = \phi_0(r) + \delta\phi(r,t); \quad \langle \delta\phi(r,t) \rangle = 0 \tag{2.34}$$

$$C(r,t) = C_0(r) + \delta C(r,t); \quad \langle \delta C(r,t) \rangle = 0 \tag{2.35}$$

$$\Sigma_f(r,t) = \Sigma_f(r) + \delta\Sigma_f(r,t); \quad \langle \delta\Sigma_f(r,t) \rangle = 0 \tag{2.36}$$

$$\Sigma_a(r,t) = \Sigma_a(r) + \delta\Sigma_a(r,t); \quad \langle \delta\Sigma_a(r,t) \rangle = 0 \tag{2.37}$$

Substituting (2.34)–(2.37) into (2.32) and (2.33), subtracting the static equations, neglecting the second-order terms such as $\delta\phi(r,t)\,\delta\Sigma_a(r,t)$, the remaining equations are amenable for a temporal Fourier transform. This makes it possible to eliminate $\delta C(r,\omega)$ by expressing it in terms of $\delta\phi(r,\omega)$, which will lead to an equation for $\delta\phi(r,\omega)$ in the form

$$D\,\nabla^2\delta\phi(r,\omega) + \left\{ \left[1 - \frac{i\omega\beta}{i\omega+\lambda} \right] \nu\Sigma_f(r) - \Sigma_a(r) - \frac{i\omega}{v} \right\} \delta\phi(r,\omega) =$$
$$\left\{ \delta\Sigma_a(r,\omega) - \left[1 - \frac{i\omega\beta}{i\omega+\lambda} \right] \delta\nu\Sigma_f(r,\omega) \right\} \phi_0(r) \equiv S(r,\omega) \tag{2.38}$$

Formally, the left-hand side (l.h.s.) of (2.38) is rather similar to that of the static equation (2.31), but it is an inhomogeneous equation. The inhomogeneous part, the right-hand side (r.h.s.), represents the perturbation, or the noise source $S(r,\omega)$, composed of the fluctuations of the macroscopic cross section, some frequency-dependent factors, and the static flux. Another difference is that, due to the Fourier transform, the equation contains complex terms, both in the noise and the noise source.

With some rearrangement, assuming now the space-independence of the static cross sections, and introducing notations that are standard in power reactor noise, (2.38) can be written in a simpler form as

$$\nabla^2\delta\phi(r,\omega) + B^2(\omega)\cdot\delta\phi(r,\omega) = \frac{S(r,\omega)}{D} \tag{2.39}$$

with

$$B^2(\omega) = B_0^2 \left(1 - \frac{1}{\rho_\infty \cdot G_0(\omega)} \right) \tag{2.40}$$

where B_0^2 is the static buckling, from (2.31) obtained as

$$B_0^2 = \frac{\nu\Sigma_f - \Sigma_a}{D} \tag{2.41}$$

$G_0(\omega)$ is the so-called zero-power reactor transfer function, defined as

$$G_0(\omega) = \frac{1}{i\omega\left(\Lambda + \dfrac{\beta}{i\omega+\lambda} \right)} \tag{2.42}$$

and further,

$$\rho_\infty = 1 - \frac{1}{k_\infty} = 1 - \frac{\Sigma_a}{\nu \Sigma_f}. \tag{2.43}$$

The noise source $S(r, \omega)$ was already defined in (2.38).

Although the frequency dependence of the amplitude and the phase of $G_0(\omega)$ is well known, for later reference we show here a plot of these quantities in Figure 2.2. The figures show the well-known property that in the so-called plateau region $\lambda \leq \omega \leq \beta/\Lambda$, the amplitude is constant, having the approximate value

$$|G_0(\omega)| \approx \frac{1}{\beta} \tag{2.44}$$

whereas the phase is close to zero, i.e. $G_0(\omega)$ can be approximated as having only real values. Such a frequency dependence will actually be seen even in the space-dependent noise.

Equation (2.39) shows a resemblance to the Langevin equation, widely used in physics for the description of the motion of a particle influenced by deterministic and random forces, although (2.39) is formulated in the frequency domain. However, the

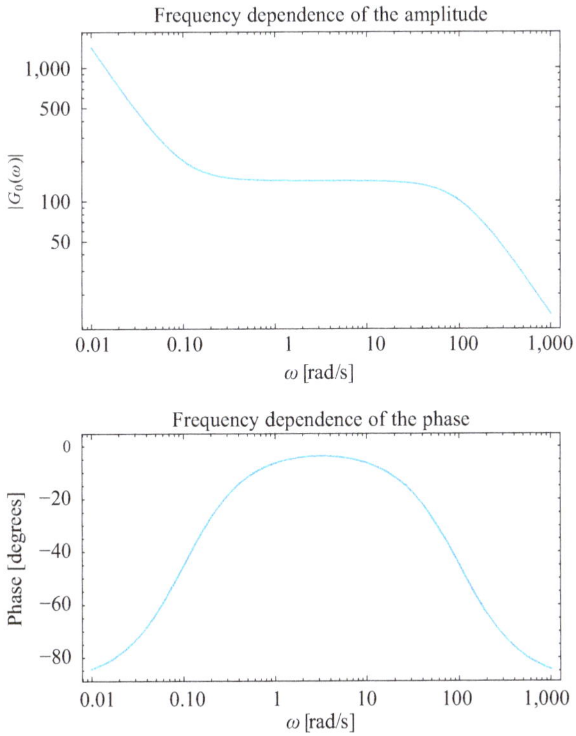

Figure 2.2   *Frequency dependence of the amplitude and the phase of the zero-power transfer function $G_0(\omega)$*

formal equivalence is obvious, since the noise source $S(r, \omega)$ on the r.h.s. is composed of the random functions $\delta \nu \Sigma_f(r, t)$ and $\delta \Sigma_a(r, t)$. In a concrete case, they will not be given by deterministic functions, rather as random processes, characterised by their auto- and cross-correlation functions. This means that the solution of the equation, $\delta \phi(r, \omega)$, will also be only determined up to its auto- and cross-power spectra. The power spectral densities can be formally obtained directly from the Fourier transforms of the functions involved by the Wiener–Khinchin theorem, so the main emphasis is on obtaining the solutions in the frequency domain. In many cases, only the time-dependence of the perturbation will be stochastic, and its space-dependence deterministic. At any rate, (2.39) is referred to as the Langevin equation, and treating power reactor noise problems through the equation the Langevin technique.

Equation (2.39) can be conveniently solved by the Green's function technique. One defines the Green's function through the equation

$$\nabla^2 G(r, r', \omega) + B^2(\omega) \cdot G(r, r', \omega) = \frac{\delta(r - r')}{D} \tag{2.45}$$

With this, the solution of (2.39) is given as

$$\delta \phi(r, \omega) = \int_{V_R} G(r, r', \omega) S(r', \omega) \, dr' \tag{2.46}$$

which is identical with (2.28).

Thus, we see that the neutron physics transfer function of the core is the Green's function of the corresponding noise equation. The properties of the induced neutron noise will therefore depend on both the reactor physical behaviour of the transfer function of the unperturbed core, as well as on the characteristics of the perturbation.

Because it leads to some confusion in the literature, it is worth mentioning here that, for simplicity, often the equation for the Green's function of one-group theory is defined as

$$\nabla^2 G(r, r', \omega) + B^2(\omega) \cdot G(r, r', \omega) = \delta(r - r') \tag{2.47}$$

This means that either the convolutional integral (2.46) has to be modified to

$$\delta \phi(r, \omega) = \frac{1}{D} \int_{V_R} G(r, r', \omega) S(r', \omega) \, dr' \tag{2.48}$$

or the noise source has to be redefined as

$$S(r, \omega) \rightarrow \frac{S(r, \omega)}{D} \tag{2.49}$$

in which case the convolutional integral (2.46) can be used unchanged. Sometimes these various definitions are mixed in one and the same paper, so it always has to be made clear which definition is used.

In a 1D system the solution of the equation for the Green's function defined by (2.45), and replacing x' with x_0, is given as

$$G(x, x_0, \omega) =$$

$$-\frac{1}{DB(\omega)\sin 2B(\omega)a} \begin{cases} \sin B(\omega)(a+x)\sin B(\omega)(a-x_0) & x \le x_0 \\ \sin B(\omega)(a-x)\sin B(\omega)(a+x_0) & x > x_0 \end{cases} \qquad (2.50)$$

In the early days of reactor kinetics and dynamics, when the computational power was much more limited than today, space–time-dependent problems were often solved with the help of the so-called reactor kinetic approximations. These approximations have their corresponding versions in the linearised, frequency-dependent noise equations. Although the kinetic approximations have played out their role from the computational point of view, they are still important in the understanding and interpretation of the structure of the induced noise and in the designing of the unfolding methods to recover the parameters of the noise source. Therefore, a very brief description will be given of the two simplest ones, which play a role in power reactor diagnostics, because they will also be needed for the terminology used later on in the methodology.

The basis of the reactor kinetic methods, here described in one-group theory for simplicity, is the factorisation of the space–time-dependent flux into an amplitude factor $P(t)$ which depends only on time, and a normalised shape function $\psi(r, t)$ as [30,31]

$$\phi(r, t) = P(t)\psi(r, t) \qquad (2.51)$$

To make the factorisation unambiguous, the normalisation condition

$$\frac{\partial}{\partial t}\int_{V_R}\phi_0(r)\psi(r, t)\,dr = 0 \qquad (2.52)$$

is used with the initial condition

$$\phi(r, t = 0) = \phi_0(r) = P_0\,\psi_0(r) \qquad (2.53)$$

assuming that the perturbation started at time $t = 0$. Choosing $\psi_0(r) = \phi_0(r)$ will yield $P_0 = 1$, although other normalisations are also possible.

In power reactor noise theory, similarly to (2.34), both the amplitude factor and the shape function are split up to static values and fluctuations with zero expectation:

$$P(t) = P_0 + \delta P(t) \qquad (2.54)$$

$$\psi(r, t) = \psi(r, t = 0) + \delta\psi(r, t) \qquad (2.55)$$

For the space–time-dependent neutron noise, this will lead to

$$\delta\phi(r, t) = \delta P(t)\cdot\psi_0(r) + P_0\cdot\delta\psi(r, t) = \delta P(t)\cdot\phi_0(r) + \delta\psi(r, t) \qquad (2.56)$$

where in the last equality we used the normalisation $P_0 = 1$. With a temporal Fourier transform, one has in the frequency domain

$$\delta\phi(r, \omega) = \delta P(\omega)\cdot\phi_0(r) + \delta\psi(r, \omega) \equiv \delta\phi(r, \omega)_{p.k.} + \delta\phi(r, \omega)_{s.d.} \qquad (2.57)$$

As the notation here indicates, the first term of the sum on the r.h.s. is called the point kinetic, or reactivity-driven term of the noise, whereas the second is called the space-dependent component. In linear theory, the point kinetic term is easily obtained as

$$\delta\phi(\mathbf{r},\omega)_{p.k.} = \rho(\omega)\, G_0(\omega)\, \phi_0(\mathbf{r}) \tag{2.58}$$

where $\rho(\omega)$ is the reactivity effect of the perturbation,

$$\rho(\omega) = \frac{\displaystyle\int_{V_R} [\delta\,\nu\Sigma_f(\mathbf{r},\omega) - \delta\,\Sigma_a(\mathbf{r},\omega)]\,\phi_0^2(\mathbf{r},\omega)\,\mathrm{d}\mathbf{r}}{\displaystyle\nu\Sigma_f\int_{V_R}\phi_0{}^2(r)\,\mathrm{d}\mathbf{r}} \tag{2.59}$$

and the zero-power transfer function $G_0(\omega)$ was defined in (2.42).

Note that the point kinetic term, in which the space and frequency (time) dependence is factorised, is also space-dependent, but its space dependence is that of the static flux; hence, if one normalises the noise with the static flux, this term becomes constant in space. In this terminology, the term "space-dependent" refers to deviations from the point kinetic behaviour, i.e. from the space-dependence of the static flux. It is also easy to see that, as a result of the normalisation condition (2.52), one has

$$\int \phi_0(\mathbf{r})\delta\psi(\mathbf{r},t)\,\mathrm{d}\mathbf{r} = 0 \tag{2.60}$$

i.e. the space-dependent component of the noise and the static flux are orthogonal.

This terminology will also be kept when using two-group theory, where the spatial structure of the noise will be more detailed. As it will be seen in the next subsection, in two-group theory, two different spatial components of the noise will appear. One of them is the equivalent of the one-group solution, changing smoothly in space, and another with a short spatial relaxation length. These will be called the global and the local components of two-group neutron noise. Alternative definitions are also in use in the literature, in which the point kinetic term is called the global one and the rest the local, but we do not adopt these. In the description that we use throughout, the global component consists of a point kinetic and a space-dependent term. In this respect, "space-dependent" refers to the deviation from point kinetics of the *global* component.

These three different components have different roles and significance in power reactor noise diagnostics. If the only goal is to detect the appearance of a perturbation (which may be the result of incipient failure), and only the frequency dependence of the perturbation is of interest, then the point kinetic component is usually the most helpful, because for perturbations with a non-negligible reactivity effect, this component is the strongest and has the longest spatial range. However, the point kinetic component does not carry any information on the position of the disturbance.

If the goal is to locate the position of a localised perturbation, then the most suitable for this purpose is the space-dependent part of the global component. It is sensitive to the position of the perturbation, and still it has a relatively long spatial relaxation. The local component has a much more pronounced space dependence and hence a sensitive dependence on the position of the perturbation. However, due to its short spatial relaxation length, it is only useful if the detector is in the vicinity of the perturbation, which is very seldom the case.

The main use of the local component is in the diagnostics of perturbations propagating with the coolant, such as vapour in a two-phase flow or bubbles in a two-component flow. To determine the transit time of the flow between two axially displaced detectors, it is necessary to use a noise component whose spatial relaxation is shorter than the distance between the detectors. The proximity to the perturbation is usually granted, since the detectors are immersed into the flow, and the perturbation, propagating with the coolant, will pass both detectors.

From the aforementioned facts, it follows that the efficiency or applicability of the various diagnostic methods depends largely on the relative weights of the point kinetic, space-dependent and local components. These, in turn, depend on the properties of both the perturbation and the properties of the core, so a general classification cannot be given. With regard to the perturbation, if the reactivity effect of the perturbation is zero, then obviously, the system will respond in a space-dependent manner, and no point kinetic contributions will exist. Such is the case of absorbers or fuel elements laterally vibrating in a flux with zero spatial derivative, or the propagating perturbations at certain frequencies. If the perturbation is homogeneous in space, then the global component of the noise will only contain the point kinetic component.

From the point of view of the system, the ratio between the point kinetic and space-dependent components of the transfer function depends primarily on the frequency of the perturbation and on the system size. For low frequencies and/or small system sizes, the point kinetic component of the Green's function dominates. This means that if the reactivity effect of the perturbation is not zero, then the point kinetic (reactivity) component of the noise will dominate. This can be seen from the expression of (2.50) for the 1D Green's function in the following way. With a simple analysis, it can be shown that when $\omega \to 0$, since

$$B(\omega) \approx B_0 \left(1 - \varepsilon(\omega)/2\right) \tag{2.61}$$

where

$$\varepsilon(\omega) = \frac{1}{\rho_\infty G_0(\omega)} \tag{2.62}$$

the leading term of $G(x, x_0, \omega)$ in $\varepsilon(\omega)$ is proportional to

$$G(x, x_0, \omega) \approx -\frac{G_0(\omega)\,\phi_0(x)\,\phi_0(x_0)}{\nu \Sigma_f} \tag{2.63}$$

where now the flux function $\phi_0(x)$ is square-normalised. Convolving this expression with the noise source $S(x, \omega)$ (taken also in the limit of $\omega \to 0$, and utilising (2.59) leads to the point kinetic term of the noise, (2.58), proving the point kinetic behaviour at low frequencies.

For higher frequencies, or larger core dimensions, the behaviour becomes increasingly space-dependent. The range of the frequencies and/or system sizes where one or the other component dominates depends on the material and geometrical properties of the core. The local component is largely insensitive to frequencies and system sizes within the practically interesting frequencies and core sizes. Its relative weight in the system transfer depends on the material properties of the core but also on the frequency, due to the frequency dependence of the amplitude of the point kinetic term.

In the analysis of the various systems, the size of the cores and the material properties are given. Most quantitative work will be made at the characteristic frequency of 10 rad/s. This frequency lies in the plateau region of most systems, which is also the characteristic range of the perturbations. Hence, the results will mostly differ due to the different sizes and different material properties of the various systems.

The point kinetic term requires only the calculation of the fluctuation of the amplitude factor, which is given by the point kinetic equation. Due to its linearised character, it is decoupled from the equation for the fluctuation of the shape function. The equation for the shape function $\delta\psi(r, t)$ is not simpler than the equation for the full noise; hence, it is simpler to obtain it by subtracting the point kinetic term from the full solution. It is in the calculation of the shape function that the kinetic approximations play a role. The point kinetic approximation means that this term is neglected, and the noise is represented only with its point kinetic term. A better approximation, which is valid for higher frequencies and/or system sizes, is the adiabatic approximation, which states that the shape function in each time instant is equal to the normalised static eigenfunction of the reactor, belonging to the instantaneous value of the perturbation at that time instant. The fluctuation of the shape function is then obtained by subtracting the static flux from it. This approximation is easy to calculate, and it is very useful in understanding the structure of the induced noise.

So far, (2.39), (2.45) and (2.46) are written as deterministic equations, and indeed they are valid also for deterministic perturbations, such as monochromatic vibrations of a component. Often it gives insight to obtain solutions for deterministic perturbations. This is because the transfer function of the core, $G(r, r', \omega)$, is independent of whether the noise source is stochastic or deterministic; since it only depends on the parameters of the undisturbed system, it is always deterministic.

It is, however, at this point that the random character of the noise source can be taken into account. Assuming that the cross sections fluctuate in a stochastic manner, then the noise source $S(r, \omega)$ will also be a random function, whereby (2.39) is indeed formally a Langevin-type equation, written in the frequency domain. Random processes cannot be characterised by individual realisations, only with statistical descriptors. To characterise a time-dependent random process, one should specify the one-point (at an arbitrary time), two-point (the joint distribution at two arbitrary times), etc., joint distributions of a process, which is impossible in a general case. However, in practice it is sufficient to treat some low-order statistical moments, similarly to the zero-power noise, where only the mean and the variance of the underlying discrete process were used.

For continuous functions, the auto- and cross-correlation and covariance are used. Since we are dealing with processes with zero mean, the correlations are equal to the covariance, which simplifies the situation. It is assumed (which is nearly exclusively always valid, or is a very good approximation) that the processes are stationary and ergodic. Stationarity means that the correlation functions, which are dependent on two time instances, depend only on the time difference. For instance, for the cross-correlation (between two spatial points) of the absorption cross-section fluctuations, one has

$$CCF_{\delta\Sigma_a}(\mathbf{r},t,\mathbf{r}',t+\tau) = \langle \delta\Sigma_a(\mathbf{r},t)\delta\Sigma_a(\mathbf{r}',t+\tau) \rangle \equiv CCF_{\delta\Sigma_a}(\mathbf{r},\mathbf{r}',\tau) \quad (2.64)$$

whereas the ergodicity means that ensemble averages can be replaced by time averages:

$$CCF_{\delta\Sigma_a}(\mathbf{r},\mathbf{r}',\tau) = \lim_{T\to\infty} \frac{1}{2T} \int_{-T}^{T} \delta\Sigma_a(\mathbf{r},t)\delta\Sigma_a(\mathbf{r}',t+\tau)\,dt \quad (2.65)$$

The auto- and cross-power spectra of stationary processes are defined as the temporal Fourier transforms of the auto- and cross-correlations. For instance, the cross-correlation of the neutron noise in two detector positions $\mathbf{r}_1$ and $\mathbf{r}_2$ is given as

$$CPSD_{\delta\phi}(\mathbf{r}_1,\mathbf{r}_2,\omega) = \int_{-\infty}^{\infty} CCF_{\delta\phi}(\mathbf{r}_1,\mathbf{r}_2,\tau)\,e^{-i\omega\tau}\,d\tau \quad (2.66)$$

Making use of the ergodic property, expressed with the time average in (2.65), that is representing $CCF_{\delta\phi}(\mathbf{r}_1,\mathbf{r}_2,\tau)$ in (2.66) with a time integral, leads to the Wiener–Khinchin theorem, which states that the auto- and cross-power spectral densities of stationary random ergodic processes can be represented as products of the Fourier transforms of the processes involved, such as

$$APSD_{\delta\phi}(\mathbf{r},\omega) \propto |\delta\phi(\mathbf{r},\omega)\delta\phi^*(\mathbf{r},\omega)| \quad (2.67)$$

and likewise

$$CPSD_{\delta\phi}(\mathbf{r}_1,\mathbf{r}_2,\omega) \propto \delta\phi(\mathbf{r}_1,\omega)\delta\phi^*(\mathbf{r}_2,\omega) \quad (2.68)$$

The proportionality factor diverges when $T \to \infty$ in (2.65), but it allows one to use it in relationships when the same operation is performed on both sides of the equation. In practical work, one only deals with time series of finite lengths, when the proportionality factor remains finite.

With the help of the Wiener–Khinchin theorem, one can derive an expression for the auto- and cross-spectra of the neutron noise in terms of the cross-spectrum of the driving force (perturbation) $S(\mathbf{r},\omega)$ as

$$CPSD_{\delta\phi}(\mathbf{r}_1,\mathbf{r}_2,\omega) =$$

$$\int_{V_R}\int_{V_R} G(\mathbf{r}_1,\mathbf{r}',\omega)\,G^*(\mathbf{r}_2,\mathbf{r}'',\omega)\,CPSD_S(\mathbf{r}',\mathbf{r}'',\omega)\,d\mathbf{r}'\,d\mathbf{r}'' \quad (2.69)$$

Note that even for the calculation of the auto-spectrum of the neutron noise, i.e. when $\mathbf{r}_1 = \mathbf{r}_2$, the cross-spectrum of the noise source is needed.

Often the space- and time-dependence of the perturbation can be separated, and at the same time the sought parameter is the position of the localised perturbation.

Such is the case with vibrations of control rods or fuel assemblies, or with channel-type thermal hydraulic instabilities. In that case the important information is in the space-dependence of the noise, which can be obtained from the equation for the space- and frequency-dependent neutron noise, without the need for turning to power spectra.

2.3.3   Two-group theory

Even in the case of simple models with homogeneous thermal cores, in certain cases the use of two-group theory is needed. One is the case of heavily reflected reactors, when the perturbation takes place close to the core boundary, or in the reflector, due to the presence of the reflector peak and the strongly deviating space dependence of the fast and thermal fluxes. From the practical point of view, this is not an important case, because most likely the few in-core detectors would be unlikely in the neighbourhood of the perturbations. Hence, this in itself would not necessitate the use of two-group theory.

A more important fact is when the local component of the noise plays a role. So far only the point kinetic and the so-called space-dependent components of the noise were mentioned, which both have relatively slow spatial relaxation. It has been known from the early days of reactor oscillator experiments that there is also a local component of the noise, which exists in the neighbourhood of the perturbation [32]. This component plays a significant role whenever propagating perturbations occur in the core, such as the two-phase flow in BWRs. It is the existence of the local component that makes it possible to determine flow velocity and other thermal hydraulic parameters of the coolant. However, only an energy-dependent theory, in which slowing down and thermal diffusion can be separated, is capable of reconstructing the local component. Two-group theory is the simplest energy-dependent theory, which can be used to quantify the local component of the noise.

As mentioned earlier, the need of using two-group theory comes to a different light when it comes to fast systems. Treating fast systems with one-group theory would give poor results even for the static calculations. A better representation is to define two groups, which are separated in energy at the threshold energy of fast fission in the fertile component of the fuel [28]. This is valid even in non-reflected homogeneous systems. Those two groups could be characterised as the fast and epithermal groups. On the other hand, they would still lead to two different roots of the characteristic equation, and hence they are still capable to treat the reflector effect and the local component of the noise.

The widespread use of two-group theory in analytical noise calculations was introduced in connection with the interpretation of in-core neutron noise in BWRs. In this process the two-group equations for thermal reactors were used. The corresponding analytical formulae will be first shown in the forthcoming for completeness. Namely, in the literature the full formulae are not available; usually they use approximations that are valid only in LWRs. The full formalism, without approximations, was used in [33], but no explicit forms of the formulae used were given. This gap will be filled up here, not the least because they are suitable for the analysis of LWRs of the Gen-III type. After that, the modifications needed for the use of two-group theory for fast systems will be given.

2.3.3.1 Two-group noise theory for thermal systems

The static equations for the direct flux read as

$$\begin{bmatrix} D_1\nabla^2 - \Sigma_1 & \nu\Sigma_{f2} \\ \Sigma_R & D_2\nabla^2 - \Sigma_{a2} \end{bmatrix} \begin{bmatrix} \phi_1(x) \\ \phi_2(x) \end{bmatrix} = 0 \tag{2.70}$$

where

$$\Sigma_1 \equiv \Sigma_{a1} + \Sigma_R - \nu\Sigma_{f1} \tag{2.71}$$

and the rest of the notations is standard. In the aforementioned, it is assumed that the system is critical, i.e. $k_{eff} = 1$. The 1D bare core has extrapolated boundaries at $x = \pm a$, and zero flux conditions will be used at the extrapolated boundaries, i.e. $\phi_i(-a) = \phi_i(a) = 0, i = 1, 2$.

The static fluxes are given as

$$\phi_1(x) = \cos(B_0 x) \tag{2.72}$$

$$\phi_2(x) = \frac{\Sigma_R}{\Sigma_{a2} + D_2 B_0^2} \cos(B_0 x) \equiv c_\mu \cos(B_0 x) \tag{2.73}$$

with $B_0 = \pi/(2a)$. Here the normalisation was used that the maximum amplitude value of the fast flux is unity. The critical buckling B_0^2 can be obtained from the material properties via (2.70) as

$$B_0^2 = -\frac{1}{2}\left(\frac{\Sigma_1}{D_1} + \frac{\Sigma_{a2}}{D_2}\right) + \frac{1}{2}\sqrt{\left(\frac{\Sigma_1}{D_1} + \frac{\Sigma_{a2}}{D_2}\right)^2 + 4\frac{\Sigma_R\nu\Sigma_{f2} - \Sigma_1\Sigma_{a2}}{D_1 D_2}} \tag{2.74}$$

The equations for the fast and thermal noises $\delta\phi_1(x,\omega)$ and $\delta\phi_2(x,\omega)$, respectively, are derived by the usual way of splitting the cross sections and the space–time-dependent flux to stationary and fluctuating parts, linearising the equations by neglecting the products of fluctuating quantities and eliminating the time derivatives and the delayed neutron precursors by temporal Fourier transform. One arrives at

$$\begin{bmatrix} D_1\nabla^2 - \Sigma_1(\omega) & \nu\Sigma_{f2}(\omega) \\ \Sigma_R & D_2\nabla^2 - \Sigma_2(\omega) \end{bmatrix} \begin{bmatrix} \delta\phi_1(x,\omega) \\ \delta\phi_2(x,\omega) \end{bmatrix} = \begin{bmatrix} S_1(x,\omega) \\ S_2(x,\omega) \end{bmatrix} \tag{2.75}$$

In the aforementioned, the following notations and definitions are used:

$$\Sigma_{f,i}(\omega) = \Sigma_{f,i}\left(1 - \frac{i\omega\beta}{\lambda + i\omega}\right); \qquad i = 1, 2 \tag{2.76}$$

$$\Sigma_1(\omega) = \Sigma_{a1} + \Sigma_R + \frac{i\omega}{v_1} - \nu\Sigma_{f1}(\omega) \tag{2.77}$$

and

$$\Sigma_2(\omega) = \Sigma_{a2} + \frac{i\omega}{v_2} \tag{2.78}$$

The noise sources $S_1(x, \omega)$ and $S_2(x, \omega)$ are defined in terms of the cross-section fluctuations and the static fluxes as follows:

$$S_1(x, \omega) = \left[\delta \Sigma_R(x, \omega) + \delta \Sigma_{a_1}(x, \omega) - \delta \nu \Sigma_{f_1}(x, \omega) \left(1 - \frac{i\omega\beta}{i\omega + \lambda} \right) \right] \phi_1(x)$$

$$- \delta \nu \Sigma_{f_2}(x, \omega) \left(1 - \frac{i\omega\beta}{i\omega + \lambda} \right) \phi_2(x) \tag{2.79}$$

and

$$S_2(x, \omega) = -\delta \Sigma_R(x, \omega)\phi_1(x) + \delta \Sigma_{a_2}(x, \omega)\phi_2(x) \tag{2.80}$$

For reasons of simplicity, in the present study the fluctuations of the diffusion coefficients were neglected. This is in accordance with standard practice, and is justified by the fact that the effect of the fluctuations of the diffusion coefficients is much smaller than that of the other cross sections [29]).

The equation for the Green's matrix that connects the noise in the fast and thermal groups with the fast and thermal noise sources can then be written as

$$\begin{bmatrix} D_1 \nabla^2 - \Sigma_1(\omega) & \nu\Sigma_{f2}(\omega) \\ \Sigma_R & D_2 \nabla^2 - \Sigma_2(\omega) \end{bmatrix} \begin{bmatrix} G_{11}(x, x', \omega) & G_{12}(x, x', \omega) \\ G_{21}(x, x', \omega) & G_{22}(x, x', \omega) \end{bmatrix}$$

$$= \begin{bmatrix} \delta(x - x') & 0 \\ 0 & \delta(x - x') \end{bmatrix} \tag{2.81}$$

Equation (2.81) can be written in a symbolic notation as

$$\hat{L}(x, \omega)\hat{G}(x, x', \omega) = \hat{I}\delta(x - x') \tag{2.82}$$

where $\hat{I}$ is the 2×2 unit matrix. From (2.81) the noise can be obtained as

$$\begin{bmatrix} \delta\phi_1(x, \omega) \\ \delta\phi_2(x, \omega) \end{bmatrix} = \int_0^H \begin{bmatrix} G_{11}(x, x', \omega) & G_{12}(x, x', \omega) \\ G_{21}(x, x', \omega) & G_{22}(x, x', \omega) \end{bmatrix} \begin{bmatrix} S_1(x', \omega) \\ S_2(x', \omega) \end{bmatrix} dx' \tag{2.83}$$

or, again in symbolic form, as

$$\delta\vec{\Phi}(x, \omega) = \int \hat{G}(x, x', \omega)\vec{S}(x'\omega) \, dx' \tag{2.84}$$

The elements of the aforementioned Green's matrix express the transfer between the noise source and the induced noise in the various groups. Thus, G_{12} connects the effect of a noise source in the thermal group at space point x' to the noise induced at point x in the fast group.‖ From here, it is seen that for the determination of the thermal noise only, one needs the second row of the Green's matrix. However, this quantity alone cannot be determined from a simple vector equation, having only two components, rather all four components of the Green's matrix need to be determined

---

‖ The matrix elements G_{ij} were denoted as $G_{j \to i}$ in [7], which is an alternative notation one often encounters in the literature for the indexing.

simultaneously. As it will be seen soon, these two components constitute a column in the adjoint Green's matrix, thus they can be obtained from one single solution of a vector equation. Due to this circumstance, using the adjoint Green's matrix, and some of its components called as "dynamic adjoints", were preferred in most literature so far. In our case, we are interested in the noise in both the fast and the thermal group, so we need the full Green's matrix anyway, and present a full solution for the Green's matrix.

The solution of (2.81) depends on the two roots[¶] $\mu^2(\omega)$ and $\nu^2(\omega)$ of the characteristic equation of (2.81), which are obtained as

$$\mu^2(\omega) = -\frac{1}{2}\left(\frac{\Sigma_1(\omega)}{D_1} + \frac{\Sigma_2(\omega)}{D_2}\right) +$$
$$\frac{1}{2}\sqrt{\left(\frac{\Sigma_1(\omega)}{D_1} + \frac{\Sigma_2(\omega)}{D_2}\right)^2 + 4\frac{\Sigma_R\nu\Sigma_{f2}(\omega) - \Sigma_1(\omega)\Sigma_2(\omega)}{D_1 D_2}} \tag{2.85}$$

and

$$\nu^2(\omega) = \frac{1}{2}\left(\frac{\Sigma_1(\omega)}{D_1} + \frac{\Sigma_2(\omega)}{D_2}\right) +$$
$$\frac{1}{2}\sqrt{\left(\frac{\Sigma_1(\omega)}{D_1} + \frac{\Sigma_2(\omega)}{D_2}\right)^2 + 4\frac{\Sigma_R\nu\Sigma_{f2}(\omega) - \Sigma_1(\omega)\Sigma_2(\omega)}{D_1 D_2}} \tag{2.86}$$

Defining

$$c_\mu(\omega) = \frac{\Sigma_1(\omega) + D_1\mu^2(\omega)}{\nu\Sigma_{f2}(\omega)} \tag{2.87}$$

and

$$c_\nu(\omega) = \frac{\Sigma_1(\omega) - D_1\nu^2(\omega)}{\nu\Sigma_{f2}(\omega)} \tag{2.88}$$

the solutions for the elements of the Green's matrix can be given as follows. Define the basic solutions $g_\mu(x, x', \omega)$ and $g_\nu(x, x', \omega)$, respectively, belonging to the two roots as

$$g_\mu(x, x', \omega) = \begin{cases} \sin[\mu(\omega)(a+x)]\sin[\mu(\omega)(a-x')]; & x < x' \\ \\ \sin[\mu(\omega)(a-x)]\sin[\mu(\omega)(a+x')]; & x > x' \end{cases} \tag{2.89}$$

and

$$g_\nu(x, x', \omega) = \begin{cases} \sinh[\nu(\omega)(a+x)]\sinh[\nu(\omega)(a-x')]; & x < x' \\ \\ \sinh[\nu(\omega)(a-x)]\sinh[\nu(\omega)(a+x')]; & x > x' \end{cases} \tag{2.90}$$

[¶]The notation ν or $\nu(\omega)$ should not be mixed up with the symbol used for the average number of neutrons in a fission event. Hopefully it is always clear from the context which one is meant. To avoid this ambiguity, in some publications the Greek symbol λ was used for the root corresponding to the local component, but this can also be mistaken for the delayed neutron decay constant.

the solutions for the four elements of the Green's matrix can be given, column-wise, as follows:

$$\begin{bmatrix} G_{11}(x,x',\omega) \\ G_{12}(x,x',\omega) \end{bmatrix} = \begin{bmatrix} 1 \\ c_\mu(\omega) \end{bmatrix} A_{\mu 1}(\omega)\, g_\mu(x,x',\omega)$$

$$+ \begin{bmatrix} 1 \\ c_\nu(\omega) \end{bmatrix} A_{\nu 1}(\omega)\, g_\nu(x,x',\omega) \tag{2.91}$$

with

$$A_{\mu 1}(\omega) = \frac{c_\mu(\omega)\,\nu\Sigma_{f2}(\omega)}{D_1^2\,\mu(\omega)\,\sin[2\,\mu(\omega)\,a]\,(\mu^2(\omega)+\nu^2(\omega))} \tag{2.92}$$

and

$$A_{\nu 1}(\omega) = \frac{c_\mu(\omega)\,\nu\Sigma_{f2}(\omega)}{D_1^2\,\nu(\omega)\,\sinh[2\,\nu(\omega)\,a]\,(\mu^2(\omega)+\nu^2(\omega))} \tag{2.93}$$

In a similar manner, for the second column of the Green's function matrix in (2.81) we obtain

$$\begin{bmatrix} G_{21}(x,x',\omega) \\ G_{22}(x,x',\omega) \end{bmatrix} = \begin{bmatrix} 1 \\ c_\mu(\omega) \end{bmatrix} A_{\mu 2}(\omega)\, g_\mu(x,x',\omega)$$

$$+ \begin{bmatrix} 1 \\ c_\nu(\omega) \end{bmatrix} A_{\nu 2}(\omega)\, g_\nu(x,x',\omega) \tag{2.94}$$

with

$$A_{\mu 2}(\omega) = \frac{\nu\Sigma_{f2}(\omega)}{D_1\,D_2\,\mu(\omega)\,\sin[2\,\mu(\omega)\,a]\,(\mu^2(\omega)+\nu^2(\omega))} \tag{2.95}$$

and

$$A_{\nu 2}(\omega) = \frac{\nu\Sigma_{f2}(\omega)}{D_1\,D_2\,\nu(\omega)\,\sin[2\,\nu(\omega)\,a]\,(\mu^2(\omega)+\nu^2(\omega))} \tag{2.96}$$

The above constitute the full solution of the two-group theory of noise in BWRs, i.e. in LWRs. Group 1 refers to the fast group and Group 2 to the thermal group. These formulae were used in many works in the past for the illustration of the existence of the local component, although with some simplifications which are allowable for LWRs, such as neglecting μ^2 in the sum $(\mu^2(\omega)+\nu^2(\omega))$ in (2.92)–(2.96).

At this point, it is worth mentioning that an alternative way of calculating the induced neutron noise is to use the adjoint Green's function, often called the "dynamic adjoint". This is in order to distinguish it from the static adjoints $\phi_{1,2}^\dagger(x)$, which are simply obtained from the static adjoint equations, obtained by transposing the matrix on the l.h.s. of (2.70). The use of the dynamic adjoint in noise calculations was introduced by van Dam [34,35], and it has certain practical advantages in noise calculations in thermal systems. These advantages have been discussed widely in the literature [36,37], out of which one will be mentioned next.

The equations for the adjoint Green's function are given by the adjoint two-group frequency-dependent diffusion operator. In two-group diffusion theory, the adjoint Green's function $\hat{\mathbf{G}}^\dagger(x, x', \omega)$ is also a 2×2 matrix, whose elements are determined by the equation:

$$\hat{\mathbf{L}}^\dagger(x, \omega)\hat{\mathbf{G}}^\dagger(x, x', \omega) = \hat{\mathbf{I}} \cdot \delta(x - x') \tag{2.97}$$

where $\hat{\mathbf{I}}$ is the 2×2 unit matrix and $\hat{\mathbf{G}}^\dagger(x, x', \omega)$ is given as

$$\hat{\mathbf{G}}^\dagger(x, x', \omega) = \begin{bmatrix} G_{11}^\dagger & G_{12}^\dagger \\ G_{21}^\dagger & G_{22}^\dagger \end{bmatrix} \tag{2.98}$$

and further

$$\hat{\mathbf{L}}^\dagger = \hat{\mathbf{L}}^T \tag{2.99}$$

where the superscript T denotes the transpose of the matrix operator. It is easy to show from the properties of the adjoint that with the help of the adjoint Green's matrix, the noise can be expressed as

$$\delta\bar{\Phi}^T(x, \omega) = \int_{V_R} \bar{\mathbf{S}}^T(x', \omega)\hat{\mathbf{G}}^\dagger(x', x, \omega)\, \mathrm{d}x' \tag{2.100}$$

where $\delta\bar{\Phi}^T(x, \omega)$ and $\bar{\mathbf{S}}^T(x', \omega)$ denote row vectors.

Equation (2.100) shows that in order to determine the fast and the thermal noise, one needs to know the first and the second column of the adjoint Green's matrix, respectively. These can be calculated from a single vector equation each, and this constitutes a certain advantage when only the noise in one of the two groups is to be determined. In the classic noise studies, only the thermal noise was of interest, which can be calculated by the second column of the adjoint Green's matrix [7]. This vector is called, somewhat misleadingly, the "adjoint function" in the literature, with its two components being called the "fast adjoint", $\Psi_1^\dagger$, and the "thermal adjoint", $\Psi_2^\dagger$:

$$\begin{bmatrix} G_{12}^\dagger(x, x', \omega) \\ G_{22}^\dagger(x, x', \omega) \end{bmatrix} \equiv \begin{bmatrix} \Psi_1^\dagger(x, x', \omega) \\ \Psi_2^\dagger(x, x', \omega) \end{bmatrix} \tag{2.101}$$

In this terminology the fast adjoint means the transfer between the *noise source in the fast group* and the thermal noise, and the thermal adjoint the transfer between the *thermal noise source* and the thermal noise. In our general notations here, $G_{ij}^\dagger$ stands for the transfer from the noise source in group i to the neutron noise in group j.

Since we calculate both the fast and the thermal noise, neither the direct nor the adjoint Green's function technique has any advantages over the other. Actually, as it is also easy to show, the direct and adjoint Green's functions obey the relationship

$$G_{ij}^\dagger(x, x', \omega) = G_{ji}(x', x, \omega) \tag{2.102}$$

Further, as it was pointed out in [38], since the dependence of the Green's function elements is identical on x and x' in a bare core, we have

$$G_{ij}^\dagger(x, x', \omega) = G_{ji}(x, x', \omega) \tag{2.103}$$

And since in the simple model considered here (a homogeneous bare core), a solution can be obtained which is analytical in both arguments x and x', the two functions are equivalent in all respects. For this reason, there is no need to show both the direct and the adjoint Green matrix elements, and in the continuation we restrict ourselves to the direct Green's functions.

It is worth to spend a few clarifying words in order to connect the terminology which will be used in the continuation, with the one dominating in the literature. This is because the latter is largely focused on the study of the noise in the thermal group only. In our case, because of including also fast systems, we have to extend the treatment also to the noise in the fast group, and this requires to widen and partly redefine the terminology. Moreover, the existing literature of two-group reactor noise is dominated by the use of the adjoint technique, which we do not implement here, rather use the direct (forward) Green's function, and to point out the existing equivalences is also important.

The first comment in this respect is that in view of (2.101) and (2.103), the component G_{21} of the direct (forward) Green's matrix is equal to the fast adjoint $\Psi_1^\dagger$, and G_{22} is equal to the thermal adjoint, $\Psi_1^\dagger$. Now in the literature, the difference $\Psi_1^\dagger - \Psi_2^\dagger$ of the fast and thermal adjoints is called the "removal adjoint" [38]. This term arises from the study of neutron noise in BWRs, related to the propagating perturbations due to the two-phase flow, because the noise in the thermal group induced by the fluctuations of the removal cross section are transferred by $\Psi_1^\dagger - \Psi_2^\dagger$. This will be shown concretely in Section 2.3.4.4. Actually, since $\Psi_1^\dagger - \Psi_2^\dagger$ yields the induced noise in the thermal group, it should be called the *thermal removal adjoint*. As mentioned before here, it is clear that in terms of the direct Green's function, the thermal removal is equal to $G_{21} - G_{22}$. Because of its practical use, it is interesting to investigate the properties of this difference. Since in our calculations we use the direct Green's matrix elements, the difference $G_{21} - G_{22}$ is called the "thermal removal adjoint", denoted by Ψ_2. Likewise, for obvious reasons, the difference $G_{11} - G_{12}$ is called the "fast removal adjoint", denoted by Ψ_1, since it yields the noise in the fast group, induced by the fluctuations of the removal cross section.

On the first sight, it might appear somewhat contradictory or illogical to call a quantity "adjoint", while it is based on quantities calculated by forward methods. However, there is a very clear purpose with introducing this terminology. This is because the word "adjoint", inspired by the phrase "removal adjoint", indicates in the continuation that it is composed by a linear combination of two of the four elements of the 2×2 Green's matrix (or the adjoint Green's matrix, for that matter). On the other hand, when the term "Green's function" is used, it always and exclusively refers to single elements of the Green's matrix. Besides, it is also a more compact term than other alternatives, such as "removal Green's function" or "removal transfer function".

The transfer between a propagating perturbation of the coolant density and the neutron noise is determined by the aforementioned defined fast and thermal removal adjoints only in water-moderated reactors. In liquid metal-cooled fast reactors, or fluoride salt-cooled thermal cores, as well as both fast and thermal MSRs, neutron moderation in the coolant plays a negligible role. Density fluctuations of the coolant will mostly affect other cross sections, primarily absorption, and in MSRs also the fission cross sections.

From the structure of the noise sources, (2.79) and (2.80), one can deduce that for such cores, the transfer between a propagating perturbation of the coolant density and the neutron noise will be determined by the transfer functions

$$\Psi_1(x, x', \omega) = \alpha_1\, G_{11}(x, x', \omega) + \alpha_2\, G_{12}(x, x', \omega) \tag{2.104}$$

for the noise in the fast group, and

$$\Psi_2(x, x', \omega) = \alpha_1\, G_{21}(x, x', \omega) + \alpha_2\, G_{22}(x, x', \omega) \tag{2.105}$$

in Group 2 (the thermal or epithermal group). The coefficients α_1 and α_2 will depend on the type of the fuel and the coolant of the core. Their calculation will be discussed in Chapters 3 and 4 in connection with the concrete reactor types. For a BWR, $\alpha_1 = 1$ and $\alpha_2 = -1$.

Because of their role in the properties of the neutron noise induced by propagating perturbations, the quantities Ψ_1 and Ψ_2 defined in (2.104) and (2.105) will be calculated and plotted for the liquid metal-cooled and molten salt reactors treated in Chapters 3 and 4. Since apart from LWRs, $\alpha_1 \neq -\alpha_2$, for all those cores they will not be called removal adjoints, rather fast and thermal "propagation adjoints". For LWRs, the term removal adjoints will be retained.

For the sake of illustration, as well as for later comparison with the findings for the next generation reactors in the following two chapters, we show here two examples of the Green's matrix elements for a Gen-II BWR, which were used in an early publication for the study of the properties of the local component of BWR in-core noise [38]. In that study, data of a commercial BWR were used, and the group constants corresponding to the three-dimensional (3D) model were corrected to include radial leakage, such that the 1D model corresponded to the axial coordinate of the core with the same height. In addition to the power reactor, also a small system was studied, which was obtained by increasing the fission cross sections by 11%. The data corresponding to the power reactor are shown in Table 2.1.

With the data mentioned here, the width of the slab, representing the height of the core, is obtained as $H = 368$ cm. In the original paper, the height was given as 372 cm.

Table 2.1   Data of a commercial BWR for a 1D model (from Reference [38])

Parameter	Fast		Thermal
D [cm]	1.709		0.5290
$\nu\Sigma_f$ [cm^{-1}]	0.004653		0.07254
Σ_a [cm^{-1}]	0.00733		0.05940
Σ_R [cm^{-1}]		0.05940	
v [cm/s]	$1.7549\ 10^7$		$3.9040\ 10^5$
β		0.007	
λ [s^{-1}]		0.1	

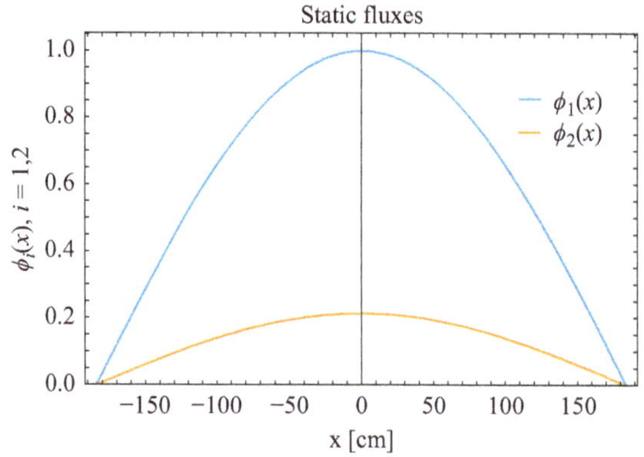

Figure 2.3   The static fluxes in a commercial BWR

The slight difference is due to the aforementioned simplifications in the derivations, which were applied in [38], but not used in our calculations.

The static fluxes in Groups 1 and 2 are shown in Figure 2.3. As usual, in commercial PWRs and BWRs the fast flux is a couple of times larger than the thermal one.

Below we show quantitative values of the space- and frequency-dependence of the amplitude and the phase of the components of the Green's function matrix, as well as those of the fast and the thermal removal adjoint, as described in the foregoing. The space dependence of the amplitude and the phase of the components of the Green's matrix for two different positions x_0 of the perturbation is shown in Figure 2.4 for a frequency of $\omega = 10$ rad/s, which is well within the plateau region. The blue and green lines stand for G_{11} and G_{12} which represent the transfer of the perturbations in the fast and the thermal groups, respectively, into the fast component of the noise. The orange and red lines stand for G_{21} and G_{22}, representing the transfer or the same perturbations to the thermal noise.

It is seen in Figure 2.4 that the global response is rather space-dependent, as expected, which is manifested by the nearly linear space dependence, which deviates from the cosine shape of the static flux significantly. The local component is visible only in G_{22}, which connects the perturbation in the thermal group to the induced noise in the thermal group. The phase is monotonically decreasing away from the perturbation, with a small local peak at the position of the perturbation,

The spatial dependence of the amplitude and the phase of the fast and the thermal removal adjoints is shown in Figure 2.5. These stand for the transfer of the fluctuations of the removal cross section (which represents a perturbation, i.e. a noise source, in both the fast and the thermal groups but with different sign), to the induced fast and thermal neutron noise, respectively. As is known from the literature, the local peak is much more pronounced in the thermal removal adjoint than in the pure component G_{22}, which is also seen from Figures 2.4 and 2.5.

The fast removal adjoint, which describes the space dependence of the noise in the fast group, was not calculated in previous works. One reason is that in Gen-II reactors, only thermal detectors are used, and hence the noise in the fast group was

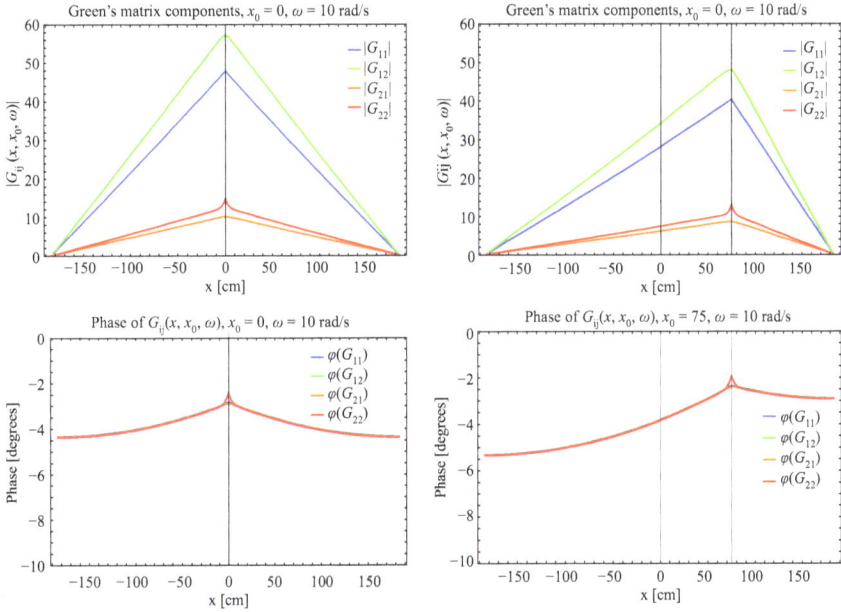

Figure 2.4   Space dependence of the amplitude and the phase of the components of the Green's function for $x_0 = 0$ (left column) and x_0 75 cm (right column), respectively, in a commercial BWR

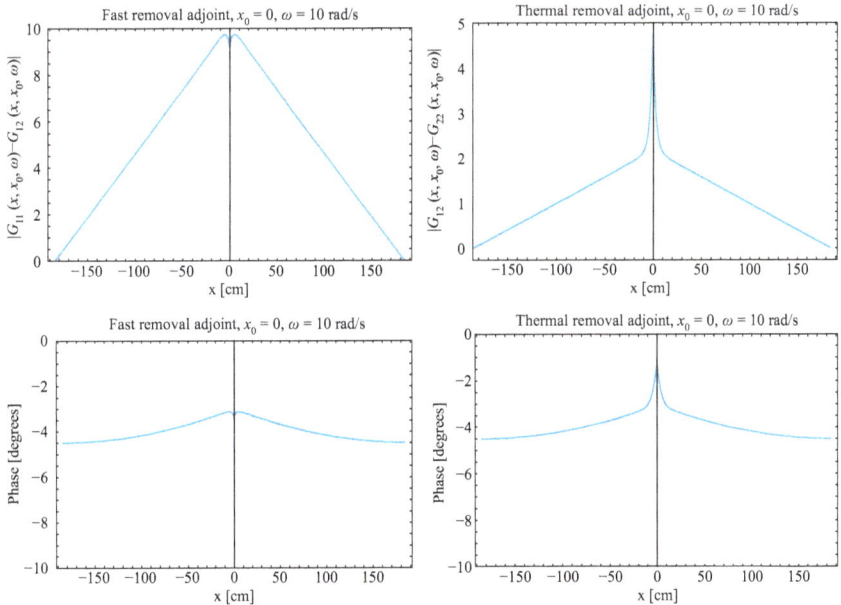

Figure 2.5   Space dependence of the amplitude (upper figures) and the phase (lower figures) of the fast (left column) and thermal (right column) removal adjoints for $x_0 = 0$ in a commercial BWR

not investigated. The situation will be different in the fast systems of Gen-IV, and for comparison it is interesting to show it here, even if for a thermal system. As the upper left plot shows, the amplitude of the fast removal adjoint shows a small local component (which is not visible in G_{11} and G_{12}). It is interesting that unlike for the thermal removal adjoint, it is "out-of-phase" with the global component, in that it appears as a small dip, in contrast to the peak present in the thermal removal adjoint. It is seen from the space dependence of the amplitude of the removal adjoints in Figure 2.5, that the possibilities of measuring propagating perturbations, such as two-phase flow, the thermal noise is much more suitable in thermal reactors than the noise in the fast group, where the local component is vanishingly small even in the removal adjoint. The spatial behaviour of the phases, although changing in space much slower, shows a similar behaviour.

The results are in agreement with those in [38] for the space dependence of the thermal adjoint. The small quantitative differences are due to the simplifications made in [38]. The fast adjoint, or all individual elements of either the direct or the adjoint Green's matrix, were not calculated there. In that publication only the components G_{21} and G_{22} were calculated, and were called the fast and thermal adjoint, respectively. The four elements of the direct and/or adjoint Green's function were calculated for a few water-moderated systems in [33], but the phases were not shown there either.

The frequency dependence of the amplitude and the phase of the four components of the Green's function are shown in Figure 2.6 for two different detector

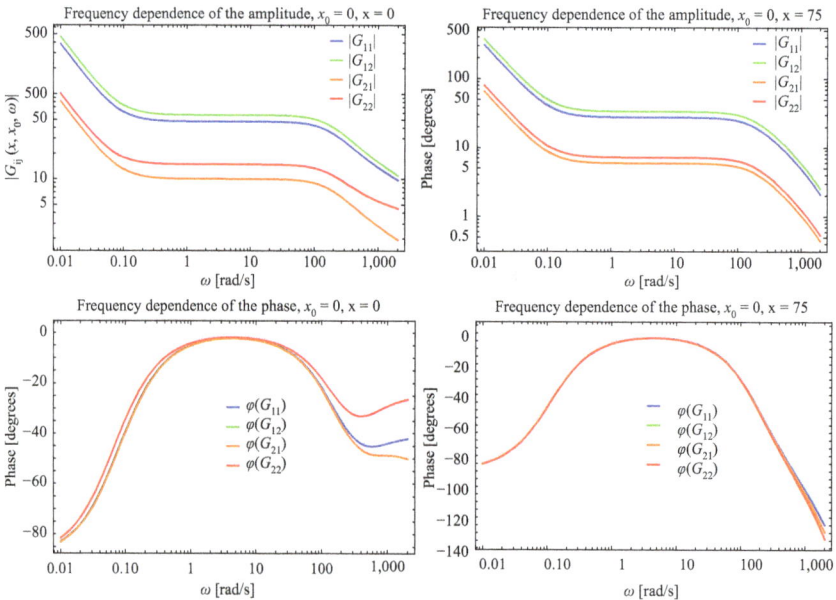

Figure 2.6   *Frequency dependence of the amplitude and the phase of the components of the Green's function for $x_0 = 0$, with $x = 0$ (left column) and $x = 75$ cm (right column), respectively, in a commercial BWR*

positions: one at the position of the perturbation (left column), and one at 75 cm, which is outside the range of the local component (right column). For a better comparison with the one-group results and with the behaviour of the zero reactor transfer function $G_0(\omega)$, the phase is shifted by $180°$. This is because the zero reactor transfer function connects the reactivity to the flux fluctuations, whereas the elements of the Green's matrix transfer the *negative* reactivity to neutron noise (cf. (2.83) and (2.79)–(2.80), which show that the G_{ij} are integrated with the fluctuations of the absorption cross sections with a positive sign, whereas for the fluctuations of the fission cross sections with a negative sign).

With this convention, the frequency dependence of both the amplitude and the phase resembles to that of the zero-power transfer function $G_0(\omega)$. There are though some deviations, which are due to the fact that the zero-power transfer function is related to the global component of the neutron noise, which does not contain the local component. The effect of the local component, whose upper break frequency is much higher than that of the global component (β/Λ) is felt only in the spatial neighbourhood of the perturbation, and beyond β/Λ. This is best seen in the phase behaviour at the location of the perturbation (lower left plot of Figure 2.6), and to a much smaller extent in the amplitude of G_{22}, at high frequencies. At the observation point x_0, 75 cm away from the perturbation, no effect is visible (right column of Figure 2.6).

It has to be added that the effect of the local component, even in the vicinity of the perturbation, is only seen in an LWR at unrealistically high frequencies, which are not encountered in practice. The only experimental observation of the effect of the upswing of the phase at high frequencies close to the perturbation was registered in a reactor oscillator experiment in the heavy water reactor (HWR) NORA in Kjeller, Norway ([40]; see also [41], Figure 9.4). This is because the break frequency of $G_0(\omega)$ in an HWR is an order of magnitude lower than in an LWR, which made it possible to notice the effect immediately above 10 rad/s. It is also seen that in the observation point 60 cm away from the perturbation, no effect of the local component is seen.

For later comparison with SMR results in Chapter 4, we also show results for the small system, analysed in [38]. This system was generated from the large commercial system by increasing the fission cross sections by 11%. The core height then became 88 cm. The same plots will be displayed as for the large commercial BWR.

The results for the space dependence of the amplitude and the phase of the components of the Green's matrix are shown in Figure 2.7. It is seen that this system behaves in a much more point kinetic way than the large commercial core. The space dependence of the components of the Green's function matrix follows very closely that of the static flux. The local peak in G_{22} is barely visible.

The spatial dependence of the amplitude and the phase of the fast and the thermal removal adjoints for the small system is shown in Figure 2.8. Interestingly, in the thermal removal adjoint, the peak is still well discernible, indicating that two phase flow diagnostics in small BWRs is still possible.

The frequency dependence of the amplitude and the phase of the four components of the Green's function for the small system are shown in Figure 2.9 for two

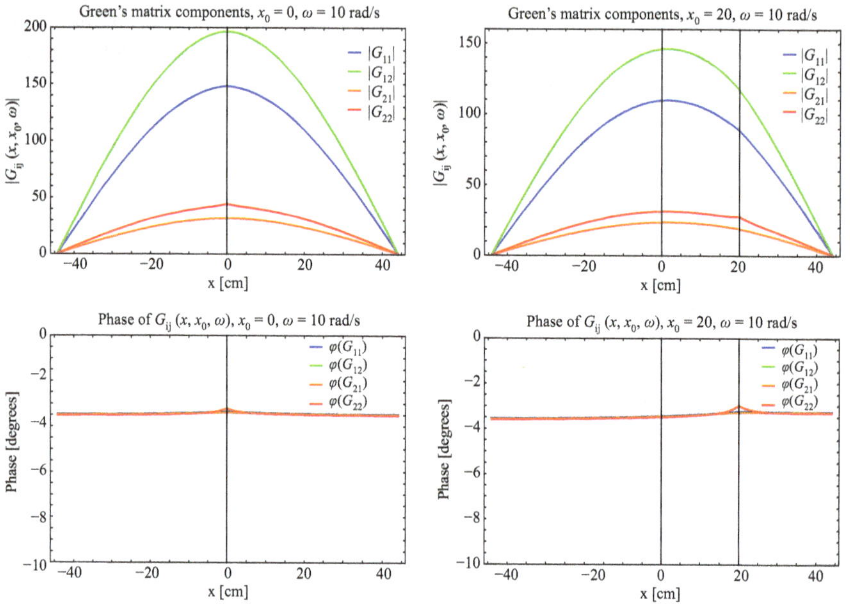

Figure 2.7   *Space dependence of the amplitude and phase of the components of the Green's function for $x_0 = 0$ (left column) and $x_0 = 20$ cm (right column), respectively, in a small BWR*

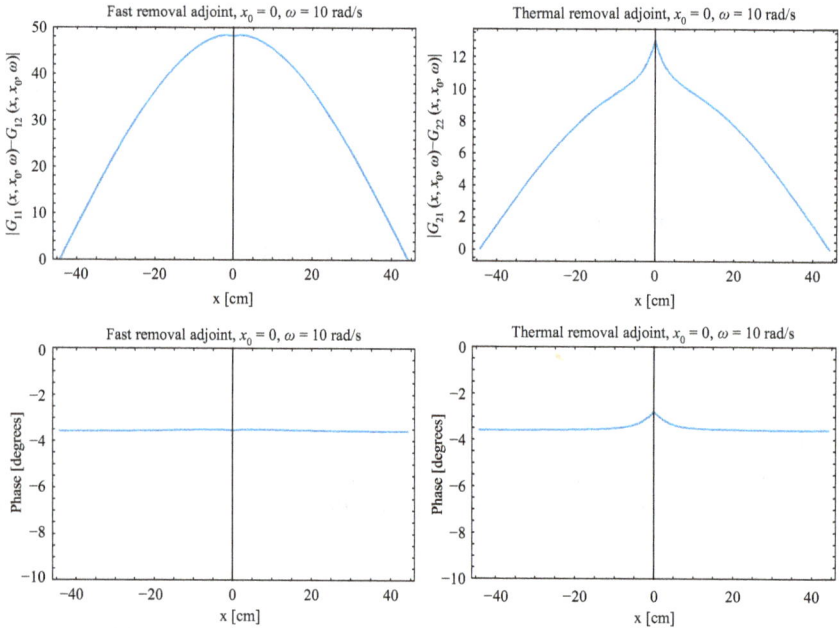

Figure 2.8   *Space dependence of the amplitude and the phase of the fast and thermal removal adjoints for $x_0 = 0$ in a small BWR*

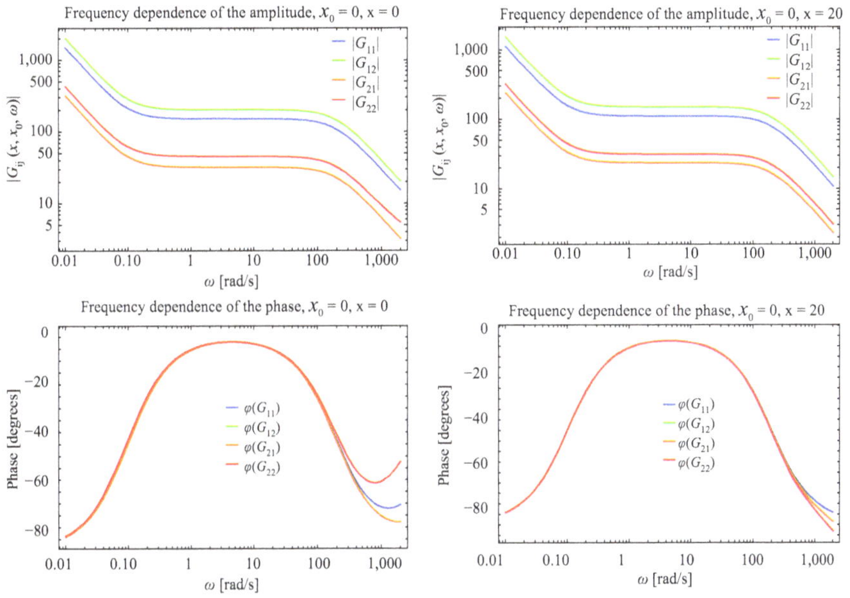

Figure 2.9 *Frequency dependence of the amplitude and the phase of the components of the Green's function for $x_0 = 0$, with $x = 0$ (left column) and $x = 15$ cm (right column), respectively, in a small BWR*

different detector positions: one at the position of the perturbation (left column), and one at $x = 20$ cm. In relative terms, i.e. in units of x/a, this is approximately to the same position inside the core as in the large system, but in absolute terms it is much closer to the perturbation. Therefore, although the local component appears with a smaller weight in the Green functions' components as in the large system, its effect still can be seen in the behaviour of the phase at high frequencies.

2.3.3.2 Two-group noise theory for fast systems

For fast reactors, which have a negligible thermal flux (see Figure 2.10), the traditional approach is not applicable. In order to use a two-group method, the model to be used has to be modified in several respects. To account for the dominantly fast neutrons in the system, the threshold energy between the slow and the fast group should be chosen much higher than in a thermal reactor. A practical suggestion is to use the threshold energy of the fissionable material, such as 1.35 MeV for ^{238}U [28]. With this choice of the energy separating Groups 1 and 2, the latter is definitely not thermal, not even "slow", but for reasons of convenience, we still often refer to Group 2 as "slow" or "epithermal".

Obviously, for this case the group constants need to be calculated in a different way when collapsing from a continuous energy or many-group transport code, and the structure of the data will be rather different from that of an LWR. The difference

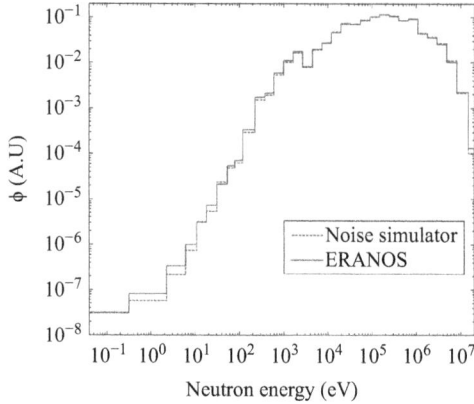

Figure 2.10   The energy spectrum of neutrons in a sodium-cooled fast reactor (SFR)

in the numerical values of the input data itself does not require the change of the formulae given in the previous subsection. There is, however, another difference, which will lead to a modification of the formulae. Namely, with such a choice of threshold energy, the fission source will appear in both groups, according to the corresponding fractions of the fission energy spectrum. This is actually valid for both the prompt and the delayed neutrons. In fact, with a high threshold, all delayed neutrons will appear in Group 2.

In order to maintain generality when allowing for the use of an arbitrary separation energy between the two groups, we assume the following input data:

χ_{p1} – fraction of the prompt neutrons that appear in Group 1;
χ_{p2} – fraction of the prompt neutrons that appear in Group 2;
χ_{d1} – fraction of the delayed neutrons that appear in Group 1;
χ_{d2} – fraction of the delayed neutrons that appear in Group 2.

In the static equations, the total spectral indices χ_1 and χ_1 will appear, which are given as

$$\chi_i = (1 - \beta)\, \chi_{p,i} + \beta\, \chi_{d,i}; \qquad i = 1, 2 \tag{2.106}$$

With these preliminaries, the static equations will read as

$$\begin{bmatrix} D_1 \nabla^2 - \Sigma_{a1} - \Sigma_R + \chi_1 \nu\Sigma_{f1} & \chi_1 \nu\Sigma_{f2} \\ \Sigma_R + \chi_2 \nu\Sigma_{f1} & D_2 \nabla^2 - \Sigma_{a2} + \chi_2 \nu\Sigma_{f2} \end{bmatrix} \begin{bmatrix} \phi_1(x) \\ \phi_2(x) \end{bmatrix} = 0 \tag{2.107}$$

Introducing the notations

$$\Sigma_1 = \Sigma_{a1} + \Sigma_R - \chi_1 \nu\Sigma_{f1} \tag{2.108}$$

$$\Sigma_2 = \Sigma_{a2} - \chi_2 \nu\Sigma_{f2} \tag{2.109}$$

$$\Sigma_{R2} = \Sigma_R + \chi_2 \nu\Sigma_{f1}, \tag{2.110}$$

Equation (2.107) can be written in a form similar to (2.70), i.e.

$$
\begin{bmatrix} D_1 \nabla^2 - \Sigma_1 & \chi_1 \nu \Sigma_{f2} \\ \Sigma_{R2} & D_2 \nabla^2 - \Sigma_2 \end{bmatrix} \begin{bmatrix} \phi_1(x) \\ \phi_2(x) \end{bmatrix} = 0 \tag{2.111}
$$

From here, one obtains for the critical buckling and the spectral index the expressions

$$
B_0^2 = -\frac{1}{2}\left(\frac{\Sigma_1}{D_1} + \frac{\Sigma_2}{D_2}\right) + \frac{1}{2}\sqrt{\left(\frac{\Sigma_1}{D_1} + \frac{\Sigma_2}{D_2}\right)^2 + 4\frac{\chi_1 \nu \Sigma_{f2} \Sigma_{R2} - \Sigma_1 \Sigma_2}{D_1 D_2}} \tag{2.112}
$$

and

$$
\frac{\phi_1(x)}{\phi_2(x)} = \frac{\chi_1 \nu \Sigma_{f2}}{\Sigma_1 + D_1 B_0^2} \tag{2.113}
$$

The equations for the noise and the noise source, obtained from the time-dependent equations after linearisation, Fourier transform and eliminating the delayed neutron precursors, can be written in a condensed form as

$$
\begin{bmatrix} D_1 \nabla^2 - \Sigma_1(\omega) & \nu \Sigma_{f2}(\omega) \\ \Sigma_{R2}(\omega) & D_2 \nabla^2 - \Sigma_2(\omega) \end{bmatrix} \begin{bmatrix} \delta\phi_1(x,\omega) \\ \delta\phi_2(x,\omega) \end{bmatrix} = \begin{bmatrix} S_1(x,\omega) \\ S_2(x,\omega) \end{bmatrix} \tag{2.114}
$$

Equation (2.114) looks formally similar to (2.75), except that now also the removal cross section became dependent on the frequency. However, the definition of the variables appearing in the two equations is rather different. In (2.114), the following notations are used:

$$
\chi_{1,2}(\omega) = \chi_{p1,2}(1 - \beta) + \chi_{d1,2}\frac{\lambda \beta}{i\omega + \lambda} \tag{2.115}
$$

$$
\Sigma_1(\omega) = \Sigma_{a1} + \frac{i\omega}{v_1} + \Sigma_R - \chi_1(\omega)\nu\Sigma_{f1} \tag{2.116}
$$

$$
\Sigma_2(\omega) = \Sigma_{a2} + \frac{i\omega}{v_2} - \chi_2(\omega)\nu\Sigma_{f2} \tag{2.117}
$$

$$
\Sigma_{R2}(\omega) = \Sigma_R + \chi_2(\omega)\nu\Sigma_{f1} \tag{2.118}
$$

and

$$
\Sigma_{f2}(\omega) = \chi_1(\omega)\Sigma_{f2} \tag{2.119}
$$

The components of the noise source are given as

$$
S_1(x,\omega) = (\delta\Sigma_{a_1}(x,\omega) + \delta\Sigma_R(x,\omega) - \chi_1(\omega)\delta\nu\Sigma_{f1}(x,\omega))\phi_1(x)
$$
$$
-\chi_1(\omega)\delta\nu\Sigma_{f_2}(x,\omega)\phi_2(x) \tag{2.120}
$$

and

$$
S_2(x,\omega) = -(\delta\Sigma_R(x,\omega) + \chi_2(\omega)\delta\nu\Sigma_{f1}(x,\omega))\phi_1(x)
$$
$$
+ (\delta\Sigma_{a_2}(x,\omega) - \chi_2(\omega)\delta\nu\Sigma_{f2}(x,\omega))\phi_2(x) \tag{2.121}
$$

With these definitions, the roots of the characteristic equation of (2.114) read as

$$
\mu^2(\omega) = -\frac{1}{2}\left(\frac{\Sigma_1(\omega)}{D_1} + \frac{\Sigma_2(\omega)}{D_2}\right)
$$
$$
+\frac{1}{2}\sqrt{\left(\frac{\Sigma_1(\omega)}{D_1} + \frac{\Sigma_2(\omega)}{D_2}\right)^2 + 4\frac{\Sigma_{R2}(\omega)\,\nu\Sigma_{f2}(\omega) - \Sigma_1(\omega)\,\Sigma_2(\omega)}{D_1 D_2}}
$$
(2.122)

and

$$
\nu^2(\omega) = \frac{1}{2}\left(\frac{\Sigma_1(\omega)}{D_1} + \frac{\Sigma_2(\omega)}{D_2}\right)
$$
$$
+\frac{1}{2}\sqrt{\left(\frac{\Sigma_1(\omega)}{D_1} + \frac{\Sigma_2(\omega)}{D_2}\right)^2 + 4\frac{\Sigma_{R2}(\omega)\,\nu\Sigma_{f2}(\omega) - \Sigma_1(\omega)\,\Sigma_2(\omega)}{D_1 D_2}}
$$
(2.123)

The coupling constants $c_\mu(\omega)$ and $c_\nu(\omega)$ are formally the same as in the LWR case, i.e.

$$
c_\mu(\omega) = \frac{\Sigma_1(\omega) + D_1\,\mu^2(\omega)}{\nu\Sigma_{f2}(\omega)}
$$
(2.124)

and

$$
c_\nu(\omega) = \frac{\Sigma_1(\omega) - D_1\,\nu^2(\omega)}{\nu\Sigma_{f2}(\omega)}
$$
(2.125)

but the frequency-dependent cross sections are now being defined differently from those for the case of the thermal reactors. With the aforementioned notations and definitions, the formulae for the components of the Green's function matrix will be formally identical with those of the case for LWRs, i.e. (2.89)–(2.96).

It can also be mentioned that certain differences in the notations notwithstanding, the aforementioned formulae for fast reactors are consistent with those of the thermal system. Substituting $\chi_{p1} = \chi_{d1} = 1$ and $\chi_{p2} = \chi_{d2} = 0$, they become identical with those for the thermal system. This is valid for both the Green's function, as well as for the noise source. This also means that one can use the same code for the calculation of the noise for both thermal and fast systems. The difference will be present only in the input data, and naturally the results need to be interpreted on the basis of the difference between the two types of systems.

One might ask how the $\delta\phi_1$ and $\delta\phi_2$, calculated by such a model of two-group neutron noise in fast systems, correspond to measured signals in fast reactors with fast and thermal detectors, or what detector responses correspond to the noise in Group 2. For the fast noise, the results of the model should be reasonably well correspond to the noise measured with e.g. a fission chamber, with the same energy threshold for fast fission of the core, which is also used in the generation of the group constants. For the epithermal noise in Group 2, the answer is not so straightforward. However, using a thermal detector with a $1/v$ cross section, it should be able to measure the epithermal noise, only with a much reduced efficiency than for thermal neutrons. This means that the ratio between the fast and epithermal noises, supplied by the model, will not be characteristic to the ratio of the amplitude of the measured signals. On the other hand, this is the case even in the traditional two-group theory of thermal systems, where the efficiency of fast and thermal neutron detectors is even more different. Correct amplitude ratios can only be calculated with the help of noise simulators in either case (thermal and fast cores, respectively), and using a large number of energy groups for fast cores.

However, the ratio of the amplitudes between the two groups is not a crucial, or even important question. Our goal here is not the exact quantitative reconstruction of the detector response, rather to characterise of the dynamic response of the various systems for the different perturbations. This latter is revealed by the spatial behaviour of the Green's functions, and that of the generated neutron noise. The character of the spatial response, on the other hand, is determined by the relative weights of the different components of the noise, primarily by the relative weight of the point kinetic (reactivity) component as compared to the space-dependent and local components. A prerequisite of getting this relative ratio correct is that the critical state of the core is correctly estimated by the two-group theory used. In this respect taking the energy separation between the groups to be equal to the threshold for fast fission in the core is the optimum choice [28]. This is the basis of the expectation that the dynamic behaviour of fast cores with the two-group model of the noise, suggested in this chapter, will be captured correctly.

As an illustration of the Green's function for a fast system, we show here results for a medium-size fast reactor. This is just a demonstration example, in order to bring out the main differences between a fast system and the thermal ones. The data were taken from a fluoride MSR model with zero fuel velocity, with a relatively soft spectrum. The separation energy between the groups was taken to be 0.5 MeV. The data used are shown in Table 2.2. In addition to the same data as for the thermal system, the prompt and delayed fission neutron spectral indices are also given.

With the aforementioned data, the critical thickness of a 1D slab would be 748 cm. To have a smaller core size, the absorption cross sections were readjusted to result in a critical slab width equal to 440 cm, and all the dynamic calculations were made with these modified group constants.

One immediate difference to the thermal systems that one can see in the data is in the kinetic parameters β and λ. The delayed neutron fraction is appreciably smaller than in the thermal reactor, whereas its decay constant is much larger. This means that the system response is faster, and its amplitude is larger, than in a thermal system.

The static fluxes in Groups 1 and 2 are shown in Figure 2.11. An immediate difference compared with the thermal systems shown before is that the flux in Group 2 is higher than in Group 1, which is opposite in thermal systems. This may appear counterintuitive; however, one has to take into account that Group 2 has a different definition and energy range than in thermal systems. The main difference is that with the energy threshold at 1.35 MeV, the group velocity v_2 in a fast system is one or two orders of magnitude higher than in a thermal system. Another, equally important difference is that, unlike in the thermal systems, a non-negligible part of the prompt, and a substantial part of the delayed fission neutrons appear directly in Group 2. These two facts together explain the reversed relationship between the amplitudes of the fluxes in the two groups.

Table 2.2   Data of a 1D model of a medium-sized
fluoride salt fast core

Parameter	Fast	Thermal
D [cm]	1.75736	1.04221
$\nu\Sigma_f$ [cm^{-1}]	0.00405111	0.00724195
Σ_a [cm^{-1}]	0.00253184	0.00748216
Σ_R [cm^{-1}]	0.0303787	
v [cm/s]	$1.460223 \cdot 10^9$	$1.04137 \cdot 10^8$
χ_p	0.879597	0.120403
χ_d	0.371764	0.628236
β	0.00308517	
λ [s^{-1}]	0.325805	

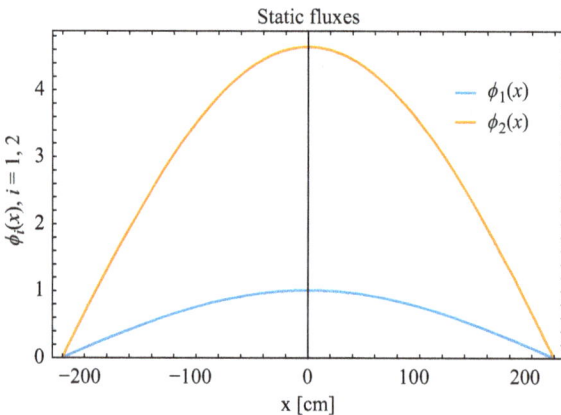

Figure 2.11   *The static fluxes in a fluoride molten salt fast core*

Quantitative results for the space and frequency dependence of the Green's function and the fast and thermal removal adjoints are shown in Figures (2.12)–(2.14), in the same order as for the thermal system. Despite the previous discussion on the difference between the removal adjoint of LWRs and the propagation adjoint of liquid metal and molten salt cores, here we still calculate the removal adjoint, i.e. using $\alpha_1 = 1$ and $\alpha_2 = -1$. Even this choice will bring out interesting differences. The proper choice of the coefficients α_i, i = 1,2, corresponding to the concrete reactor types will be made in Chapters 3 and 4 when discussing the various cores.

The space dependence of the amplitude and the phase of the components of the Green's matrix for two different positions x_0 of the perturbation is shown in Figure 2.12. One can immediately see some basic differences as compared to a large thermal system. Despite the larger size of the fast core, the space dependence is less pronounced than for the commercial BWR, and more similar to a point kinetic behaviour. Accordingly, the amplitude of the local component is much smaller.

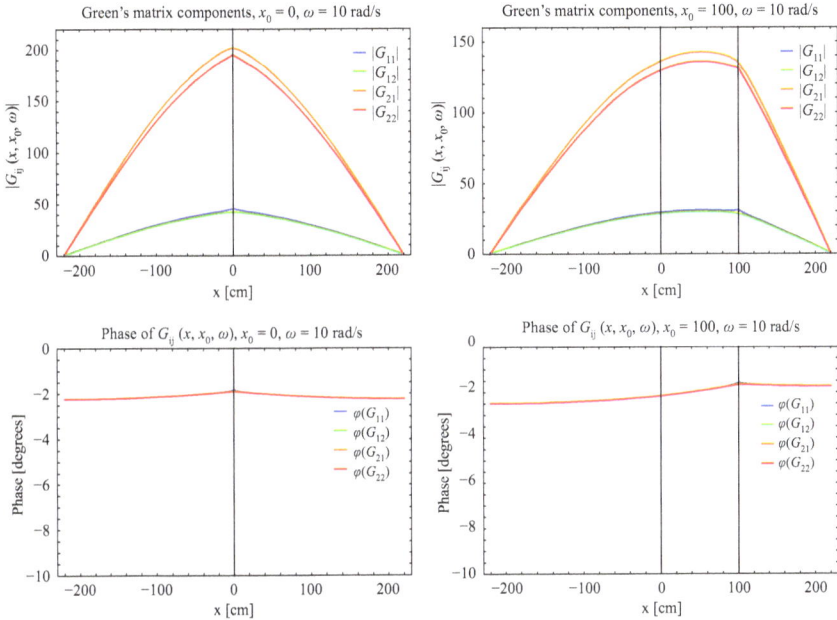

Figure 2.12   *Space dependence of the amplitude and the phase of the components of the Green's function for $x_0 = 0$ (left column) and x_0 100 cm (right column), respectively, for a medium-sized fluoride salt fast core*

A more substantial difference is that the role of the groups, Groups 1 and 2, is swapped. In the thermal system, the amplitude of the fast Green's function components G_{11} and G_{12} was significantly larger than that of the thermal components G_{21} and G_{22}. Here it is the other way round, although with the remark that for the fast system, Groups 1 and 2 are not equivalent with the fast and thermal groups of a thermal system. The swapping concerns also the local component; whereas for the thermal system it appears in G_{22}, in the present fast system it appears in G_{11}. The reasons for this are closely related to those for the static flux, described before.

The faster dynamics of the system is also reflected in the fact that the phase is everywhere closer to zero than in the corresponding spatial points of the thermal system. From the practical point of view this is not significant; the space dependence of the phase of the Green's function is too weak in either case that it should be useful for diagnostical purposes. The information in the phase will be more pronounced for the noise induced by vibrating components (fuel or control rods), as it will be seen soon.

Further new characteristics are revealed in the removal adjoints. This is a conceptual case because, as mentioned earlier, in a metal cooled core or in an MSR, the transfer of the propagating density fluctuations is transferred by the propagation adjoints, and not by the removal adjoints. Those will be calculated for the concrete cases treated in Chapters 2 and 3. Here, the removal adjoints for a fast system are

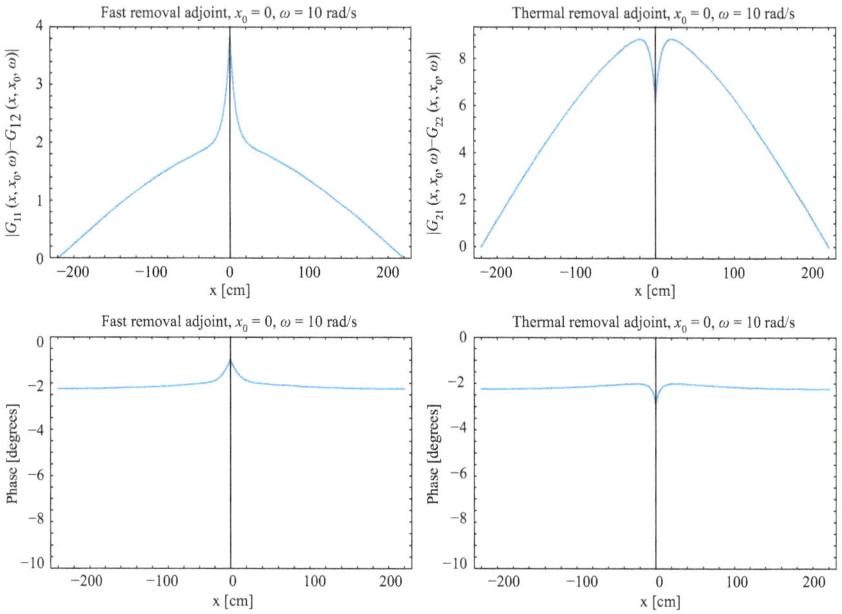

Figure 2.13   Space dependence of the amplitude and the phase of the fast and thermal removal adjoints for $x_0 = 0$ for a medium-sized fluoride salt fast core

shown only to bring out certain further differences between the thermal and fast systems, which give some insight.

The amplitude and phase of the fast and thermal adjoints are shown in Figure 2.13. The swapping of the characteristics is obvious also here, namely that it is the fast removal adjoint that has the local peak, and the thermal removal has the local dip. The phase behaviour shows a similar difference between the two systems.

There is one additional difference, which arises from the changed characteristics of the amplitudes. Namely, it is not only that the amplitude relationships between the fast and the thermal components of the Green's matrix are reversed, but also the internal amplitude relationships between G_{11} and G_{12} on the one side, and those between G_{21} and G_{22}, are also reversed. As a consequence, while in Figure 2.5 the phase of $G_{11} - G_{12}$ as well as $G_{21} - G_{22}$ were shown, respectively, in Figure 2.13 the phases of $G_{12} - G_{11}$ as well as $G_{22} - G_{21}$ are shown. A peculiar characteristic is that although the local component appears with a very small weight in G_{11}, it has a quite large relative weight in the fast removal adjoint, due to the fact that the amplitudes of the global part of G_{11} and G_{12} are much closer to each other than that of G_{21} and G_{22} in a thermal system. However, it is good to keep in mind that for fast systems, it is not the removal adjoint which is relevant, rather the propagation adjoint, which will not have the same amplifying effect on the local component as the removal adjoint.

The frequency dependence of the amplitude and the phase are shown in Figure 2.14. Whereas the form of the curves resembles qualitatively to those of

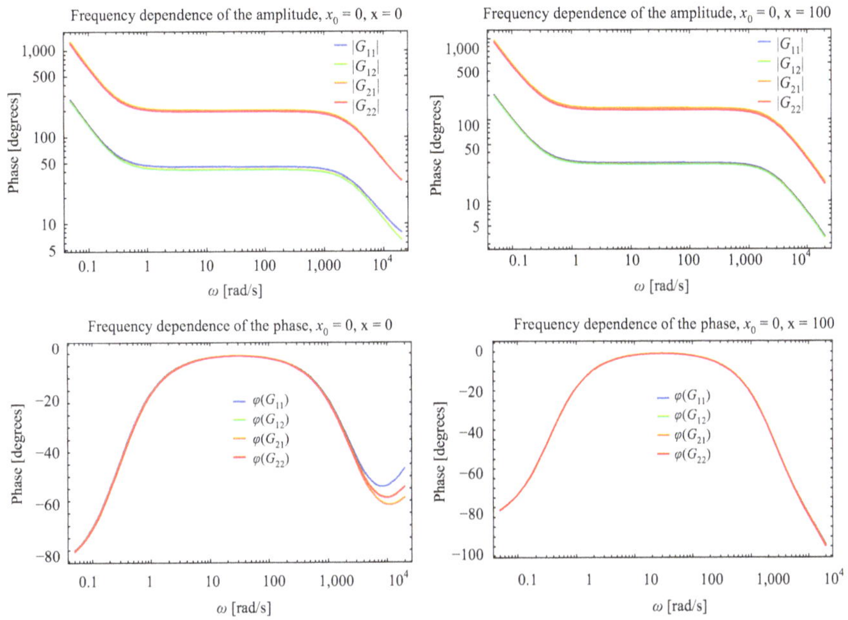

Figure 2.14   *Frequency dependence of the amplitude and the phase of the components of the Green's function for $x_0 = 0$, with $x = 0$ (left column) and $x = 60$ cm (right column), respectively, in a conceptual fluoride salt fast core*

the thermal system, a significant shift of the lower and upper break frequencies to higher frequencies can be observed. The shift of the lower break frequency is a consequence of the larger delayed neutron decay constant λ, whereas the higher upper break frequency is due to the much larger neutron velocities in both groups. The final consequence is that the plateau region, where the amplitude is constant and the phase is close to zero, is much wider than the plateau in light water systems (not to mention heavy water systems, where the upper break frequency is even much lower than in light water systems). This means that the simplification of the noise equations by neglecting the imaginary part of the noise and its frequency dependence, which makes it possible to Fourier-invert the arising results in the frequency domain to time domain, will have a wider applicability in conceptual studies.

One final, more subtle difference is that the absolute value of the Green's functions in this conceptual fluoride salt fast core is about a factor 4 larger than in the large BWR, despite the larger size of the fast core. This is essentially due to the smaller value of the delayed neutron fraction β. As was seen in the simple one-group treatment, the point kinetic response of the core is proportional to the zero reactor transfer function $G_0(\omega)$, whose approximate value at plateau frequencies is equal to $1/\beta$. A smaller delayed neutron fraction hence leads to the increase of the amplitude of the noise, which is beneficial from the point of view of noise diagnostics. This aspect is encountered also when we discuss MSRs with moving fuel, since in those reactors a fraction of the delayed neutrons is lost because they decay in the outer loop, outside the core.

2.3.4   Diagnostics of perturbations

There is a significant expertise what regards the various types of noise sources in Gen-II reactors, namely what type of anomalies can arise and what are the possibilities for their detection and quantification. Gen-IV reactors and SMRs might have new types of anomalies and corresponding diagnostic needs, but since there is no operational experience with those reactors yet, we need to restrict ourselves to the type of perturbations we are aware of in the present fleet of reactors. These type of noise sources, with corresponding anomalies, may occur in the majority of the next generation systems too. Even for the possible new, hitherto not known type of anomalies, the response of the core to the known types of perturbations may give some guidance how to detect and quantify those new types, and what the chances of successful diagnostics are.

Some of the known types of noise sources do not require knowledge the neutronic transfer function of the system, and hence neither solving of corresponding diffusion or transport equations. Such perturbations are

- vibrations of the core barrel in PWRs; on a more general note, deformations of the core, such as the "flowering effect" in SFRs;
- baffle jetting in PWRs;
- thimble tube vibrations in PWRs;
- vibrations and impacting of detector tubes in BWRs;
- monitoring BWR instability;
- determination of the MTC.

The last two out of this list are not directly anomalies, rather examples of the use of neutron noise analysis for determining operational parameters with non-intrusive methods, and the emphasis is on monitoring so that the corresponding parameters (the decay ratio and the MTC) do not shift into undesired values. Although none of the aforementioned problems are directly based on knowledge of the reactor transfer function, its knowledge still helps when elaborating the signal analysis methods for evaluating noise measurements for the problems.

In this book, we concentrate on three major types of perturbations, which are generic to the existing LWRs. We then in Chapters 3 and 4 examine the response of the core of several of the next generation systems to these perturbations, and even give an assessment of the possibility of their diagnosis, i.e. quantification. These three perturbation types are as follows:

- an absorber of variable strength
- a vibrating fuel pin or an absorber rod
- propagating perturbations of the density of the coolant.

The treatment of these perturbations and the possibility of their diagnostics will be given in the continuation, and illustrated on the BWR system presented in the fore-going. The modelling of these perturbation is based on constructing a simple model of the spatial and frequency dependence of the cross-section fluctuations that represent the perturbation, which can be used in the analytical model. Simple models of all three of the aforementioned types have long been used in the literature, but mostly in a one-group treatment. The actual magnitude of the cross-section fluctuations is uninteresting, the diagnostics is always based on the relationships between two or more detectors, which are invariant to a scaling of the strength of the cross-section fluctuations. This procedure is quite straightforward in a one-group model, in particular when only the fluctuation of only one cross section is taken into account. In a two-group treatment, and in particular when the fluctuations of several cross sections need to be taken into account, then the relative strength of the cross-section perturbations between the different cross sections, and even for any single cross section, the relative strengths between the fast and the slow group need also to be fixed.

In general, the calculation of the relative magnitude of the response of the cross sections to a certain perturbation, such as change of temperature or density of the coolant, would require involved assembly-level transport or Monte Carlo calculations. To simplify the task, we assume that the relative amplitudes or the cross-section fluctuations are proportional to the stationary values of the cross sections involved. The absolute amplitude is left as a free parameter, and hence the calculated noise is presented in arbitrary units. This, however, does not affect the spatial or frequency dependence of the noise, on which the diagnostics is based.

2.3.4.1   Variable strength absorber

For localised perturbations, such as the variable strength absorber, it is customary to use the so-called Feinberg–Galanin model, which assumes that the absorber can be represented by a spatial Dirac-delta function [4]. In one-group theory, one assumes that there is a thin absorber rod in the homogeneous system at position x_p, which can be described as

$$\Sigma_a(x) = \gamma_0 \, \delta(x - x_p) \tag{2.126}$$

Here γ_0 is the so-called Galanin constant. The perturbation then consists of the fact that the Galanin factor becomes time-dependent, and performs small fluctuations around the equilibrium value with a time-dependent amplitude $\gamma(t)$, i.e.

$$\Sigma_a(x, t) = (\gamma_0 + \gamma(t)) \, \delta(x - x_p) \tag{2.127}$$

from which it follows that in the frequency domain one has

$$\delta\Sigma_a(x, \omega) = \gamma(\omega) \, \delta(x - x_p) \tag{2.128}$$

For simplicity, it is usually assumed that the effect of the static rod on the criticality equations can be neglected, because they give only a second-order contribution to the noise, hence for both the critical flux and the calculation of the elements of the Green's matrix, the expressions derived for the homogeneous system without the static rod can be used.

At a given frequency, the spatial dependence of the neutron noise by a variable strength absorber, which is represented by infinitely thin absorber, is directly related to the elements of the Green's function, i.e. the neutronic transfer function of the system. Therefore, physical realisations of such an absorber, called "the reactor oscillator", were used in the past for experimentally determining the space- and frequency dependence of the transfer function [41]. A reactor oscillator is a well-known object, which does not require any diagnostics. Perturbations of the type that can be modelled by an absorber of variable strength are local flow channel instabilities, as the one that occurred in the Swedish BWR Forsmark-1 in 1996–97 [21]. Such an instability appears in BWRs due to unseated fuel elements. The oscillation of the flow usually occurs at a given frequency, and the diagnostic task is to determine the position of the instability, i.e. to identify the unseated fuel element.

For this reason, the frequency dependence of the fluctuating factor $\gamma(\omega)$ can be neglected, because it will be taken at a given single frequency. However, in two-group theory, we need to specify two different factors, which will be denoted as γ_{a1} and γ_{a2}. As mentioned before here, a suitable choice of these factors is

$$\gamma_{a1} = \frac{\Sigma_{a1}}{\Sigma_{a1} + \Sigma_{a2}}; \qquad \gamma_{a2} = \frac{\Sigma_{a2}}{\Sigma_{a1} + \Sigma_{a2}} \qquad (2.129)$$

It is reasonable to assume that neither the fission nor the removal cross sections are affected. Although this is straightforward for the traditional reactor oscillator, it is not obvious for the flow channel instabilities in BWRs, where the removal cross section is also perturbed. Accounting for also the fluctuations of the removal cross section would be completely straightforward, but it only would change the ratio between the noise in the fast and the slow group. It will have larger effect when it comes to propagating perturbations, as it will be seen later.

With (2.129), one has

$$\overrightarrow{\delta\Sigma_a}(x, t) = \begin{bmatrix} \gamma_{a1} \\ \gamma_{a2} \end{bmatrix} \delta(x - x_p) \qquad (2.130)$$

and thus the noise sources in Groups 1 and 2 have the form

$$S_1(x, \omega) = \gamma_{a1}\, \delta(x - x_p)\, \phi_1(x) \qquad (2.131)$$

$$S_2(x, \omega) = \gamma_{a2}\, \delta(x - x_p)\, \phi_2(x) \qquad (2.132)$$

Finally, the induced neutron noise in Groups 1 and 2 is given as

$$\delta\phi_1(x, \omega) = \gamma_{a1}\, G_{11}(x, x_p, \omega)\phi_1(x_p) + \gamma_{a2}\, G_{12}(x, x_p, \omega)\phi_2(x_p) \qquad (2.133)$$

and

$$\delta\phi_2(x, \omega) = \gamma_{a1}\, G_{21}(x, x_p, \omega)\phi_1(x_p) + \gamma_{a2}\, G_{22}(x, x_p, \omega)\phi_2(x_p) \qquad (2.134)$$

It is seen in the formulae here, that there is no visible difference whether the noise source components belong to a thermal or a fast system. The difference for these two systems is embedded solely in the elements G_{ij} of the Green's matrix. The difference between the expression of the perturbation whether it concerns a thermal or a fast system will be seen in the case of a vibrating fuel pin.

An illustration of the neutron noise induced by a variable strength absorber is shown in Figure 2.15 for the large BWR for two different absorber positions x_p. Similarly to the individual components of the Green's matrix, the amplitude and the phases are shown separately.

Not surprisingly, the space dependence of the noise is very similar to that of the Green's functions themselves, as seen in Figure 2.4, since they were also very similar to each other, and the noise is a linear combination of these components in

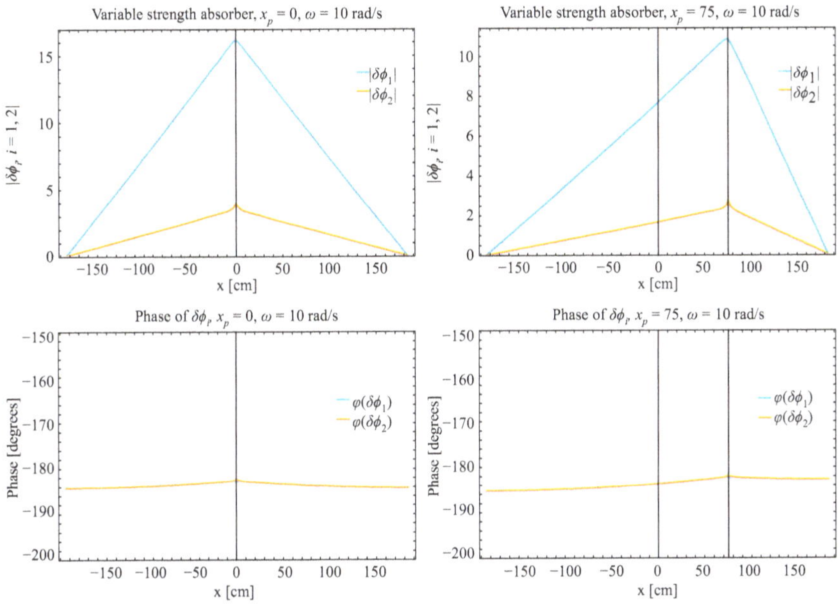

Figure 2.15   *Space dependence of the amplitude and the phase of the noise in the two energy groups, induced by an absorber of variable strength for $x_p = 0$ (left column) and $x_p = 75$ cm (right column), respectively, in a commercial BWR*

both groups. The small local component is present only in the noise in the thermal group. Since the induced noise is out of phase with the oscillation of the absorber strength, plus the physical phase delay of the transfer, the phase is below $-\pi$ (180°).

The phase changes very moderately throughout the core, when plotted with a scale between -200 and -150 degrees, and obviously it is not useful for the localisation of the noise source. The amplitude, on the other hand, shows a significant deviation from the point kinetic behaviour, which makes it a promising tool for the localisation of the position of the absorber. In the present 1D model, one way of investigating the sensitivity of the localisation procedure is to check the sensitivity of the relationships between the amplitude and phase of the signals of two detectors at fixed positions. Naturally, in a real case this is a 2D problem, where at least three detector signals are needed, and the procedure of the localisation is performed based on 2D neutronic calculations and with machine learning methods. Our goal here is not to suggest a concrete procedure for the localisation in a real case, rather to compare the possibilities of localisation between the different reactor systems.

In a practical case, it is not the Fourier transforms of the time signals that are used, rather the auto power spectral density (APSD) and the cross power spectral density (CPSD) of the two detectors. These spectral quantities can be obtained from the frequency-dependent signals according to the Wiener–Khinchin theorem as

$$APSD_{\delta\phi}(x, \omega) = |\delta\phi(x, \omega)|^2 \qquad (2.135)$$

and

$$\text{CPSD}_{\delta\phi_1,\delta\phi_2}(x_1, x_2, \omega) = \delta\phi(x_1, \omega)\, \delta\phi(x_2, \omega)^* \tag{2.136}$$

The aforementioned general definitions here apply to the noise in both energy groups separately.

The localisation is then performed from the signals of two detectors, one close to the left boundary ($x_1 = -0.9a$) and the other close to the right boundary ($x_2 = 0.9a$). Since the neutron noise in any given point x will also be a function of the position x_p of the absorber, we can indicate it explicitly. Denoting

$$\delta\phi_{i,x_j}(x_p) \equiv \delta\phi_i(x_j, x_p, \omega) \tag{2.137}$$

where $i = 1, 2$ stands for the energy group and $j = 1, 2$ for the detector position, we define the amplitude localisation function $\Delta_i(x_p)$ in group i as

$$\Delta_i(x_p) = \frac{\text{APSD}_{i,x_1}(x_p)}{\text{APSD}_{i,x_2}(x_p)} \tag{2.138}$$

From this ratio the scaling factor, which represents the global amplitude of the perturbation, disappears, only the relative contributions from the two groups remain, which are reasonably well approximated with the ratio of the static values.

In a similar manner, the localisation based on the phase differences between the two detectors in energy group i will be based on the phase localisation function $\theta_i(x_p)$. This is defined as

$$\theta_i(x_p) = \text{Arg}\left\{\text{CPSD}_{i,x1,x2}(x_p)\right\} = \text{Arg}\left\{\delta\phi_{i,x1}(x_p)\, \delta\phi_{i,x2}(x_p)^*\right\} \tag{2.139}$$

It is obvious that also this quantity is insensitive to the unknown scaling factor which is present in the amplitude of the cross-section perturbations.

The dependence of the amplitude and phase localisation functions $\Delta_i(x_p)$ and $\theta_i(x_p)$, respectively, on the position of the absorber of variable strength, is shown in Figure 2.16. For better visibility, the ratio of the APSDs is shown with a logarithmic scale.

The figure shows that the ratio of the APSDs of two peripherally placed neutron detectors depends relatively strongly on the position of absorber, which means that the chances of locating it are good. This is mostly due to the fact that a large commercial reactor is considered here, in which the space-dependent effects are more pronounced than in a small system, which behaves much more point kinetically. The localisation curves show no difference between the fast and the thermal detectors, except when the perturbation is close enough to one of the detectors, that the effect of the local component can be felt. This is again the consequence of the fact that both the static fluxes and the components of the Green's function have similar spatial dependence for the two groups. Hence both fast and thermal detectors are equally suitable for the task, although in thermal reactors the larger efficiency of thermal detectors makes them much better suited for the task.

The lower figure also shows that although the phase of the CPSD displays a slight linear dependence on the position of the absorber, the overall change in the phase is too small to be useful. This will be even more so in small systems, which behave in a more point kinetic way.

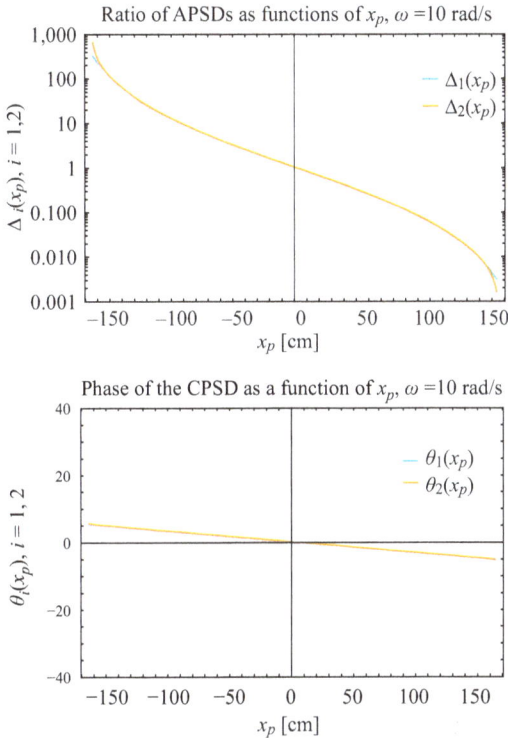

Figure 2.16   *Dependence of the amplitude and phase localisation functions on the position of the variable strength absorber in a commercial BWR*

The situation in a realistic case, which requires a 2D treatment, is less favourable than shown here. In 2D, the Green functions decay faster as a function of the distance from the perturbation than in an equivalent 1D model. In principle this should be favourable, because it means stronger space dependence, and thus larger deviation from point kinetics. The problem is rather that only detectors relatively close to the perturbation can extract the noise due to this particular perturbation from the general background noise. This was seen in the case of the localisation of the channel instability in Forsmark-1 [21]. Hence for a successful localisation it is not sufficient to have detectors placed peripherally. This again underlines the need of good instrumentation in next generation systems.

2.3.4.2   Vibrating fuel pin

Vibrations of core internals, notably control rods and fuel assemblies or fuel pins are one of the basic noise sources and related diagnostic concerns in PWRs. It is also one of the prominent examples where the applicability of the neutron noise-based localisation procedure was proved at a power plant during operation [20,27].

Flow induced vibrations of fuel pins or subassemblies can be expected to occur also in liquid metal cooled Gen-IV reactors, both in lead cooled fast reactors [42] as well as in sodium cooled ones [43]. Some LFR designs will contain wire-wrapped fuel elements, to maintain the radial gap in the rod bundle. Such fuel rods will be prone to turbulent flow induced vibrations. Although the study in [42] found that under normal conditions the amplitude of these vibrations will be stable and would not lead to impacting of the fuel pins, the task of diagnostics will be to monitor whether anomalous strong vibrations occur which might lead to impacting and pin-to-pin contact. In sodium cooled fast reactors it is the slender structure of the subassembly which is prone to flow induced vibrations through the sodium flow, as well as the possibility of so-called leakage flow [43].

In this section, we discuss the neutron noise caused by the vibrations of a thin fuel rod and the possibilities of its localisation. Treating a vibrating absorber, such as a control rod or control pin, goes in a completely analogous manner as that of the variable strength absorber, using the Feinberg–Galanin model. The difference is in the spatial form of the perturbation and of course in the cross sections affected.

We assume that the static fuel pin is described by a spatial Dirac-delta function at position x_p, which is the equilibrium (static) position, around which the vibrations take place. The perturbation is represented by the movement of the fuel pin around its equilibrium position with a random amplitude $\varepsilon(t)$. Following the same strategy as with the absorber of variable strength, one can write [4]

$$\overrightarrow{\delta \Sigma_f}(x, t) = \begin{bmatrix} \gamma_{f1} \\ \gamma_{f2} \end{bmatrix} \left[\delta(x - x_p - \varepsilon(t)) - \delta(x - x_p) \right] \tag{2.140}$$

Here, similarly to the case of the absorber of variable strength, the relative amplitudes γ_{f1} and γ_{f2} are defined as

$$\gamma_{f1} = \frac{\nu \Sigma_{f1}}{\nu \Sigma_{f1} + \nu \Sigma_{f2}}; \qquad \gamma_{f2} = \frac{\nu \Sigma_{f2}}{\nu \Sigma_{f1} + \nu \Sigma_{f2}} \tag{2.141}$$

Actually, the vibration of a fuel pin also incurs the vibration of the absorption cross sections, but we only consider the fluctuation of the fission cross sections. Accounting for absorption would possibly change the ratio between the magnitudes of the noise in the two groups, but not their spatial dependence.

Assuming that the vibration amplitude $|\varepsilon(t)|$ is very small compared to the size of the system, one can perform a one-term Taylor expansion in (2.140), after which a temporal Fourier transform yields the perturbation in the fission cross sections in the form

$$\overrightarrow{\delta \Sigma_f}(x, \omega) = -\varepsilon(\omega) \begin{bmatrix} \gamma_{f1} \\ \gamma_{f2} \end{bmatrix} \delta'(x - x_p) \tag{2.142}$$

With this, the noise sources take the form

$$S_1(x, \omega) = \varepsilon(\omega) \chi_1(\omega) \delta'(x - x_p) \left[\gamma_{f1} \phi_1(x) + \gamma_{f2} \phi_2(x) \right] \tag{2.143}$$

$$S_2(x, \omega) = \varepsilon(\omega) \chi_2(\omega) \delta'(x - x_p) \left[\gamma_{f1} \phi_1(x) + \gamma_{f2} \phi_2(x) \right] \tag{2.144}$$

Introducing the notation

$$\Phi(x) = \gamma_{f1} \phi_1(x) + \gamma_{f2} \phi_2(x) \tag{2.145}$$

using (2.143) and (2.144) in (2.83) leads to

$$\delta\phi_1(x,\omega) = -\varepsilon(\omega)\Big[\chi_1(\omega)\left\{G'_{11}(x,x_p,\omega)\Phi(x_p) + G_{11}(x,x_p,\omega)\Phi'(x_p)\right\}$$
$$+\chi_2(\omega)\left\{G'_{12}(x,x_p,\omega)\Phi(x_p) + G_{12}(x,x_p,\omega)\Phi'(x_p)\right\}\Big] \quad (2.146)$$

and

$$\delta\phi_2(x,\omega) = -\varepsilon(\omega)\Big[\chi_1(\omega)\left\{G'_{21}(x,x_p,\omega)\Phi(x_p) + G_{21}(x,x_p,\omega)\Phi'(x_p)\right\}$$
$$+\chi_2(\omega)\left\{G'_{22}(x,x_p,\omega)\Phi(x_p) + G_{22}(x,x_p,\omega)\Phi'(x_p)\right\}\Big] \quad (2.147)$$

As mentioned here, the spatial derivatives of the Green's functions have to be taken with respect to the second argument at the equilibrium fuel rod position x_p.

The amplitude and the phase of the neutron noise in a large commercial BWR (divided by the vibration amplitude $\varepsilon(\omega)$), induced by the vibrations of a fuel rod at two different equilibrium rod positions are shown in Figure 2.17.

The spatial structure of the amplitude and the phase of the neutron noise induced by vibrating absorbers and fuel rods has been extensively discussed in the literature [32,44–46], hence only some general points will be taken up here. The main point is that the structure of the noise is a result of the interplay between the point kinetic (reactivity driven), the (global) space-dependent and the local terms. While the point

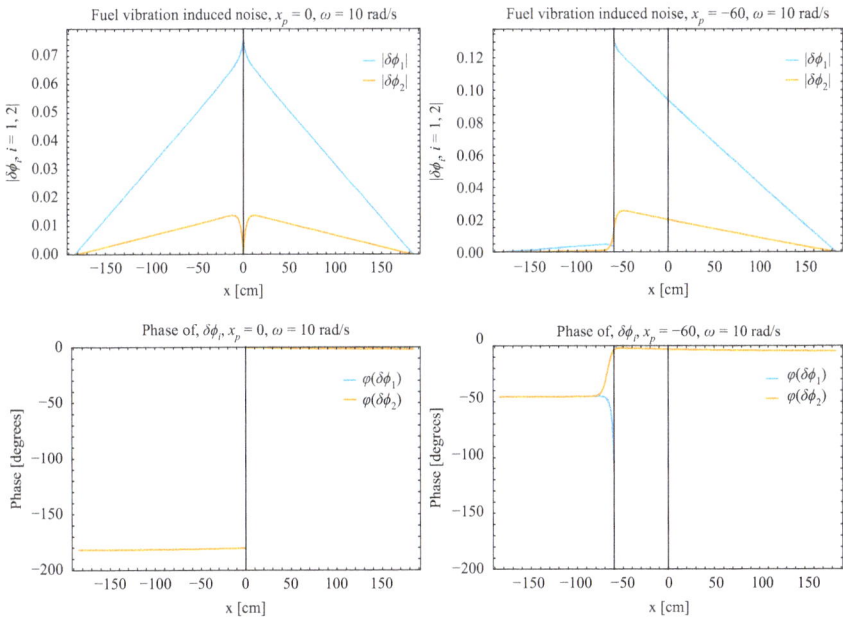

Figure 2.17 *Space dependence of the amplitude and the phase of the noise in the two energy groups, induced by a vibrating fuel rod, for $x_p = 0$ (left column) and $x_p = -60$ cm (right column), respectively, in a commercial BWR*

kinetic response is in-phase everywhere in the core, the space-dependent and local components are roughly out of phase at the two sides of the vibrating component. For a fuel pin left from the centre of the core, which is vibrating in a positive flux gradient, the reactivity noise is in-phase with the vibrations, whereas the other two components are in-phase to the right of the vibrating fuel and out-of-phase on the left to the fuel pin. This positive and negative interference, respectively, explains the difference (discontinuity) of the amplitude of the noise at the two sides of the vibrating rod.

For the central absorber, the reactivity component in first order of the vibration amplitude is zero; hence, the phase is opposite at the two sides – 0 on the right half and $-180°$ on the left side of the core. In the amplitude, a small local peak is seen in the noise in the fast group, whereas the presence of the local component leads to a peculiar behaviour in the noise in the thermal group. The reasons are explained in [46], the essence is the fact that the thermal noise is zero at the position of the fuel pin because in a thermal reactor all fission neutrons are born in the fast group, i.e. $\chi_2(\omega)$ is zero. The absence of the local peak in the thermal noise is due to the fact that the local peak is present only in G_{22}, but as is seen from (2.147), this does not enter the expressions, since $\chi_2(\omega) = 0$. This will be certainly different for fast systems, where $\chi_2(\omega)$ is not zero, due to the different choice of the energy threshold between the groups. This will be seen clearly in the forthcoming.

From the aforementioned, it is also seen that the spatial dependence of the vibration induced noise is a much more sensitive function of the position of the absorber than for the variable strength absorber, at least within a certain region. This is not due to the transfer properties of the system, which are the same for both perturbations, rather to the way the perturbation induces the noise. The noise in a given point is a result of the interference between the point kinetic component on one hand, and the global and the local components on the other. Out of the latter two, the local component can be decoupled from the discussion since the localisation in a practical cases will be made by detectors further away from the vibrating fuel element, except in a few exceptional cases. The interference between the point kinetic and space-dependent components, regarding both the amplitude and the phase, is controlled by the relative amplitude of the reactivity component. This latter, due to the strong space dependence of the gradient of the static flux, is a sensitive function of the equilibrium position of the fuel rod. This gives a much better possibility for the localisation of a vibrating fuel or absorber pin than that of a perturbation whose strength fluctuates but not its position, such as an absorber of variable strength, or a local thermohydraulic channel instability. Localisation of a vibrating component is therefore more likely to be possible even in small systems.

The feasibility of localisation can be investigated with the same amplitude and phase localisation functions $\Delta_i(x_p)$ and $\theta_i(x_p), i = 1, 2$ functions, (2.138) and (2.139), respectively, as in case of the absorber of variable strength. The localisation curves for the large commercial BWR are shown in Figure 2.18.

The dependence of the amplitude localisation function on the position of the fuel pin is still quite strong, but more complicated than for the variable strength absorber. Similarly to the latter, in a logarithmic scale it is an odd function, due to the symmetry

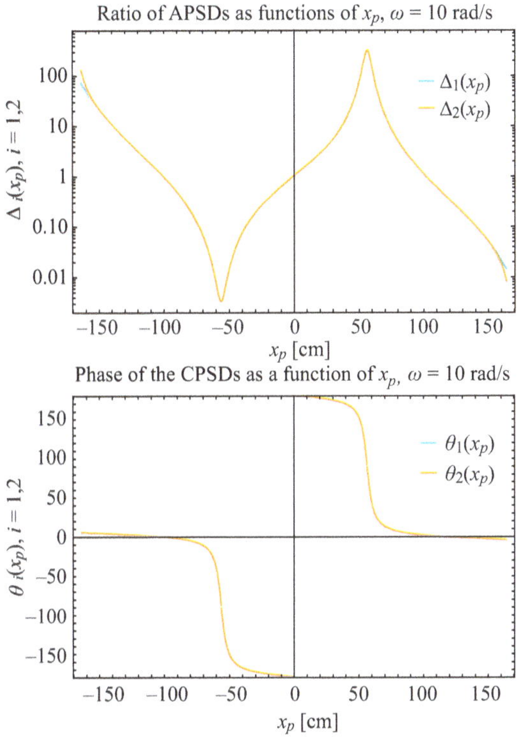

Figure 2.18   Dependence of the amplitude and phase localisation functions on the position of the vibrating fuel rod in a commercial BWR

of the system. However, in contrast, it is not a monotonic function of the position of the fuel pin. The minima and maxima corresponds to the positions of the fuel pin where the negative interference between the reactivity term and the space-dependent term is maximal, such that the APSD of one of the detector signals is close to zero. The position of the minimum at −60 cm and the maximum at 60 cm confirms well with the upper right plot of Figure 2.17, which shows that at x = −60 cm, the noise is close to zero everywhere left to the fuel pin, and hence also at the detector at x_1, which is close to the left boundary of the core.

The non-monotonic behaviour also means that the localisation from the amplitude localisation curve alone is not unambiguous, for certain values of $\Delta(x_p)$ there are three possible positions x_p which yield the same value. To some extent this ambiguity can be resolved with the help of the phase, which is much more sensitive to the position of the fuel pin, than it was in the case of the variable strength absorber. Based on the figures, the ambiguity cannot be fully eliminated. However, the consequence is only that the procedure can point out possibly two probable positions, out or which the relevant one can be selected by other methods. Also, in a 2D setting, which is the relevant case in practice, three or more detectors will be used, which may reduce the ambiguity; further, due to the faster spatial relaxation of the noise in a real

2D problem, detectors only will convey information about the closest to them. Such questions can be investigated in a realistic case with numerical noise simulators.

It is remarkable that the localisation curves shown in Figure 2.18 show considerable resemblance to the one-group results to those in Figures 7(a)–(c) of [44]. The reason why this is interesting is that the calculations in [44] concern the noise induced by a vibrating absorber, whereas the ones shown in this Section concern a vibrating fuel pin. Both the reactivity term and the space-dependent term of the noise are in opposite phase between these two type of perturbations. However, interestingly, the fact that for a given position of the vibrating component, on which side of the vibrating perturbation the reactivity and space-dependent terms have constructive or destructive interference, does not depend on the type of the vibrating component. And since the structure of the localisation curves is determined largely by the dependence of the magnitude of the interference, the localisation curves are the same whether one considers a vibrating absorber or a fuel pin. The difference in the vibration noise manifests itself in the very minor contribution of the local component, seen here, and its much larger magnitude for the vibrating absorber, which was shown in [46]. The local component, in its turn, does not affect the form of the localisation curves except in the neighbourhood of the detectors.

2.3.4.3   Vibrating absorber

Since, as it is obvious, in the MSRs the case of a vibrating fuel rod is not relevant, in those we investigate the noise induced by a vibrating absorber. Although the formulae, shown next shortly, are quite similar, the situation is still somewhat different. Especially in thermal reactors, vibrations of a thermal absorber lead to a large peak in the noise in the thermal group. This was demonstrated in [46], and it was noticed already in the pioneering experiments with the historic Clinton Pile** oscillator experiments, and treated first theoretically in [32].

The calculation of the neutron noise by a vibrating absorber goes on the same lines as for the vibrating fuel pin. Assuming that only the absorption cross section is perturbed, the perturbation can be described, analogously to (2.140) as

$$\overrightarrow{\delta\Sigma_a}(x,t) = \begin{bmatrix} \gamma_{a1} \\ \gamma_{a2} \end{bmatrix} \left[\delta(x - x_p - \varepsilon(t)) - \delta(x - x_p) \right] \tag{2.148}$$

The relative amplitudes γ_{1a} and γ_{2a} were defined in (2.129). Then, assuming again the smallness of the vibration amplitude $|\varepsilon(t)|$, one performs again a one-term Taylor expansion in (2.148), after which a temporal Fourier transform yields the perturbation in the vibrating absorption cross sections in the form

$$\overrightarrow{\delta\Sigma_a}(x,\omega) = -\varepsilon(\omega) \begin{bmatrix} \gamma_{a1} \\ \gamma_{a2} \end{bmatrix} \delta'(x - x_p) \tag{2.149}$$

The noise sources will read as

$$S_1(x,\omega) = -\varepsilon(\omega)\,\gamma_{a1},\delta'(x - x_p)\,\phi_1(x) \tag{2.150}$$

$$S_2(x,\omega) = -\varepsilon(\omega)\,\gamma_{a2}\,\delta'(x - x_p)\,\phi_2(x) \tag{2.151}$$

**The Graphite Reactor, also called the X-10 reactor, in Oak Ridge National Laboratory (ORNL). It was shut down permanently in 1963 and was designated as a National Historic Landmark in 1965.

and the noise in the two groups is obtained as

$$\delta\phi_1(x,\omega) = \varepsilon(\omega)\left[\gamma_{a1}\left\{G'_{11}(x,x_p,\omega)\phi_1(x_p) + G_{11}(x,x_p,\omega)\phi'_1(x_p)\right\}\right.$$

$$\left. + \gamma_{a2}\left\{G'_{12}(x,x_p,\omega)\phi_2(x_p) + G_{12}(x,x_p,\omega)\phi'_2(x_p)\right\}\right] \quad (2.152)$$

and

$$\delta\phi_2(x,\omega) = \varepsilon(\omega)\left[\gamma_{a1}\left\{G'_{21}(x,x_p,\omega)\phi_1(x_p) + G_{21}(x,x_p,\omega)\phi'_1(x_p)\right\}\right.$$

$$\left. + \gamma_{a2}\left\{G'_{22}(x,x_p,\omega)\phi_2(x_p) + G_{22}(x,x_p,\omega)\phi'_2(x_p)\right\}\right] \quad (2.153)$$

As before, the spatial derivatives of the Green's functions have to be taken with respect to the second argument at the equilibrium fuel rod position x_p.

2.3.4.4   Propagating perturbations

Neutron noise due to propagating perturbations occurs in all liquid cooled reactors, but the main interest in it has been primarily due to the use of in-core neutron noise in BWRs. In the early stages of power reactor noise diagnostics it was noted that one can extract information about some properties of the two-phase flow, primarily the transit time of the flow between axially placed in-core neutron detectors in the same detector tube [47]. The phase of the CPSD between two detectors showed a linear dependence on the frequency, the slope being equal to the transit time of the flow between the two detectors. It was recognised that the reason why determination of the transit time is possible was the existence of the local component of the noise.

Although the existence of such a local component was shown already in the pioneering work of Weinberg and Schweinler [32] with one-group theory with a slowing down kernel, the formalism was involved and not optimal for diagnostic purposes. It was shown by Kosály [48] that a simple two-group theory is sufficient for a transparent and practical reconstruction of the local component, which is suitable for diagnostic purposes. This is also one incentive for using two-group theory extensively in this monograph.

Propagating perturbations occur also in PWRs. These are due to density fluctuations of the coolant due to inlet temperature fluctuations, and represent a significantly smaller magnitude of the perturbation. As a rule, these are concealed by the background noise or noise from other perturbations. Nevertheless, these relatively small fluctuations of the neutron flux still play a role. Partly, with advanced signal processing techniques, which can suppress the relative weight of the local component, it was possible to determine the velocity of the coolant in a PWR by in-core neutron detectors [49,50]. And partly, the neutron noise induced by the temperature fluctuations in the coolant can be used for the determination of the MTC with noise methods [7].

Propagating perturbations will with all likelihood also play a role in the noise diagnostics of next generation nuclear systems. Some of the planned SMRs will be of the BWR type, such as the BWRX-300, developed by GE Hitachi Nuclear Energy. In these reactors application of in-core neutron noise for monitoring two-phase flow parameters will be equally relevant than in the existing BWRs. In the fast reactors of the Gen-IV type, the molten lead and sodium cooled reactors will have a high heat capacity, in which temperature and density fluctuations will play a similar role than in PWRs. The effect of those temperature fluctuations in a hard spectrum will of course have a smaller effect on the neutron noise.

More importantly, it is anticipated that two phase, or at least two-component flow may also occur in the metal coolant of fast reactors. This is for technological reasons, such as homogenisation, or removal of noble gases (He, Xe, etc.), primarily from MSRs. There is a large body of investigations of appearing a vapour phase in metal cooled reactors, summarised in [51], which also lists several methods of diagnosing and quantifying bubbles in liquid metal. However, there is no mentioning of the possibility of using neutron noise at all for this purpose. Therefore, we investigate also the possibilities of detecting the presence of two-phase flow in liquid metal cooled reactors and MSRs.

In the classic problem of BWR in-core noise diagnostics, it is customary to assume that the presence of bubbles in the water coolant primarily affects the removal cross section. With only the removal cross section perturbed, this leads to the concept of the fast and thermal "removal adjoint". Here, we keep a broader scope, and allow for the perturbation of both the absorption and the removal cross section, again with the relative weight of the individual perturbations being equal to the relative magnitude of the individual static cross sections. This approach is reasonable for water or liquid metal cooled reactors with solid fuel.

MSRs are a class for themselves from the point of view of propagating perturbations. Apart from the possibility of the appearance of bubbles, propagating perturbations with a detectable effect on the neutron noise can be expected in MSRs with both a thermal and a fast spectrum. Inlet temperature fluctuations in the fluoride salt of the MSR affect the fuel directly, unlike in a PWR were a good part of the temperature fluctuations of the coolant is transferred to the fuel by heat conduction. Changes in the fission cross sections will represent a more significant perturbation than density fluctuations in molten lead or sodium. In addition, any non-homogeneity in the distribution of the fuel in the molten salt constitutes a direct perturbation which will propagate through the core.

On the other hand, a moving fuel means that, unlike in reactors with a solid fuel the delayed neutron precursors will also move, and will decay at a point different from where they were born, even outside the core. This fact changes the physics and the dynamics of MSR which calls for a modification of the underlying description. A full analytical description in the same framework with the same transparency as the one used so far is not possible for the MSR, and hence some simplifications and approximations are necessary. The diagnostics of MSRs will be treated separately at a later section of this chapter.

As is customary for treating propagating perturbations, in the 1D model one treats the axial dependence of the reactor along the z-coordinate, where the core boundaries are at $z = 0$ and $z = H$. The propagating character of the perturbation is described such that the fluctuations entering the reactor at $z = 0$ will propagate unchanged along the axial height of the core. This is expressed by the relationship that for any affected cross section one has

$$\delta\Sigma(z,t) = \delta\Sigma\left(z = 0,\ t - z/v\right),\tag{2.154}$$

where v is the velocity of propagation[††]. After a temporal Fourier transform this has the form

$$\delta\Sigma(z,\omega) = \delta\Sigma(0,\omega)\, e^{-\frac{i\omega z}{v}} \tag{2.155}$$

It is usually assumed that the inlet temperature fluctuations are white noise, i.e. $\delta\Sigma(0,\omega) = \text{const}$. With the assumption that absorption cross sections in both groups plus the removal cross sections are perturbed, the noise sources in the two groups will be given as follows:

$$S_1(z,\omega) = [\delta\Sigma_{a1} + \delta\Sigma_R]\,\phi_1(z)\, e^{-\frac{i\omega z}{v}} \tag{2.156}$$

and

$$S_2(z,\omega) = -\delta\Sigma_R\,\phi_1(z)\, e^{-\frac{i\omega z}{v}} + \delta\Sigma_{a2}\,\phi_2(z)\, e^{-\frac{i\omega z}{v}}$$
$$= [-\delta\Sigma_R + c_\mu\delta\Sigma_{a2}]\,\phi_1(z)\, e^{-\frac{i\omega z}{v}} \tag{2.157}$$

Assuming that the amplitude of the cross-section fluctuations are proportional to their static values, and introducing a normalisation w.r.t. to the sum of the cross sections, noise sources can be written as

$$\begin{bmatrix} S_1(z,\omega) \\ S_2(z,\omega) \end{bmatrix} = \begin{bmatrix} \alpha_1 \\ \alpha_2 \end{bmatrix} \phi_1(z)\, e^{-\frac{i\omega z}{v}} \tag{2.158}$$

If only the absorption and the removal cross sections are affected by the density fluctuations of the coolant, the coefficients α_1 and α_2 are defined as

$$\alpha_1 = \frac{\Sigma_{a1} + \Sigma_R}{\Sigma_{a1} + \Sigma_{a2} + \Sigma_R}; \qquad \alpha_2 = \frac{c_\mu\Sigma_{a2} - \Sigma_R}{\Sigma_{a1} + \Sigma_{a2} + \Sigma_R} \tag{2.159}$$

Actually, as it will be seen later, parameters α_i, $i = 1,2$ depend on the type of the reactor considered. In graphite moderated systems the slowing down happens in the graphite, hence the removal cross section is not affected by the density fluctuations of the coolant. For the eight different reactor types considered, there will be five different definitions of these parameters. In the gas cooled fast reactor the effect of the density fluctuations of the coolant are negligible, hence the effect of propagating perturbations will not be considered.

By using (2.158) in (2.83) leads to the expressions for the noise in the two groups as

$$\delta\phi_1(z,\omega) = \int_0^H \Psi_1(z, z_0, \omega)\,\phi_1(z)\, e^{-\frac{i\omega z_0}{v}}\, dz_0 \tag{2.160}$$

and

$$\delta\phi_2(z,\omega) = \int_0^H \Psi_2(z, z_0, \omega)\,\phi_1(z)\, e^{-\frac{i\omega z_0}{v}}\, dz_0 \tag{2.161}$$

[††]Not to be mixed up with the one-speed neutron velocity.

Here, using the terminology defined previously, the propagation adjoints Ψ_1 and Ψ_2 are given as

$$\Psi_1(z, z_0, \omega) = \alpha_1 G_{11}(z, z_0, \omega) + \alpha_2 G_{12}(z, z_0, \omega) \tag{2.162}$$

and

$$\Psi_2(z, z_0, \omega) = \alpha_1 G_{21}(z, z_0, \omega) + \alpha_2 G_{12}(z, z_0, \omega) \tag{2.163}$$

Thus it is seen that in the case of the fluctuations of more cross sections than only the removal one, Ψ_1 and Ψ_2 take over the role of the fast and the removal adjoints, defined for thermal LWRs. Indeed, if only the fluctuations of the removal cross section are accounted for, then from (2.159) one has

$$\alpha_1 = 1; \qquad \alpha_2 = -1 \tag{2.164}$$

and hence Ψ_2 is exactly identical with the thermal adjoint of [38], even though it is composed from the elements of the direct Green's matrix. The approximation (2.164) is a rather good one for LWRs, but definitely does not hold for liquid metal cooled fast reactors. The structure of the propagation adjoints of (2.162) and (2.163) might therefore be significantly different from that of the fast and thermal adjoint of LWRs. When investigating the dynamic transfer properties of all non-LWR-type cores, the removal adjoints (2.162) and (2.163) will be plotted, since it is their structure, in particular the magnitude of the local components, that will decide the possibilities of diagnosing propagating perturbations.

A way of quantifying this possibility is to investigate the phase of the CPSD between two axially placed detectors as a function of the frequency. Assuming two axially displaced detectors at positions z_1 and z_2, this is given as

$$\varphi(\omega) = \text{Arg}[\text{CPSD}_{z_1,z_2}(\omega)] = \text{Arg}[\delta\phi_1(z_1, \omega)\,\delta\phi_1(z_2, \omega)^*] \tag{2.165}$$

If the contribution from the local component is sufficiently high, then the phase is a linear, or quasi-linear function of the frequency, and its slope is proportional to the transit time of the propagation between the two detectors. If the global or point kinetic term dominates, the phase is close to zero, and no linear behaviour can be seen. The practicality of diagnosing the propagating perturbations will be determined by the linearity of the phase.

An illustration is shown in Figure 2.19. The figure shows the dependence of the phase as a function of frequency for the large commercial BWR and the small system, both with the fast and the thermal neutron noise. The data for the calculations were taken from [38], and the results for the phase obtained from the thermal neutron noise show a good agreement with those in Figures 4 and 6 in [38], respectively. It is interesting that the phase of the cross spectrum between two fast neutron detectors follows that of the thermal detectors quite well up to about 10 Hz, despite that the local component is much less noticeable in the fast propagation adjoint than in the thermal one, and it is not a peak, rather a dip. For the small system, in which the local component is smaller for both the fast and the thermal adjoint, above 1 Hz the phase from the thermal neutron noise shows still a good agreement with the theoretical phase $\varphi = -2\pi f\tau$, where $\tau = (z_2 - z_1)/v$ is the transit time. The phase from the fast neutron noise, although globally showing a linearly decreasing behaviour, behaves

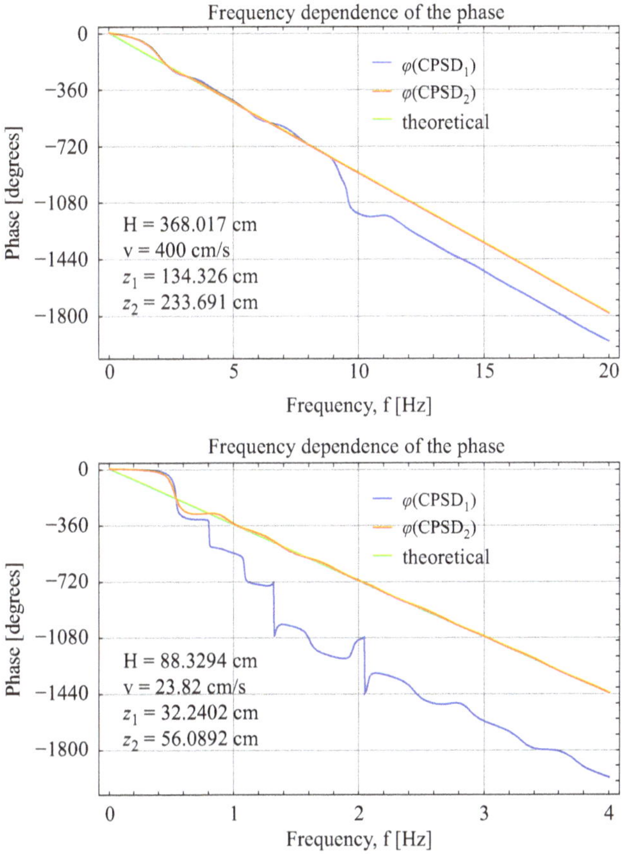

Figure 2.19   Dependence of the phase of the cross-correlation on frequency between two axially displaced detectors for propagating perturbations in a large (top) and a small (bottom) BWR

rather irregularly, and is not suitable for the determination of the transit time. One has to add that the plots shown here are idealised theoretical ones, exempt from all extra noise contained in the signals from other sources.

The aforementioned plots are based on calculations for light water BWRs, where the perturbation was only attributed to the fluctuations of the removal cross sections, which yield $\alpha_1 = 1$ and $\alpha_2 = -1$. If the fluctuations of other cross sections need also be taken into account, that will influence the amplitude and phase of the propagation adjoints Ψ_1 and Ψ_2, and hence in the end also the possibility of using the in-core neutron noise to characterise the propagating perturbations. Such effects will be discussed in connection with the concrete treatment of the various Gen-IV reactors and SMRs in the next two chapters, using the tools developed in the foregoing.

Some of the planned Gen-IV reactors and SMRs will be MSRs. As it was already mentioned, propagating perturbations are expected to play a major role in the noise diagnostics of those cores. However, MSRs have a significantly different physics and dynamics, due to the movement of the fuel and the delayed neutron precursors. The underlying equations describing the static and dynamic behaviour of MSRs are different from those used so far. In order to study the dynamics of MSRs, the theory needs to be modified. Because of these differences, in the next section, we develop a simplified theoretical framework to analyse the dynamic properties of MSRs. The model and the method elaborated in the next section are used for the analysis of the MSRs in the next two chapters.

2.3.5   *Noise source unfolding*

In the foregoing, the emphasis was on the calculation of the neutron noise by the three main perturbation types that have been the most commonly occurring in LWRs, and to some extent also in HWRs. The objective was to investigate the possibilities of diagnosing these types of perturbations in the planned Gen-IV and SMR systems, with their neutronic transport properties being different from those of recent reactors. As it was already illustrated in the previous sections, and will be further pursued in the next two chapters, general conclusions can be drawn on the chances, i.e. the efficiency and accuracy of the diagnostics, i.e. unfolding the relevant information, such as the position of a perturbation, form these investigations. However, it has not been discussed so far how such a quantitative diagnostics can be performed, there were no algorithms suggested for them. In other words, unfolding of source parameters from the induced noise, in knowledge of the transfer properties of the system, is an inverse task, whereas so far we have only dealt with the direct task, i.e. how to calculate the noise induced by a given perturbation. There are no standard methods of how to solve an inverse task, and the solution method is different from case to case. Therefore, in this section, we give a brief overview of how the unfolding of the parameters of the mentioned perturbations can be made, with some outlook of possible special circumstances for Gen-VI systems.

With regard to the solution methods of the inverse task, there is in principle no difference between the present Gen-II and Gen-III reactors and the next generation nuclear systems. On the other hand, significant developments took place in this respect with the entering of machine learning methods into the solution of inverse tasks. These developments will directly be made use of in the next generation systems. Since these methods are not yet widely used in current systems either, it is worth giving a summary of the status of these here.

Unfolding of the different perturbations requires different methods, as they also have different level of sophistications. We start with the simplest, the diagnostics of the propagating perturbations, where no sophisticated unfolding methods are needed, at least as long as only measurement of the transit time of the perturbation between two detectors is concerned. This is because no spatial aspects and corresponding treatment of spatial transfer properties need to be considered. The effect of the perturbation is measured in the radial core position where it occurs, and axially it propagates between the two detectors with known positions. The transit time of the propagation between the two detectors is determined from the slope of the phase of the CPSD as a function of the frequency (or, alternatively, from the peak of the temporal cross-correlation function, CCF(τ), see later), and this only requires signal processing methods, but no complicated algorithm. The success of the unfolding, i.e. determination of the transit time, depends on the goodness of the linearity of the phase, which at most requires some curve fitting procedure.

The goodness of the slope, i.e. its linearity, depends on the relative contribution of the local component (and its sufficiently fast spatial decay). This, in turn, depends partly on the transfer properties of the system, and partly on the nature of the perturbation, i.e. which cross section is affected the most by the perturbation. It is the first of these two questions which will be investigated in the next two chapters for a number of selected Gen-IV and SMR types. The second question, i.e. the variety of the possible propagating perturbations, and their quantitative effect on the cross sections, is much more difficult to quantify and investigate at the present state or matters, and its investigation is outside the scope of this book.

In short of knowledge on the possible character of perturbations in the next generation systems, one is left to guesses. One can though anticipate with some certainty, that the effect of propagating perturbations will not be as strong in the Gen-IV systems as in the classical case of two-phase flow in BWRs. Two-phase flow in LWRs represents primarily a strong perturbation of the removal cross section, and as it was seen, fluctuations of the removal cross section amplifies the relative weight of the local peak of the neutronic transfer.

Neither of these advantageous circumstances can be expected in metal cooled fast reactors or MSRs. Even the possible presence of voids, which is anticipated [51], will only affect the absorption cross section that will constitute a significantly smaller perturbation. In MSRs, void will affect also the fission cross sections, which is a stronger perturbation, but still without the same amplifying effect as the perturbation of the removal cross section that, by being transferred through the difference between two elements of the Green's matrix, will experience a stronger local component. In addition, as we will see, in certain thermal MSRs the same fortuitous combination of the Green's function components takes place as the one that amplifies the local component in a BWR.

A crude attempt was made in the next two chapters to account for the effect of possible perturbations on the cross sections involved in Gen-IV systems. The relative weight of the perturbation of the various cross sections, representative of the various reactors, was though rather ad-hoc, and not made with advanced assembly homogenisation codes. This work will have to be done when design details of the new systems becomes more concrete and public.

If the perturbation is not represented by voids, its effect will be even weaker. Temperature fluctuations of the liquid metal coolant will affect the reaction rates due to spectral effects, but his will be rather subtle, and are expected to be much smaller in fast reactors than in LWRs. But even in the latter, inlet temperature fluctuations in a PWR have a too small induced neutron noise to be useful for determination of the transit time.

For the determination of the transit time from neutron noise induced by temperature fluctuations of the coolant, one might take resource to advanced signal processing methods to improve the situation. Such a method was elaborated originally by Nishihara [52]. Nishihara's method was inspired by the so-called "cepstrum" technique, developed for seismic analysis, although his method is free from the approximations involved in the cepstrum method. The method is based on the peak of the cross-correlation between the two detector signals, but instead of calculating it as an inverse Fourier transform of the cross spectrum $CPSD_{12}(\omega)$, it calculates a function $\gamma_{12}(\tau)$ as the inverse Fourier transform of the CPSD divided by the square root of the APSDs of the two signals:

$$\gamma_{12}(\tau) = FFT^{-1} \left\{ \frac{CPSD_{12}(\omega)}{\sqrt{APSD_1(\omega)\,APSD_2(\omega)}} \right\} \tag{2.166}$$

This method was simplified and used for PWRs in [49,50]. Instead of using the peak of the aforementioned defined $\gamma_{12}(\tau)$ function, they used the so-called impulse response function (IRF). The IRF is defined as the inverse Fourier transform of the CPSD between the two detectors, divided by the APSD of one of the detectors:

$$IRF(\tau) = FFT^{-1} \left\{ \frac{CPSD(\omega)}{APSD(\omega)} \right\} \tag{2.167}$$

Dividing the CPSD with the autospectrum of one of the detectors, or with the square root of the product of the APSDs of the two detectors largely reduces the weight of the global component, primarily that of the reactivity or point kinetic component, and sharpens the otherwise wide peak of the CCF, making it suitable for the accurate determination of the transit time even for small perturbations with a large global component. An illustration is shown in Figure 2.20, showing the original cross-correlation function, as well as the IRF [50].

For completeness it is worth mentioning that coolant flow velocity measurements are possible also with other methods than neutron noise. Some of these are mentioned in Chapter 8, such as with ultrasonic flowmeters or with correlating signals of thermocouples. A particular possibility for liquid metal cooled reactors is to

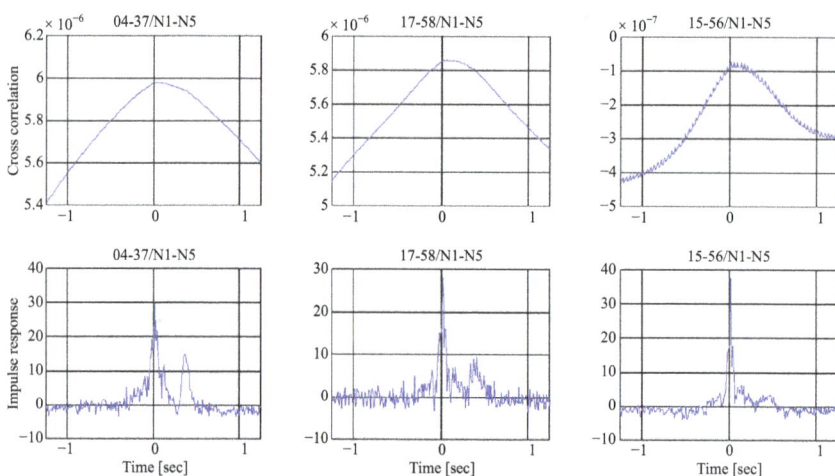

Figure 2.20   Measured cross-correlation (top) and impulse response (bottom) functions between detectors with a transport time of approximately 0.4 s. From Reference [50].

use electromagnetic measurement methods, such as the direct current or the eddy current flowmeters. However, the use of these methods for in-core measurements is problematic, they can best be used on ex-core flow loops.

In the cases mentioned so far, no algorithmic unfolding was necessary other than advanced signal processing. Estimating the transit time assumes only the existence and sufficient weight of the local component, but no knowledge of its functional form or spatial range ("field of view") is necessary to know. This is the case also if the objective is not a transit time, just the detection of incipient boiling or unexpected occurrence of voids for any reason. The situation becomes different if more involved type of diagnostics is needed, such as determining the void fraction from in-core neutron noise measurements. This is a relevant, and to this date not fully solved question even in BWRs, which has been discussed in the literature [53,54]. Recent work shows that by knowing the range (field of view) of the local component, and the void velocity at the detector position, then the void fraction, in particular in the case of low void content with sparse bubbles, can be determined [55]. This is now a substantially more involved unfolding question. Apart from the need of knowing the spatial structure of the local component, the determination of the void velocity at the detector position is not trivial either, at least if the void fraction has an axial profile, due to which the velocity is also axially dependent. The transit time is an integral of the reciprocal of the axially dependent void velocity. To extract the velocity profile, access to several transit time is necessary, usually four independent transit times [56]. Work is going on regarding this method for BWRs, and it can also be relevant for metal cooled and MSR-type Gen-IV reactors.

More involved unfolding procedures are also necessary for the diagnostics of localised noise sources inside the core, such as localisation of a variable strength absorber (such as the case of a channel-type instability), or a vibrating fuel pin/assembly or a control rod. In such cases an algorithm is necessary to unfold the position and possible other attributes of the perturbation, which is based on the measured signal and the theoretical relationship between the perturbation and the induced noise, which in turn also requires the knowledge of the neutronic transfer function.

The localisation curves shown in the previous section, which were meant to indicate the sensitivity of the amplitude ratios and phases of the CPSD between two detectors, could give an impression that measured values of these could be used to find the location of the perturbation from the position of the value of the localisation curves which agrees with the measured values. In reality the situation is much more complicated. Both of these inverse problems are 2D, i.e. the location of the suspected channel instability or vibrating rod needs to be searched on the whole horizontal cross section of the core. This means not only that at least three detectors are needed, but also that both that the calculation of the transfer function, as well as the expression for the induced noise, becomes more complicated. This latter is particularly valid for the vibration problem, where the perturbation itself becomes 2D, having a larger degree of freedom than in the simple 1D case [57]. Under such circumstances, construction of an unfolding algorithm becomes much more difficult.

Until relative recently, unfolding methods were based on simple analytical models. This is partly because an analytical solution of the direct task, which is only possible in simple models, expedites the elaboration of an inversion algorithm. And partly, because calculation of the noise in other than simple analytical models was not possible. No numerical codes were developed which could calculate the space–frequency-dependent neutron noise with the same spatial resolution in two-group diffusion theory in real inhomogeneous cores as the ICFM codes do for the static flux. In addition, a numerical modelling of the various perturbations, which in an analytical model can be represented by a Dirac-delta function that cannot directly be applied in numerical methods, was also needed. These circumstances restricted the unfolding of the noise source to the use of simple models of bare homogeneous cores. This fact notwithstanding, even such simple models were used with success for locating both a channel-type instability in a BWR [21], and an excessively vibrating control rod in a PWR [20]. This is because, in contrast to the ICFM codes, the demands on the accuracy of the calculations, as well as the dependence on the microstructure of the flux and the boundary conditions, are much milder for noise problems than in a criticality calculation.

The last two decades or so have seen a major breakthrough in this area, from which the diagnostics of Gen-IV reactors and SMRs will highly benefit. Namely, diagnostics and noise source unfolding became possible in real inhomogeneous systems. This is the result of two circumstances. One is that machine learning methods, primarily the use of ANNs, relieved, and in fact eliminated completely the need for analytical solutions of the direct task in order that the inverse task should be possible to perform. ANNs only need training patterns in pure numerical form, which can be generated by numerical solution of the direct task for a large number of cases. The second is that in the meantime, ICFM-quality codes, called "noise simulators", have been developed for the calculation of the noise in real inhomogeneous systems, together with methods of numerical simulations of the perturbations with high fidelity. Chapter 5 gives a description of how these codes are work, and also gives a survey about the international state of the art in this field.

The use of ANNs in reactor diagnostics and monitoring was pioneered largely by Bob Uhrig in the early 1990s [58] (for an overview, see also [24]). In the first period, mostly conceptual studies were made on simulated signals. One of the very early practical applications is from 1996, performed on previously measured data related to an anomaly in a reactor during operation, an excessively vibrating control rod [27]. In this application the data on which the ANN was trained were still generated by a very simple homogeneous 2D one-group diffusion model in the so-called power reactor approximation, in which the frequency-dependent buckling $B^2(\omega)$ is neglected in the noise equation. The success of this procedure was expedited by the fact that the ANN had only to find one out of the possible locations of 7 control rods in a control rod bank.

Although ANNs got started to be increasingly used in various "direct diagnostic" tasks, in which no neutronic transfer properties were needed (such as determining the decay ratio in BWR instability [59], determining the axial position of partially inserted control rods [60], identifying two-phase flow regimes from neutron radiography images [61], etc.), the integration of ANNs and advanced noise simulators took place only recently in the frame of the Euratom-supported coordinated research project CORTEX (see [23] and the references therein). The CORTEX project clearly showcased the power of combining advanced noise simulators with machine learning methods. Apart from the capability of treating real inhomogeneous cores with high fidelity, there is also an added advantage. In the algorithm-based parametric unfolding methods, the type of algorithm to be used depends on the type of the perturbation. A different algorithm has to be used for the localisation of a channel-type instability [21] and vibrating absorber rod [27]. Hence, after discovering an anomaly from the peaks in the power spectra, first an expert judgement is needed about the kind of the anomaly, and accordingly a decision as to which algorithm has to be used for the unfolding. With machine learning methods, one can develop deep learning methods, including convolutional or recurrent networks, which can to both jobs: both to determine the type of perturbation, as well as to select the appropriate network for the unfolding. It can even give warning in case of anomalous signals where the type of perturbation cannot be identified in terms of the known anomalies.

This extended functionality of the machine learning methods will prove particularly useful for the diagnostics of next generation of nuclear systems, because there is still a huge lack of experience on the possible type of perturbations or anomalies that can occur. The use of advanced noise simulators, combined with deep learning methods will therefore receive and increased significance in the surveillance and diagnostics of next generation nuclear systems.

2.3.6   Molten salt reactors

For the investigation of MSRs, one has to develop a simple analytical model of MSR in one dimension and in one- and two-group theories, similarly to the one used for traditional reactor in the previous sections. Such models were elaborated by several authors [62–71], not to mention the numerous publications using various kinetic approximations, such as the adiabatic and quasistatic approximations, as well as reduced order models. However, none of these was used for the type of analysis of the MSRs for general type of perturbations, i.e. not only for propagating perturbations, rather for all the others which together are the main focus of this book. Hence here, similarly to the analytical noise theory for traditional systems, elaborated in the previous part of this chapter, we describe both the one-group and two-group theories of neutron noise in MSR. This two-step approach justified by the fact that, similarly to the traditional case, the one-group description gives physical insight that is used to interpret the two-group results. This insight is even more significant here than for traditional cores because no closed analytical solution of the noise equations is possible to obtain for finite fuel velocities in two-group theory. Hence a simplification, based on the prompt recirculation approximation and related reasoning will be used, whose validity can be investigated in a one-group model, in which the case of finite fuel velocities is tractable. This way the analysis in one-group theory is used to justify the use of the same approximation in two-group theory.

The significance of this approximation arises also from another circumstance when using a simple 1D model for MSR. Namely as it was seen in the previous sections in this chapter, to handle some of the perturbations, such as the method of locating vibrating fuel elements or absorbers, the only modelled dimension of the system must be the horizontal one, with a slab with thickness of the diameter of the core. However, in reactors with a moving fuel, the effects of the movement of the precursors is only possible if the selected coordinate in the 1D model is the axial one. This means that for the MSR, in principle only the propagating perturbations can be handled, since they also take place in the axial direction. For the handling of the other perturbations, a 2D model would be necessary. Fortunately, as we shall see it soon, with the aforementioned approximation, it is possible to resolve this problem by still using a 1D model. Since this approximation, and its proof is not available in the literature, it will be derived here through a concise, but complete description of both the one-group and two-group cases.

2.3.6.1   One-group theory of thermal MSRs

We assume a homogeneous unreflected slab reactor with the axial dimension along the z axis. The flow of the fuel is also assumed to take place along this direction. The extrapolated system boundaries will lie at $z = 0$ and $z = H$, where the flux is supposed to vanish. This is the usual choice of the coordinate system when the propagation of the coolant (in BWRs) or the fuel (in MSRs) is described. However, for practical reasons, occasionally we also will switch to the x-coordinate system, when the extrapolated boundaries of the system will lie between $x = -a$ and $x = a$. The fuel, flowing axially with a constant velocity u, will be lead back from the core outlet at $z = H$ to the core inlet at $z = 0$ through loop of length L. The total length T of the recirculation will be thus $T = H + L$. The corresponding transit times will be denoted as $\tau_c = H/u$ for the passage through the core, $\tau_L = L/u$ in the external loop, and $\tau = \tau_c + \tau_L$ for the total recirculation time.

Assuming a uniform fuel velocity which is the same in the core and in the outer loop is only used for convenience, in reality this will not be the case. But this simplification will not lead to any discrepancy of the model. The different velocity in the outer loop can be simply compensated by redefining a fictitious length of the outer loop, which gives the same recirculation time as the real loop with the real fuel velocity.

The equations of the space and time dependence of the neutron flux and the precursors are read as

$$\frac{1}{v}\frac{\partial \phi(z,t)}{\partial t} = D\nabla^2 \phi(z,t) + \left[\nu\Sigma_f(z,t)(1-\beta) - \Sigma_a(z,t)\right]\phi(z,t) + \lambda C(z,t)$$

(2.168)

$$\frac{\partial C(z,t)}{\partial t} + u\frac{\partial C(z,t)}{\partial z} = \beta\nu\Sigma_f(z,t)\phi(z,t) - \lambda C(z,t) \qquad (2.169)$$

The flux is assumed to disappear at the extrapolated boundaries:

$$\phi(0,t) = \phi(H,t) = 0 \qquad (2.170)$$

For the MSR, it is necessary to specify boundary conditions for the delayed neutron precursors, whose density will not vanish either at the outlet or at the inlet, due to their streaming. The condition one can set is that the number density of delayed neutron precursors at the inlet is the same as that at the outlet with a time $\tau_L = L/u$ earlier, but decreased by the decay during this time:

$$C(0,t) = C(H, t - \tau_L)e^{-\lambda\tau_L} \qquad (2.171)$$

The assumption of zero flux at the extrapolated boundary and a non-zero density of delayed neutron precursors is somewhat non-physical. However, it was shown by considering logarithmic boundary conditions for the flux at the physical boundary, that this approximation gives physically meaningful results [65].

The static equations read as

$$DV^2\phi_0(z) + (\nu\Sigma_f(1 - \beta) - \Sigma_a(z))\,\phi_0(z) + \lambda C_0(z) = 0 \tag{2.172}$$

and

$$u\frac{dC_0(z)}{dz} = \beta\nu\Sigma_f\phi_0(z) - \lambda C_0(z) \tag{2.173}$$

with the stationary boundary conditions

$$C_0(0) = C_0(H)e^{-\lambda\tau_L} \tag{2.174}$$

Equations (2.172) and (2.173) show that for the MSR, the delayed neutron precursors do not disappear from the static equations. This is an indication that the MSR equations have a higher dimensionality than for the traditional reactors.

Similarly to the traditional systems, the equation for the precursor density can be solved to express $C_0(z)$ with $\phi_0(z)$, and one can derive an integro-differential equation for solely the flux. By integrating (2.173) one obtains

$$C_0(z) = e^{-\frac{z\lambda}{u}}\frac{\beta\nu\Sigma_f}{u}\left(\frac{1}{e^{\lambda\tau} - 1}\int_0^H e^{\frac{z'\lambda}{u}}\phi_0(z')dz' + \int_0^z e^{\frac{\lambda z'}{u}}\phi_0(z')dz'\right) \tag{2.175}$$

where it is seen that only the full recirculation time τ appears, which is somewhat counter-intuitive. As it was discussed in the literature [63], the first term on the r.h.s. of (2.175) corresponds to the contribution from the precursors of all previous recirculations when the precursors passed through the core completely, whereas the second term is from the most current passage of precursors from the inlet to the point z.

Inserting (2.175) into (2.172) leads to a single integro-differential equation for the static flux in the form

$$DV^2\phi_0(z) + (\nu\Sigma_f(1 - \beta) - \Sigma_a)\,\phi_0(z) + e^{-\frac{z\lambda}{u}}\frac{\lambda\beta\nu\Sigma_f}{u}$$

$$\left(\frac{1}{e^{\lambda\tau} - 1}\int_0^H e^{\frac{z'\lambda}{u}}\phi_0(z')dz' + \int_0^z e^{\frac{\lambda z'}{u}}\phi_0(z')dz'\right) = 0 \tag{2.176}$$

As it is seen, due to the indefinite integral in the last term, the characteristic equation is of third order, as opposed to the case of the traditional systems with the corresponding equation being a pure second-order differential equation, with a characteristic equation of second order. However, with a third-order equation, a fully analytical solution is still possible [65,72], and it is obtained as follows. By differentiating once with respect to z and rearranging, one obtains a simple linear differential equation with constant coefficients in the form

$$\phi_0'''(z) + \frac{\lambda}{u}\phi_0''(z) + B_0^2\phi_0'(z) + \left(\frac{\lambda}{u}B_0^2 + C\right)\phi_0(z) = 0 \tag{2.177}$$

or, equivalently,

$$\phi_0'''(z) + \frac{\lambda}{u}\phi_0''(z) + B_0^2\phi_0'(z) + B_S^2\phi_0(z) = 0 \tag{2.178}$$

with

$$B_0^2 = \frac{\nu\Sigma_f(1-\beta) - \Sigma_a}{D} \tag{2.179}$$

$$C = \frac{\lambda\beta\nu\Sigma_f}{uD} \tag{2.180}$$

and

$$B_S^2 = \frac{\lambda\left(\nu\Sigma_f - \Sigma_a\right)}{uD} \tag{2.181}$$

Here, B_S^2 is equal to λ/u times the static buckling of an equivalent traditional reactor with static fuel. The characteristic equation is a third-order polynomial in the variable k (inverse length) obtained as

$$k^3 + \frac{\lambda}{u}k^2 + B_0^2k + B_S^2 = 0 \tag{2.182}$$

Hence the characteristic equation has three roots. One can show that one root is real, and the other two are complex conjugates. In view of this, the solution can be written in the form

$$\phi_0(z) = A_1 e^{\alpha z}\sin(\beta z) + A_2 e^{\alpha z}\cos(\beta z) + A_3 e^{\gamma z} \tag{2.183}$$

where α, β and γ are related to the real and imaginary parts of the three roots. Of the three unknown coefficients, two can be eliminated by using the boundary conditions of vanishing flux at $z = 0$ and $z = H$, whereas substituting the resulting form, containing only one undetermined coefficient, will supply the criticality condition. Once the flux is found in an analytical form, the precursor density can easily be derived through (2.175).

By this way, analytical solutions can be obtained for both the static flux, and for the neutron noise, or for the Green's function of the system. For the sake of later reasoning, we show here some quantitative results. These conceptual calculations were made with data taken from [62]. These correspond to a traditional U-fuelled thermal power reactor (Table 2.3). This is only for purposes of demonstration; later on calculations will be made by data corresponding to real MSRs.

Calculations were made for several different fuel velocities, by accounting for an extrapolation distance of 10 cm. Results for the static flux regarding the dependence of its shape on the fuel velocity are shown in Figure 2.21. The variation of the spatial

Table 2.3   Parameters of the one-group MSR model used in the calculations

H [cm]	L [cm]	D [cm]	Σ_a [cm^{-1}]	$\nu\Sigma_f$ [cm^{-1}]	β	λ [s^{-1}]	v [cm/s]
300	400	0.33	0.01	0.0100362	0.0065	0.1	$1.532 \cdot 10^6$

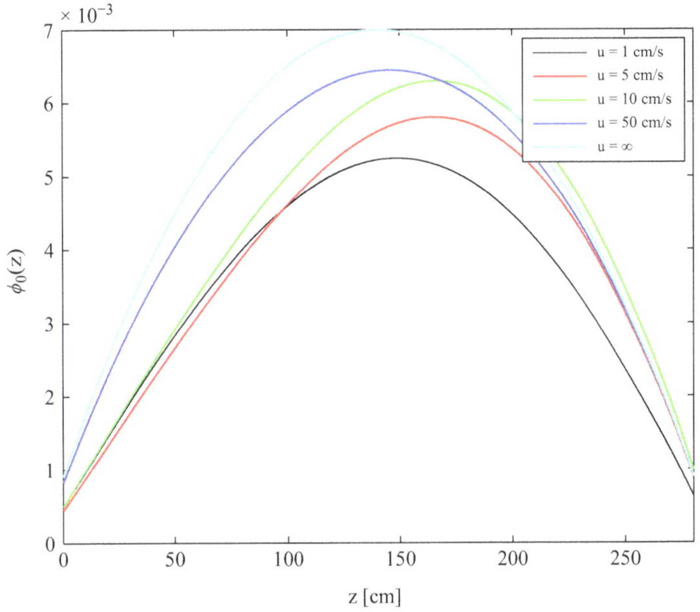

Figure 2.21   The static neutron flux for several fuel velocities

distribution of the precursors is more significant, but it is not interesting for our reasoning, and hence will not be shown here. The figure shows that the initally symmetric
cosine-shaped static flux at $u = 0$ starts to get distorted with increasing fuel velocities,
the maximum being shifted towards the core exit. However, this dependence is not
monotonic; at higher velocities, the maximum starts shifting back towards the centre.
At high velocities, for which $\tau \cdot \lambda \ll 1$, the flux shape becomes symmetric again
around half height of the core. It is seen that this happens already at the relatively
moderate fuel velocity of $u = 50$ cm/s.

Physically, this is due to the fact that the contribution from the multiply recirculated precursors starts to dominate over the ones which are generated in the current
cycle. This corresponds to the dominance of the third term of (2.176) over the last one
on the l.h.s. This lends the possibility to introduce an approximation by neglecting the
last term, which leads to significant simplifications. Namely, the characteristic equation becomes of second order, which simplifies the formalism substantially. Since
this approximation can be formally derived from (2.176) by letting $u \to \infty$, in the
previous publications this approximation was called the case of "infinite fuel velocity" [63,64,70]. Although this terminology will also be kept to some extent here,
especially when it makes notations simpler, but in general here we prefer to call it the
"prompt recirculation approximation". This is to emphasise that the approximation
is sufficiently good even at finite fuel velocities, and to avoid conceptual difficulties
such as the fuel velocity being larger than the neutron velocity, etc.

In the continuation, we therefore use the approximation of prompt recirculation,
both in the static and the dynamic case, both in one-group and two-group theories.

In two-group theory, this is the only way of getting simple analytical solutions. The suitability of the approximation can be investigated in the one-group case, where analytical solutions can be obtained also with finite fuel velocities.

The static equation in the prompt recirculation approximation has the form

$$D\nabla^2\phi_0(z) + (\nu\Sigma_f(1-\beta) - \Sigma_a)\,\phi_0(z) + \frac{\beta\nu\Sigma_f}{T}\int_0^H \phi_0(z')dz' = 0 \tag{2.184}$$

which can be rewritten as

$$\nabla^2\phi_0(z) + B_0^2\,\phi_0(z) + \frac{\eta_0}{D}\int_0^H \phi_0(z')dz' = 0 \tag{2.185}$$

with

$$B_0^2 = \frac{\nu\Sigma_f(1-\beta) - \Sigma_a}{D}$$

$$\eta_0 = \frac{\nu\Sigma_f\beta}{D} \tag{2.186}$$

Since the last term of (2.185) does not depend on z and hence can be treated as a constant, the solution can be searched in the form of the full solution of the homogeneous solution and a particular solution of the inhomogeneous equation. From the boundary conditions and from the condition that the full solution needs to fulfil (2.185), one obtains the static flux as

$$\phi_0(x) = A\left[\cos(B_0\,x) - \cos(B_0\,a)\right] \tag{2.187}$$

where A is an arbitrary constant, and the criticality equation in the form

$$B_0^2\cos(B_0\,a) + \frac{2a\,\eta_0}{T}\cos(B_0\,a) - \frac{2\eta_0}{TB_0}\sin(B_0\,a) = 0 \tag{2.188}$$

A quantitative comparison between the cases of a traditional reactor ($u = 0$) and the prompt recirculation approximation ($u = \infty$) is shown in Figure 2.22 in the x-coordinate system, with $x = 0$ being at the centre of the core, i.e. at half height. The data in Table 2.3 correspond to a critical system with $u = 0$. The core with the prompt recirculation ($u = \infty$) has the same size and was made critical by adjusting the cross sections through the criticality equation (2.188). One can see that, as expected, the two fluxes are both symmetric, and quite similar to each other, although the delayed neutron precursor distributions (not shown here) are quite different for the two cases.

We will now calculate the Green's function of the dynamic problem both for finite fuel velocities, as well as in the prompt recirculation approximation. First, from (2.168) and (2.169), assuming for simplicity the fluctuations of the absorbing cross sections only, the linearised equations for the fluctuations of the flux and the precursor density are obtained as

$$\frac{1}{v}\frac{\partial\delta\phi(z,t)}{\partial t} = D\nabla^2\delta\phi(z,t) + (\nu\Sigma_f(1-\beta) - \Sigma_a)\,\delta\phi(z,t)$$
$$+ \lambda\delta C(z,t) - \delta\Sigma_a(z,t)\phi_0(z) \tag{2.189}$$

$$\frac{\partial\delta C(z,t)}{\partial t} + u\frac{\partial\delta C(z,t)}{\partial z} = \beta\nu\Sigma_f\delta\phi(z,t) - \lambda\delta C(z,t) \tag{2.190}$$

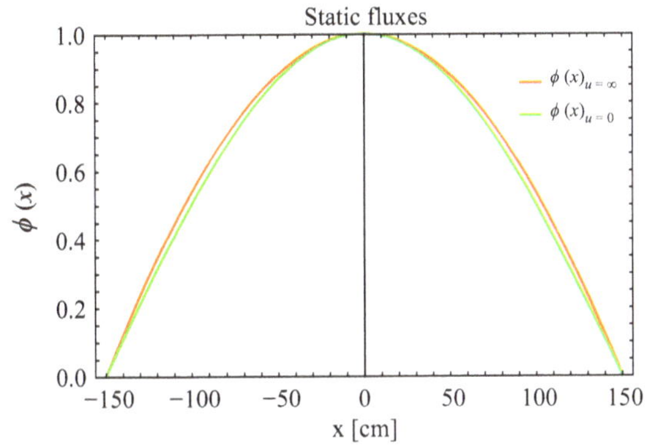

Figure 2.22   Comparison of the static fluxes between a traditional core ($u = 0$ with the data of Table 2.3) with an equivalent MSR with $u = \infty$

with the boundary condition of vanishing flux fluctuations at the extrapolated boundaries and the condition for δC in the form

$$\delta C(0, t) = \delta C(H, t - \tau_L)e^{-\lambda \tau_L} \tag{2.191}$$

After a temporal Fourier transform, integrating (2.190) to express $\delta C(z, \omega)$ with $\delta \phi(z, \omega)$, one arrives at the noise equation for $\delta \phi(z, \omega)$ as

$$\nabla^2 \delta \phi(z, \omega) + B^2(\omega)\, \delta \phi(z, \omega) + e^{-\frac{(\lambda+i\omega)z}{u}} \frac{\lambda \eta_0}{u}$$

$$\times \left(\frac{1}{e^{(\lambda+i\omega)\tau} - 1} \int_0^H e^{\frac{(\lambda+i\omega)z'}{u}} \delta\phi(z', \omega)dz' + \int_0^z e^{\frac{(\lambda+i\omega)z'}{u}} \delta\phi(z', \omega)dz' \right) \tag{2.192}$$

$$= \frac{\delta \Sigma_a(z, \omega)\phi_0(z)}{D}$$

where

$$B^2(\omega) = \frac{\nu \Sigma_f(1 - \beta) - \Sigma_a - \dfrac{i\omega}{v}}{D} = B_0^2 - \frac{i\omega}{vD} \tag{2.193}$$

is the dynamic buckling, and η_0 was defined in (2.186). The Green's function of (2.192) obeys the equation

$$\nabla^2 G(z, z_0, \omega) + B^2(\omega)\, G(z, z_0, \omega) + e^{-\frac{(\lambda+i\omega)z}{u}} \frac{\lambda \eta_0}{u}$$

$$\times \left(\frac{1}{e^{(\lambda+i\omega)\tau} - 1} \int_0^H e^{\frac{(\lambda+i\omega)z'}{u}} G(z', z_0, \omega)\,dz' + \int_0^z e^{\frac{(\lambda+i\omega)z'}{u}} G(z', z_0, \omega)\,dz' \right)$$

$$= \delta(z - z_0) \tag{2.194}$$

This equation was solved with the methods described in [63] for various fuel velocities with the data of Table 2.3. The results for $z_0 = H/2$ are shown in Figure 2.23 for several different fuel velocities.

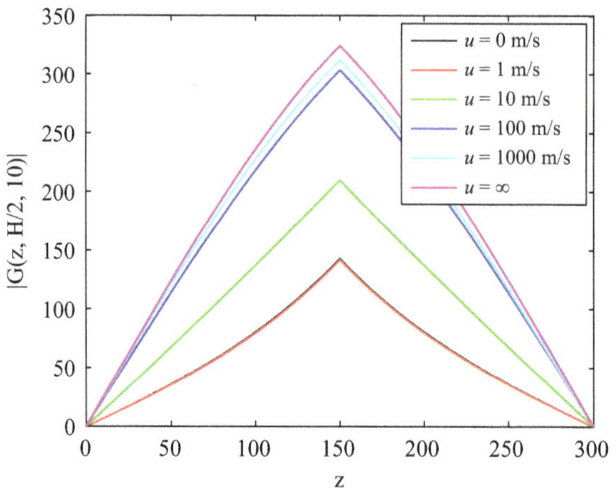

Figure 2.23   *The space dependence of the Green's function in a large system for several different velocities from $u = 0$ to $u = \infty$ with $z_0 = H/2$ and $\omega = 10$ rad/s*

The figure shows that for increasing fuel velocities, the character of the Green's function changes monotonically. For $u = 0$ (the lowermost curve), i.e. for the case as a traditional system, corresponding to the fact that it is a large power reactor, the shape of the Green's function shows a considerable space dependence, i.e. deviation from the point kinetic behaviour, which has the shape of the static flux. With increasing fuel velocities, the originally concave shape first turns to linear space dependence, after which the shape becomes convex. This means that with the moving fuel, the behaviour changes towards a more point kinetic response.

As discussed in the literature, this is for two reasons. One is that with increasing fuel velocity, an increasingly larger fraction of delayed neutron precursors decays outside the core, hence the effective delayed neutron fraction decreases. As it is seen from (2.44), at the plateau frequency of 10 rad/s, a decrease of β leads to the increase of the magnitude, and hence the contribution from, the point kinetic component. The other effect leading to a more point kinetic-type behaviour is the strengthening of the spatial coupling of the fission chains by the movement of the precursors, by the fact that precursors generated at one point will decay at another spatial point. As we see it soon, out of these two effects, the decrease of the effective delayed neutron fraction is the one that is dominating.

Figure 2.23 also shows that the convergence to the results of the prompt recirculation case, i.e. $u = \infty$, takes place at higher fuel velocities than for the static flux. Although the results are not universal, since the ratio of the length of the external loop and the core axial dimension also play a role, still from this figure with the data of LWRs, it seems that the case closest to the practical applications, with $u = 10$ m/s, is in between the case of static fuel and the MSR with prompt recirculation. Actually, the results obtained by the assumption of prompt recirculation can also be considered as a measure of the largest effect the movement of the fuel can have on the dynamic response of the system, and is therefore still a useful tool.

The equation for the Green's function with the prompt recirculation approximation can be derived from the noise equation (2.192), by performing the limit $u \to \infty$ and $\tau \to 0$ leading to

$$D\nabla^2 \delta\phi(z,\omega) + \left(\nu\Sigma_f(1-\beta) - \Sigma_a - \frac{i\omega}{v} \right) \delta\phi(z,\omega)$$

$$+ \frac{\lambda\beta\nu\Sigma_f}{(\lambda+i\omega)T} \int_0^H \delta\phi(z',\omega)\,dz' = \delta\Sigma_a(z,\omega)\phi_0(z) \qquad (2.195)$$

From this, the equation for the Green's function of (2.195) corresponding to the definition (2.47), with the shift of the spatial variable from z to x reads as

$$\nabla^2 G(x,x_0,\omega) + B^2(\omega)G(x,x_0,\omega) + \frac{\eta(\omega)}{T} \int_{-a}^{a} G(x,x_0,\omega)dx' = \delta(x-x_0) \qquad (2.196)$$

where

$$\eta(\omega) = \frac{\lambda\eta_0}{\lambda+i\omega} \quad \text{with} \quad \eta_0 = \frac{\nu\Sigma_f\beta}{D} \qquad (2.197)$$

the latter already having been defined in (2.186).

The last term on the l.h.s. of (2.196) is still independent of x and hence can be considered as constant, hence the solution can still be sought as a sum of the solution of the homogeneous and the inhomogeneous equations. This is though not identical with the solution of the static equation, rather it is a modification of the method of solving for the collided and uncollided flux in transport or diffusion theory. The inhomogeneous equation consists of (2.196) without the integral term, i.e. without the last term on the l.h.s., whereas the homogeneous equation consists of (2.196) without the Dirac delta on the r.h.s. but with the integral term included on the l.h.s., where the integral term is over the full solution. Both the inhomogeneous and the homogeneous solutions have to fulfil the boundary conditions separately.

As is described in [63], the solution of the inhomogeneous equation is formally identical with that of the traditional reactors, but with different parameters, whereas the space-dependence of the homogeneous solution is the same as what was called the inhomogeneous solution for the static flux. Since in this case the inhomogeneous equation contains a term which is not proportional with the sought solution, there is no undetermined factor left, and the result is given as

$$G(x,x_0,\omega) = -\frac{\eta(\omega)\varphi(x)\varphi(x_0)}{TK(\omega)B^2 \cos Ba}$$

$$-\frac{1}{B\sin 2Ba} \begin{cases} \sin B(a-x_0)\sin B(a+x) & x < x_0 \\ \sin B(a+x_0)\sin B(a-x) & x > x_0 \end{cases} \qquad (2.198)$$

where $B \equiv B(\omega)$,

$$\varphi(x) = \cos(Bx) - \cos(Ba) \qquad (2.199)$$

and the factor $K(\omega)$ is defined as

$$K(\omega) = B^2 \cos(Ba) + \frac{2a\eta(\omega)}{T}\cos(Ba) - \frac{2\eta(\omega)}{TB}\sin(Ba) \qquad (2.200)$$

Note that $K(0) = 0$ because the r.h.s. of (2.200) reduces to the criticality equation One notes the symmetry in x and x_0: the fact that (2.195) is self-adjoint is reflected in the Green's function.

The second term in the Green's function is formally identical with the Green's function of the traditional systems, where the first term is not present. For traditional reactors, the point kinetic behaviour can be derived from the properties of this second term (which is the only one appearing in the solution) as ω tends to zero, because the term $\sin(2Ba)$ tends to zero. In the present case, $\sin(2Ba)$ does not tend to zero when ω tends to zero, and correspondingly, the point kinetic behaviour will not be related to the second term, rather it will arise from the first one. Namely, taking the limit $\omega \to 0$, the second term of (2.198) will remain finite, whereas the first will diverge since $K(\omega) \to 0$, as mentioned here before. Hence one has

$$G(x, x_0, \omega) \sim \frac{\eta_0 \varphi(x) \varphi(x_0)}{T K(\omega) B_0^2 \cos B_0 a} \tag{2.201}$$

when $\omega \to 0$. Since the Green's function is now factorised in frequency and space dependence, and the space dependence is further factorised into a product of the static flux at x and x_0, respectively, it is shown that an MSR with the prompt recirculation approximation also shows point kinetic behaviour at low frequencies. The analysis can be made similarly with considering systems of decreasing size instead of frequency, with a similar conclusion.

At this point it is interesting to make a comparison between the Green's function of a traditional system, (2.50), with that of an equivalent MSR in the prompt recirculation approximation, (2.198) for various frequencies, with the same system size. Here again the data from Table 2.3 were used, which describe a critical system with static fuel. The MSR in the prompt recirculation approximation with the same system size had the buckling adjusted with the criticality equation (2.188).

The space-dependence of the amplitude of the Green's functions of an MSR and a traditional system of power reactor size, for $x_0 = 0$ and for various frequencies, can be seen in Figure 2.24. At very low frequencies (Figure 2.24(a)), the spatial form of the two Green's functions is the same, being equal to that of the static flux. This is a consequence of the fact that both systems behave in a point kinetic manner. At the same time, the amplitude of the Green's function of the MSR is much larger than that of the traditional system. The reason for this is that, as discussed earlier (see (2.44) and the discussion around it), the amplitude of the point kinetic component, which is dominating at these frequencies, is proportional to $1/\beta$. Although both systems have the same β value, the effective delayed neutron fraction is much smaller in the MSR than in a corresponding traditional reactor, due to the fact that a large portion of precursors decays outside the core in the MSR. For the MSR in the prompt recirculation approximation, a rough estimate is that the effective delayed neutron fraction β_{eff} can be estimated as

$$\beta_{eff} \approx \beta \frac{H}{L+H} = \beta \frac{H}{T} \tag{2.202}$$

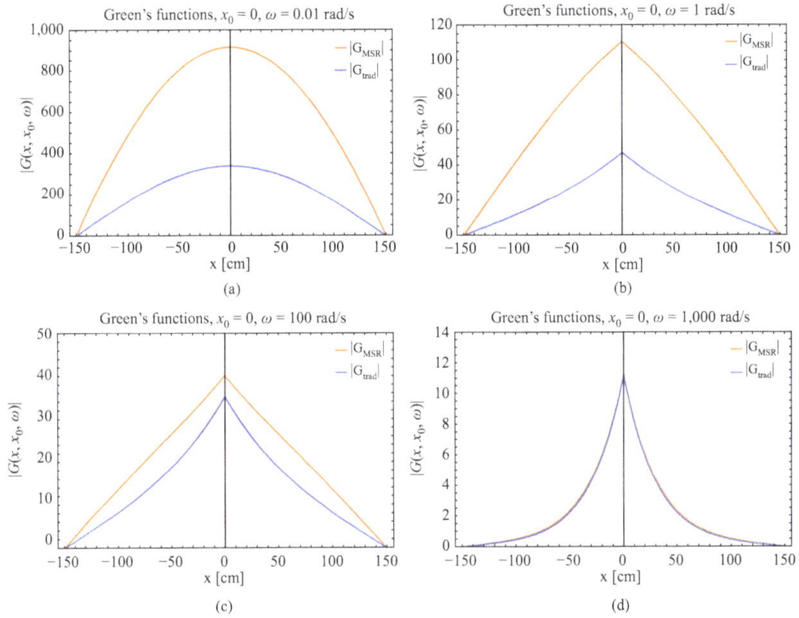

Figure 2.24 Comparison between Green's functions for a traditional reactor with zero fuel velocity and an MSR with the prompt recirculation approximation, for various frequencies in a large system ($H = 300$ cm). (a) $\omega = 0.01$ rad/s; (b) $\omega = 1$ rad/s; (c) $\omega = 100$ rad/s; (d) $\omega = 1000$ rad/s

For the case here, $H = 300$ cm and $L = 400$ cm, one obtains that the amplitude of the Green's function of an MSR is over 2 times larger than that of the corresponding traditional system, which is confirmed by Figure 2.24(a) (and even by Figure 2.24(b)).

From a practical point of view, the significance of this fact is that at low frequencies, and even at plateau frequencies for smaller systems, the amplitude of the noise, induced by perturbations in the core, will be larger for systems with a smaller effective delayed neutron fraction. Apart from MSRs which have a smaller effective delayed neutron fraction due to decay outside the core, Gen-IV reactors fuelled with thorium or plutonium will have a smaller fraction of delayed neutrons per fission than the traditional LWRs. This might be helpful from the practical point of view of diagnosing various perturbations.

Figure 2.24 also shows how the point kinetic behaviour gradually disappears and the system response to localised perturbations becomes localised for increasing frequencies. It is also seen that the MSR retains a point-kinetic behaviour for larger frequencies than a traditional reactor. At very high frequencies (Figure 2.24(d)), where only the prompt neutron behaviour determines the transfer properties, the response of the two systems becomes identical.

These results are consistent with the ones shown in Figure 2.23, for a given frequency but varying fuel flow velocity, and hence decreasing the effective delayed neutron fraction. These results all point to the fact that the main reason for the difference in the dynamic system response, expressed by the transfer function, are due to the difference in the effective delayed neutron fraction, and to a much less degree to the movement of the precursors inside the core. As it was mentioned earlier already in the literature, the fact that delayed neutron precursors decay at a point different from the position where they were born, enhances the spatial neutronic coupling, enhancing point kinetic behaviour, and also has an effect of the effective delayed neutron fraction by the fact that the importance of the position of decay may be different from that of the birth of the precursor. According to the aforementioned results, the dominating fact is the decrease of the effective delayed neutron fraction, and that it can be estimated from the geometry of the core and the external loop by (2.202).

This suggests the idea that a qualitative assessment of the dynamic behaviour of a molten salt system can be effectively estimated by the transfer function of a traditional system, but using a reduced delayed neutron fraction. This assumption can actually be investigated by making the same calculation as in Figure 2.24, but using the effective delayed neutron fraction by the geometrical data of the system. The results are shown in Figure 2.25.

These results show that the shape of the transfer functions for a traditional core with an effective delayed neutron fraction and that of an MSR in the prompt

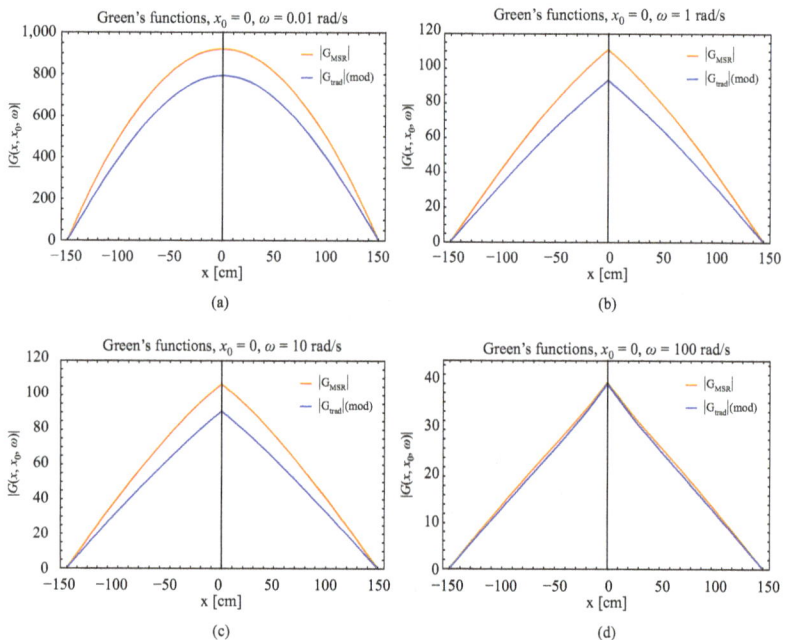

Figure 2.25   *Comparison between Green's functions for a traditional reactor with zero fuel velocity but reduced effective delayed neutron fraction, and an MSR with the prompt recirculation approximation, for various frequencies in a large system ($H = 300$ cm). (a) $\omega = 0.01$ rad/s; (b) $\omega = 1$ rad/s; (c) $\omega = 100$ rad/s; (d) $\omega = 1000$ rad/s*

recirculation approximation are very similar. The amplitudes of the traditional system are somewhat lower, but this actually even enhances the usefulness of the approximation of using the traditional transfer function with reduced effective delayed neutron fraction. Namely, as Figure 2.23 shows, the amplitude of the Green's function with a finite velocity are also lower than that of the prompt recirculation approximation.

This shows that the dynamical response of an MSR can be estimated to a very good approximation by using zero fuel velocity, i.e. the traditional Green's function, but with the reduced effective delayed neutron fraction. This approach has several advantages. For one thing, it solves the dilemma that in a 1D model, the selected dimension has to be the direction of the fuel flow, to describe properly the MSR characteristics. This would have excluded the possibility of selecting the dimension transversal to the flow, to describe certain perturbations. By using the traditional Green's function, the direction of the flow of the fuel does not have any significance, one can select also a direction perpendicular to the flow to calculate the noise of vibrations in an MSR. Another advantage is that when calculating the effect of propagating perturbations, as is shown in (2.160) and (2.161), calculation of the noise requires an integral of the noise source describing a propagating perturbation with the Green's function. Whereas the analytical form of the Green's function of the MSR in the prompt recirculation approximation is not much more complicated than that of a traditional system, still it is simpler to use the formula already calculated for traditional reactors.

As it will be seen in the next subsection, the conclusions drawn in the present one-group treatment will also be applicable in two-group theory, which is the one used in the evaluation of the Gen-IV reactors of the MSR type in Chapters 3 and 4.

2.3.6.2   Two-group theory of MSR

As mentioned earlier, no analytical solution of the MSR equations exists in two-group theory for finite velocities. Analytical formulae can be derived in the prompt recirculation approximation. This approach will be pursued in this section, and a comparison with traditional solid fuel reactors will be performed, similarly as in the previous section in one-group theory. The purpose is to show that the grounds for the conclusion that the neutronic transfer properties of the MSRs can be approximated by the equations of tradition systems with a reduced effective delayed neutron fraction hold also in two-group theory.

Although for the traditional systems, a two-group theory of the noise in systems with a fast spectrum was elaborated, here we keep to thermal MSRs only. A two-group theory of fast MSRs in the prompt recirculation approximation is doable, but it is beyond the scope of this book. We assume that the conclusions drawn from the analysis of thermal MSRs are also applicable to fast MSRs. Since the conclusion is that the formalism of traditional reactors can be used with some modifications, the modified traditional two-group theory for fast reactors is used in the analysis of fast MSRs.

The two-group theory of MSRs in the prompt recirculation approximation was elaborated in [64]. Nevertheless, the derivation (even if not in all details), and in particular the final formulae, will be reproduced here. This is because the final formulae are not given in [64], and in addition, the published formulae contain several typos (sign errors). Our purpose, like in the preceding sections, is to supply both the formulae, as well as the input data, for the readers, so that they can also repeat the calculations (or do calculations for other systems), if they wish.

The condensed form of the static equations for the thermal MSR in the prompt recirculation approximations reads as

$$
\begin{bmatrix} D_1 \nabla^2 - \Sigma_1 & \nu\Sigma_{f2}(1-\beta) \\ \Sigma_R & D_2\nabla^2 - \Sigma_{a2} \end{bmatrix} \begin{bmatrix} \phi_1(x) \\ \phi_2(x) \end{bmatrix} = \begin{bmatrix} -\dfrac{\beta\,\nu\Sigma_{f2}}{T} \displaystyle\int_{-a}^{a} \phi_2(x')\,dx' \\ 0 \end{bmatrix}
$$

(2.203)

where now Σ_1 is defined as

$$\Sigma_1 = \Sigma_{a1} + \Sigma_R \tag{2.204}$$

whereas Σ_2 is defined as

$$\Sigma_2 = \Sigma_{a2} \tag{2.205}$$

The solution is obtained with the same methodology as in the one-group case: the r.h.s. of the equation can be considered as constant, hence the solution can be sought in the form of the sum of solution of the homogeneous equation (with a zero r.h.s.) and a particular solution of the inhomogeneous equation. The solution which fulfils the boundary conditions at $x = \pm a$ is given as

$$
\overrightarrow{\phi_0}(x) = A \begin{bmatrix} 1 \\ c_\mu \end{bmatrix} (\cos(\mu x) - \cos(\mu a)) + B \begin{bmatrix} 1 \\ c_\nu \end{bmatrix} (\cosh(\nu x) - \cosh(\nu a))
$$

(2.206)

where

$$
c_\mu = \frac{\Sigma_R}{\Sigma_2 + D_2\mu^2} = \frac{\Sigma_1 + D_1\mu^2}{\nu\Sigma_{f2}(1-\beta)} \tag{2.207}
$$

and

$$
c_\nu = \frac{\Sigma_R}{\Sigma_2 - D_2\nu^2} = \frac{\Sigma_1 - D_1\nu^2}{\nu\Sigma_{f2}(1-\beta)} \tag{2.208}
$$

Both forms of the coupling factors c_μ and c_ν were written out explicitly, because both are used in manipulating and compacting the various formulae, such as the criticality equation.

Here, μ and ν are the positive roots of the characteristic equation which are given as

$$
\mu^2 = -\frac{1}{2}\left(\frac{\Sigma_1}{D_1} + \frac{\Sigma_2}{D_2}\right) + \frac{1}{2}\sqrt{\left(\frac{\Sigma_1}{D_1} + \frac{\Sigma_2}{D_2}\right)^2 - 4\frac{\Sigma_1\Sigma_2 - \Sigma_R\nu\Sigma_f(1-\beta)}{D_1 D_2}} \tag{2.209}
$$

and

$$\nu^2 = \frac{1}{2}\left(\frac{\Sigma_1}{D_1} + \frac{\Sigma_2}{D_2}\right) + \frac{1}{2}\sqrt{\left(\frac{\Sigma_1}{D_1} + \frac{\Sigma_2}{D_2}\right)^2 - 4\frac{\Sigma_1\Sigma_2 - \Sigma_R\nu\Sigma_f(1-\beta)}{D_1 D_2}} \qquad (2.210)$$

Putting (2.206) into the second equation of (2.203) yields the relationship between the coefficients A and B as

$$B = \frac{\mu^2 c_\mu \cos(\mu a)}{\nu^2 c_\nu \cosh(\nu a)} A \qquad (2.211)$$

whereas using the first equation of (2.203) gives the criticality condition

$$\cos(\mu a) D_1 \mu^2 \left(\frac{c_\mu}{c\nu} - 1\right) + \frac{2\beta\nu\Sigma_{f2} c_\mu}{T}\left\{\frac{\sin(\mu a)}{\mu}\right.$$
$$\left. + \cos(\mu a)\left[\frac{\mu^2}{\nu^3}\tanh(\nu a) - a\left(1 + \frac{\mu^2}{\nu^2}\right)\right]\right\} = 0 \qquad (2.212)$$

This expression is both more compact (by utilising (2.207) and (2.208)) than the criticality equation (19) in [64], as well as it is free from the sign errors present in the latter.

For a quantitative illustration, similarly to the one-group case, the data of a light water-moderated core are used. For the two-group case, we use the data of the BWR core of Section 2.3.3, Table 2.1. The critical size of the core is $H = 2a = 368$ cm with static fuel. The corresponding MSR in the prompt recirculation approximation was made critical with the same size by using the criticality equation (2.212) to adjust the fission cross sections to make the system critical. The static fluxes for the two systems are shown in Figure 2.26. Unlike in the one-group case, no difference between the static fluxes is visible. The figure also shows that the contribution of the ν-terms, the second term on the r.h.s. of (2.206), is negligible.

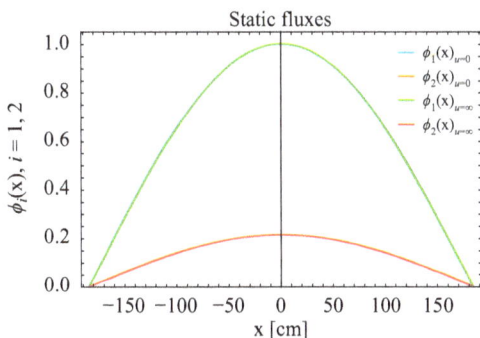

Figure 2.26   Comparison of the static fluxes between a traditional core ($u = 0$ with the data of Table 2.1) and an equivalent MSR with $u = \infty$

The equation for the space- and frequency-dependent Green's function matrix is given as

$$
\begin{bmatrix} D_1\nabla^2 - \Sigma_1(\omega) & \nu\Sigma_{f2}(1-\beta) + \dfrac{\beta\nu\Sigma_{f2}}{T}\dfrac{\lambda}{\lambda+i\omega}\displaystyle\int_{-a}^{a}\mathrm{d}x' \\[2mm] \Sigma_R & D_2\nabla^2 - \Sigma_2(\omega) \end{bmatrix} \times
$$

$$
\begin{bmatrix} G_{11}(x,x_0,\omega) & G_{12}(x,x_0,\omega) \\[2mm] G_{21}(x,x_0,\omega) & G_{22}(x,x_0,\omega) \end{bmatrix} = \begin{bmatrix} \delta(x-x_0) & 0 \\[2mm] 0 & \delta(x-x_0) \end{bmatrix}
$$

(2.213)

with

$$
\Sigma_1(\omega) = \Sigma_{a1} + \Sigma_R - \nu\Sigma_{f2}(1-\beta) + \frac{i\omega}{v_1}
\tag{2.214}
$$

and

$$
\Sigma_2(\omega) = \Sigma_{a2} + \frac{i\omega}{v_2}
\tag{2.215}
$$

and the integral has to be taken with respect to the first argument of the Green's function.

The solution goes along the same lines as for the one-group case, namely that the full Green's function matrix is given as the sum of the solution of the homogeneous equation, $\hat{G}^{inh}$, where the integral in the first row of (2.213) is omitted, and that of the solution of the homogeneous equation, $\hat{G}^h$, where the identity Dirac-delta matrix on the r.h.s. is replaced by zero, but the integral term in (2.213) is accounted for. Again, the solution of the inhomogeneous equation is formally identical with that of the traditional systems, the only difference is that it is used with different values of the parameters. The solution of the columns for the homogeneous solution are formally the same as that of the static equation. The details of the calculations can be found in [64], here we only quote the full solution.

The roots of the characteristic equation are given as

$$
\mu^2(\omega) = -\frac{1}{2}\left(\frac{\Sigma_1(\omega)}{D_1} + \frac{\Sigma_2(\omega)}{D_2}\right) +
$$

$$
\frac{1}{2}\sqrt{\left(\frac{\Sigma_1(\omega)}{D_1} + \frac{\Sigma_2(\omega)}{D_2}\right)^2 - 4\frac{\Sigma_1(\omega)\Sigma_2(\omega) - \Sigma_R\,\nu\Sigma_{f2}(1-\beta)}{D_1 D_2}}
\tag{2.216}
$$

and

$$
\nu^2(\omega) = \frac{1}{2}\left(\frac{\Sigma_1(\omega)}{D_1} + \frac{\Sigma_2(\omega)}{D_2}\right) +
$$

$$
\frac{1}{2}\sqrt{\left(\frac{\Sigma_1(\omega)}{D_1} + \frac{\Sigma_2(\omega)}{D_2}\right)^2 - 4\frac{\Sigma_1(\omega)\Sigma_2(\omega) - \Sigma_R\,\nu\Sigma_{f2}(1-\beta)}{D_1 D_2}}
\tag{2.217}
$$

The coupling constants $c_\mu(\omega)$ and $c_\nu(\omega)$ are given by

$$c_\mu = \frac{\Sigma_R}{\Sigma_2(\omega) + D_2\mu^2(\omega)} = \frac{\Sigma_1(\omega) + D_1\mu^2(\omega)}{\nu\Sigma_{f2}(1-\beta)} \tag{2.218}$$

and

$$c_\nu = \frac{\Sigma_R}{\Sigma_2(\omega) - D_2\nu^2(\omega)} = \frac{\Sigma_1(\omega) - D_1\nu^2(\omega)}{\nu\Sigma_{f2}(1-\beta)} \tag{2.219}$$

With these, the two columns ($j = 1, 2$) of the inhomogeneous Green's matrix are given as

$$\begin{bmatrix} G_{1j}^{inh}(x, x_0, \omega) \\ G_{2j}^{inh}(x, x_0, \omega) \end{bmatrix} = \begin{bmatrix} 1 \\ c_\mu(\omega) \end{bmatrix} A_{\mu j}(\omega)\, g_\mu(x, x_0, \omega)$$

$$+ \begin{bmatrix} 1 \\ c_\nu(\omega) \end{bmatrix} A_{\nu j}(\omega)\, g_\nu(x, x_0, \omega) \tag{2.220}$$

where the basic solutions $g_\mu(x, x_0, \omega)$ and $g_\nu(x, x_0, \omega)$ were defined in (2.89) and (2.90). The coefficients $A_{\mu j}(\omega)$ and $A_{\nu j}(\omega)$, $j = 1, 2$ are given by

$$A_{\mu 1}(\omega) = \frac{c_\mu(\omega)\, \nu\Sigma_{f2}(1-\beta)}{D_1^2\, \mu(\omega)\, \sin[2\,\mu(\omega)\, a]\, (\mu^2(\omega) + \nu^2(\omega))} \tag{2.221}$$

$$A_{\nu 1}(\omega) = \frac{c_\mu(\omega)\, \nu\Sigma_{f2}(1-\beta)}{D_1^2\, \nu(\omega)\, \sinh[2\,\nu(\omega)\, a]\, (\mu^2(\omega) + \nu^2(\omega))} \tag{2.222}$$

$$A_{\mu 2}(\omega) = \frac{\nu\Sigma_{f2}(1-\beta)}{D_1 D_2\, \mu(\omega)\, \sin[2\,\mu(\omega)\, a]\, (\mu^2(\omega) + \nu^2(\omega))} \tag{2.223}$$

and

$$A_{\nu 2}(\omega) = \frac{\nu\Sigma_{f2}(1-\beta)}{D_1 D_2\, \nu(\omega)\, \sin[2\,\nu(\omega)\, a]\, (\mu^2(\omega) + \nu^2(\omega))} \tag{2.224}$$

For the formulae of the homogeneous component of the Green's matrix, we define the functions

$$\varphi_\mu(x, \omega) = \cos[\mu(\omega)\, x] - \cos[\mu(\omega)\, a] \tag{2.225}$$

and

$$\varphi_\nu(x, \omega) = \cosh[\nu(\omega)\, x] - \cosh[\nu(\omega)\, a] \tag{2.226}$$

as well as the integrals $\tilde{g}_\mu(x_0, \omega)$ and $\tilde{g}_\nu(x_0, \omega)$ of $g_\mu(x, x_0, \omega)$ and $g_\nu(x, x_0, \omega)$, respectively, w.r.t. their first arguments:

$$\tilde{g}_\mu(x_0, \omega) = \int_{-a}^{a} g_\mu(x, x_0, \omega)\, dx = \frac{2\, \sin[\mu(\omega)\, a]\, \varphi_\mu(x_0, \omega)}{\mu(\omega)} \tag{2.227}$$

and

$$\tilde{g}_\nu(x_0, \omega) = \int_{-a}^{a} g_\nu(x, x_0, \omega)\, dx = -\frac{2\, \sinh[\nu(\omega)\, a]\, \varphi_\nu(x_0, \omega)}{\nu(\omega)} \tag{2.228}$$

With these notations, the two columns ($j = 1,2$) of the homogeneous Green's function matrix are obtained as

$$
\begin{bmatrix} G^h_{1j}(x, x_0, \omega) \\ G^h_{2j}(x, x_0, \omega) \end{bmatrix} =
$$

(2.229)

$$
\begin{bmatrix} 1 \\ c_\mu(\omega) \end{bmatrix} A_j(x_0, \omega)\, \varphi_\mu(x, \omega) + \begin{bmatrix} 1 \\ c_\nu(\omega) \end{bmatrix} B_j(x_0, \omega)\, \varphi_\nu(x, \omega)
$$

where

$$
A_j(x_0, \omega) = \frac{\beta \nu \Sigma_{f2}}{K(\omega)\, T} \frac{\lambda}{\lambda + i\omega} \Big\{ c_\mu(\omega) A_{\mu j}(\omega)\, \tilde{g}_\mu(x_0, \omega) +
$$

(2.230)

$$
c_\nu(\omega) A_{\nu j}(\omega)\, \tilde{g}_\nu(x_0, \omega) + \Big\}
$$

$$
B_j(x_0, \omega) = \frac{\mu^2(\omega)\, c_\mu(\omega)}{\nu^2(\omega)\, c_\nu(\omega)} \frac{\cos(\mu(\omega)\, a)}{\cosh(\nu(\omega)\, a)} A_j(x_0, \omega)
$$

(2.231)

and

$$
K(\omega) = \cos[\mu(\omega)\, a]\, D_1\, \mu^2(\omega) \left[\frac{c_\mu(\omega)}{c_\nu(\omega)} - 1 \right] + \frac{2\,\beta\,\nu\Sigma_{f2}\, c_\mu(\omega)}{T} \frac{\lambda}{\lambda + i\omega}
$$

$$
\times \left\{ \frac{\sin[\mu(\omega)\, a]}{\mu(\omega)} + \cos[\mu(\omega)\, a] \left[\frac{\mu^2(\omega)}{\nu^3(\omega)} \tanh[\nu(\omega)\, a] - a\left(1 + \frac{\mu^2(\omega)}{\nu^2(\omega)}\right) \right] \right\}
$$

(2.232)

The full Green's function matrix consists of the sum of the inhomogeneous and the homogeneous Green's matrices. Similarly to the one-group case, (2.198), the asymptotic behaviour of the Green's function matrix for $\omega \to 0$ is determined by the homogeneous solution, through the fact that the $K(\omega)$ of (2.232) reverts to the criticality equation, leading to point kinetic behaviour.

Here, everything is specified for the calculation of the elements of the Green's function matrix for an MSR in the prompt recirculation approximation. For a quantitative illustration, the aforementioned BWR data which were used in the static calculations will be utilised. The Green's functions were calculated both with static fuel ($u = 0$), further as an MSR in the prompt recirculation approximation, and finally with static fuel, but reduced delayed neutron fraction. The main purpose is to confirm that the hypothesis that an MSR with finite fuel velocities in the practical range can be reliably modelled with a core with static fuel, but reduced delayed neutron fraction, similarly to the same analysis that was made in the one-group case, Figures 2.24 and 2.25, also holds in two-group theory.

Due to the fact that the Green's matrices have four elements, it is not practical to show the Green's functions for the traditional and the MSR case in one plot. Instead, the space-dependence of three versions of the Green's functions are shown separately, for three different frequencies, namely for $\omega = 0.01$, 10 and 1000 rad/s, respectively. The three variants are the traditional core with solid fuel, the MSR version in the prompt recirculation approximation, and finally the traditional Green's functions, but calculated with a reduced effective delayed neutron fraction.

The results are shown in Figures 2.27–2.29. In the calculations with the prompt recirculation approximation, and also in the traditional core with a reduced delayed neutron fraction, the results depend on the ratio of the length of the full recirculation loop and the core height. This incurs a certain arbitrariness in the results, since these data had to be chosen without having access to such design data of the planned MSRs. In the present calculations, the core height H was 368 cm, and the total length of the recirculation loop was chosen as $T = 900$ cm.

The results for $\omega = 0.01$ rad/s and 1000 rad/s are of less practical interest, they are displayed here primarily to show that the tendencies and relationships between the traditional cores with solid fuel, the MSR in the prompt recirculation approximation and the modified traditional system with a reduced delayed neutron fraction all show the same tendencies with frequency as in the one-group case. At very low frequencies (Figure 2.27), much below the lower break frequency of the zero-power reactor transfer function $G_0(\omega)$ at $\lambda = 0.1$ rad/s, all Green's functions show a point kinetic behaviour, with the amplitude of the Green's functions being higher with moving fuel, the highest being in the prompt recirculation approximation. This ratio is more

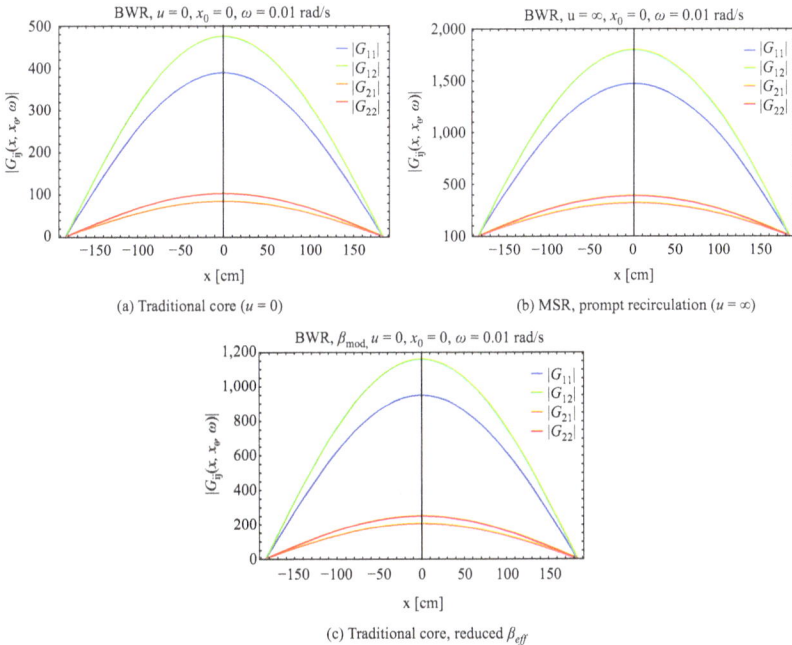

Figure 2.27   *Comparison between Green's functions for (a) a traditional reactor with zero fuel velocity, (b) an MSR with the prompt recirculation approximation, and (c) a traditional core with reduced effective delayed neutron fraction with data of a commercial BWR at $\omega = 0.01$ rad/s*

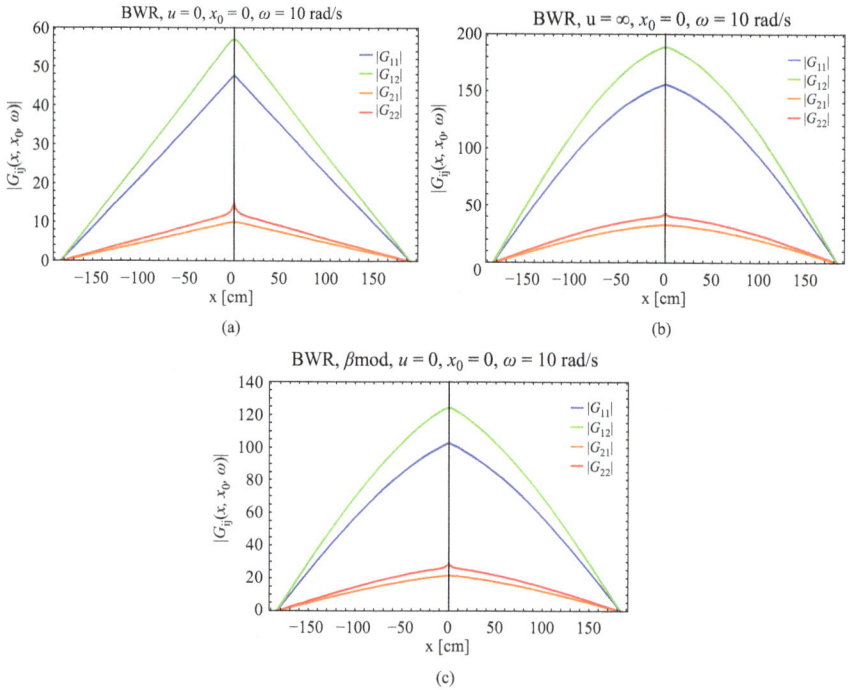

Figure 2.28   *Comparison between Green's functions for (a) a traditional reactor with zero fuel velocity, (b) an MSR with the prompt recirculation approximation, and (c) a traditional core with reduced effective delayed neutron fraction with data of a commercial BWR at $\omega =$ 10 rad/s*

than 3, larger than in the one-group case, which is partly due to the fact that a larger T/H ratio is used here.

At the other extreme, at a frequency much larger than the upper break frequency of $G_0(\omega)$ (Figure 2.29), all three cases show very similar results, in complete agreement with the findings of the one-group calculations. The practically interesting case is when ω lies at the plateau, which is shown in Figure 2.28 ($\omega = 10$ rad/s). Whereas the ratio of the amplitudes between the MSR with prompt recirculation (Figure 2.28(b)) and the traditional core (Figure 2.28(a)) is approximately the same as with the very low frequency, the spatial dependencies are rather different. The MSR with the prompt recirculation approximation shows a relatively small deviation from the point kinetic behaviour, whereas the traditional system exhibits the quite marked spade dependence and large deviation from point kinetics, which is characteristic to power reactors at the plateau frequencies. The traditional system, with an effective delayed neutron fraction scaled down with the factor H/T ((Figure 2.28(c)), still shows much milder space dependence and more point kinetic-like behaviour, at the same time exhibiting a smaller amplitude.

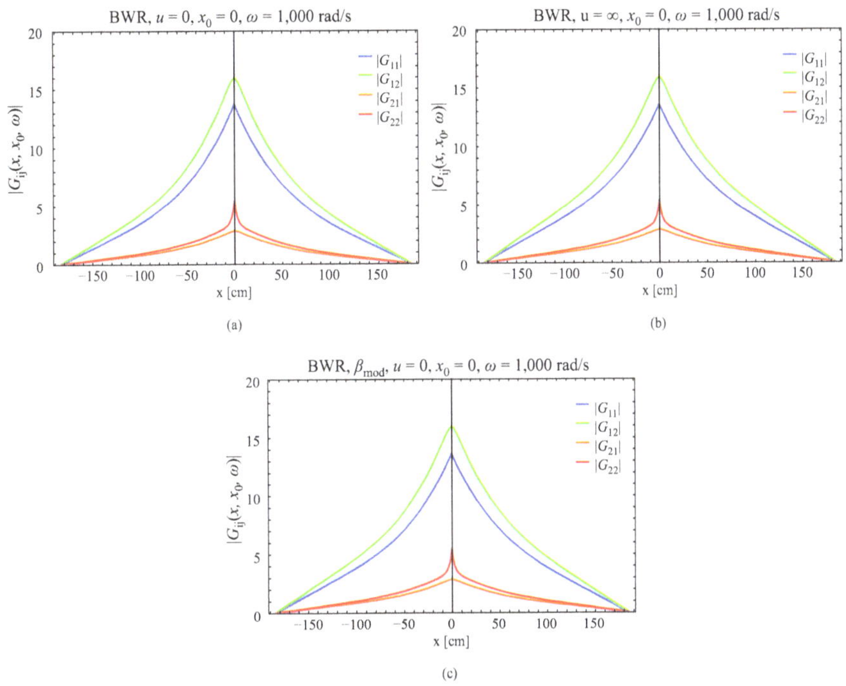

Figure 2.29 *Comparison between Green's functions for (a) a traditional reactor with zero fuel velocity, (b) an MSR with the prompt recirculation approximation, and (c) a traditional core with reduced effective delayed neutron fraction with data of a commercial BWR at $\omega = 1000$ rad/s*

A similar analysis for the same three types of assumptions, i.e. a traditional reactor with original and reduced effective delayed neutron fraction and for an MSR operating in the prompt recirculation approximation was made for the MSDR, a large thermal MSR, with the nuclear data corresponding to a real MSR. This reactor will be analysed in detail in the next chapter, here it is only used to further confirm the hypothesis about the suitability of modelling an MSR as a conventional reactor with reduced delayed neutron fraction. Although the spatial response of the MSDR is different from that of the BWR, the tendencies with varying frequencies, and the relationship between the three different conceptual modes of operation were the same. These facts are in good agreement with the conclusion which was drawn in the one-group case. Namely, that since the amplitude and the spatial character of an MSR with finite velocity in the practical velocity range is "in between" a traditional reactor ($u = 0$) and an MSR in the prompt recirculation case ($u = \infty$) (see Figure 2.23), the transfer properties (Green's functions) of the traditional system with reduced delayed neutron fraction capture these characteristics very well.

Therefore, instead of using the much more involved MSR equations, in the assessment of the dynamic response of MSR cores, the transfer functions (Green's functions) of a traditional reactor with stationary fuel can be used, with the delayed neutron fraction downscaled in proportion of the core to the total loop length. In the evaluation of the MSR-type Gen-IV reactors and SMRs, this strategy will be used. Although the aforementioned analysis was made for a thermal MSR, by extrapolation we assume that the validity of the approximations also holds for the cores with a fast spectrum. This is important, since the two-group theory of fast MSRs has not been elaborated even in the prompt recirculation approximation.

2.3.7 The use of gamma detector signals

One particular aspect of the diagnostics of next generation systems is the type of instrumentation that will be available. Reactor diagnostics has been so far relied on the analysis of neutron detector signals, either in the pulse counting mode for reactivity measurements in low power systems, or in the current mode for power reactor diagnostics. This is also the main emphasis of this book.

However, it might be interesting to consider alternative possibilities, notably to use signals of gamma ray detectors. Gamma photons accompany all nuclear processes in the core, and in particular they are also produced in fission, moreover with a much larger multiplicity (average number of photons per fission is more than 7) than neutrons. Although gamma photons do not undergo multiplication by themselves, yet due to the fact that new gamma photons are generated at each fission event, gamma photons actually carry the same information about the statistics of the development of fission chains as the neutrons. This fact has been recognised and utilised already in nuclear safeguards measurements, where the multiplication and the measurement circumstances are simpler than in reactor measurements.

There has been an interest in using gamma counts instead of neutron pulse counts for reactivity measurements in low power systems already in 1966 [73]. The renewed interest and recent developments in zero-power noise will be described, with some development of the underlying theory, in Chapter 6, Section 6.4.

With regard to power reactor noise diagnostics, the advantages of gamma-sensitive detectors is not obvious in in-core measurements. The higher multiplicity of gamma photons per fission is counteracted by the large uncorrelated background of delayed gammas from fission products. Some attempts were made with in-core gamma detectors which will be described in this section, but as a rule the use of gamma detectors for power reactor noise analysis has not yet been developed.

One aspect of Gen-IV reactors and SMRs that may revive interest in gamma measurements is that in-core nuclear instrumentation may not be available in several of the new designs. The tight design of the fuel pitch in sodium and lead cooled cores, as well as the compact design of SMRs may exclude the possibility of using in-core detectors. Then, gamma detectors have the advantage that they may be more effective than neutron detectors, due to the smaller attenuation and longer spatial correlation range of gamma detectors. Therefore, in the rest of this section, we summarise the available experience with regard to gamma measurements for power reactor noise diagnostics.

One of the first attempts on gamma measurements for power reactor noise was reported by Kenney [74,75] in the early noise meetings, such as the Florida conference in 1967 [3] and at the Japan–US Seminar in 1968 [76]. At the latter conference, Osborn discussed the feasibility and possible advantages of gamma measurements for noise analysis [77]. At that time, one main incentive for gamma measurements was the difficulty of neutron measurements in a high gamma background, such as shut-down reactivity measurements. Although Gen-IV reactors may also have a higher gamma background than LWRs, this aspect is not crucial any longer, with the development of high temperature fission chambers and Campbelling techniques [78].

The high gamma background is actually disadvantageous for gamma measurements as well, because most of it is not due to fission gammas, rather to the delayed gammas from fission products. However, most of the latter can be eliminated by energy discrimination, because the energy of fission gammas is higher. A threshold of 4 MeV was suggested in the aforementioned publications. Another possibility is to use two-detector correlations instead of one detector, because the uncorrelated background is largely eliminated in a two-detector measurement.

Osborn's main point for the advantages of using gamma detectors is their efficiency in ex-core measurements, in that they are effective for correlation measurements at distances larger from the core than neutron detectors, and their field of view into the core is larger than that of neutron detectors. Osborn attempts to support this statement with a simple transport model of neutrons and gammas, where both species obey a transport equation, in which the neutron reactions serve as the source term for the gammas. The treatment is rather cursory and the conclusions are rather qualitative than quantitative.

The question of the use of gamma detection for power reactor noise diagnostics was taken up by Kostic and Seifritz in 1971 [79]. This was still more of a conceptual study than aiming for practical applications, and the main goal was to substantiate Osborn's conclusions. They considered an infinite homogeneous system, and introduced the space- and time-dependent gamma Green's function. It is defined as the gamma response at point R at time t, due on one fast neutron introduced into the system at point r' at an earlier time t'. Only prompt gammas are considered, and they are produced by thermal neutrons through fission and (n, γ) capture processes. The generated gammas propagate from their source with an attenuation kernel. The gamma source $S_\gamma(r, t, E)$ of gamma photons with energy E at point r, and time t, due to an injected neutron as point r' is given through the neutronic transfer function $G_n(|r - r'|, t - t')$ as

$$S_\gamma(r, t, E) = Q(E) \, G_n(|r - r'|, t - t')$$ (2.233)

with $Q(E)$ being the intensity of gamma photons generated with energy E by the neutron reactions. The gamma Green's function is then given as

$$G_\gamma(|R - r'|, t - t', E) = \int_{V_r} d^3 r \, K(\mu|R - r|) \, S_\gamma(|r - r'|, t - t', E)$$ (2.234)

where $K(\mu|R - r|)$ is the attenuation kernel with attenuation coefficient μ. In a 3D infinite homogeneous system it is given as

$$K(\mu|R - r|) = \frac{e^{-\mu|R-r|}}{4 \pi \, |R - r|^2}$$ (2.235)

It is to be noted that, unlike in [73,77], the gamma propagation is assumed to be instantaneous, in view of the high speed of the gamma photons. This simplifies the treatment significantly. A similar concept will be used in Chapter 6, Section 6.4 for the derivation of the gamma-counting-based Feynman-alpha formula. In the present case, it has the further advantage that it makes the conversion of the formulae to the frequency domain simpler, and it is sufficient to specify the neutronic transfer function only in the frequency domain.

For this latter, in [79] a similar analytical form was assumed as for the gamma attenuation, although with a complex frequency-dependent exponent $\kappa(\omega)$. Both one- and two-group representations of the neutronic Green's function were considered. In the two-group case, out of the four components of the Green's matrix, only the transfer between a fast source and a thermal generated neutron had to be considered, since the gamma photons were assumed to be generated by thermal neutrons only. Having both the gamma attenuation kernel and the neutronic transfer function specified, quantitative studies were made with the data of LWRs. In particular, the spatial cross-correlation function (CCF) was calculated between two points, to compare the spatial relaxation of the correlations for neutrons and gammas.

According to the expectations, it was found that if the gamma attenuation is strong, then the spatial range of the correlations is very close between neutrons and photons. On the other hand, with a weak gamma attenuation, the correlation range of photons was found larger than for neutrons, basically because the photons carried further the existing correlations until they got scattered or absorbed. By this way Kostic and Seifritz confirmed Osborn's claim that "gamma detectors see a larger volume of the reactor, and are hence less sensitive to spatial effects." The latter statement though does not follow from the former, and it has to be proven separately.

It is interesting that when using the two-group neutronic transfer function, the authors arrive to the same two terms and roots of the characteristic equation as the local and global terms discussed in the foregoing. Based on the quantitative values of the roots, which determine the spatial relaxation, the authors use the terminology "macro-scale" and "micro-scale". They are nevertheless not interested in the "micro-scale" component, because their goal is to analyse the long range spatial correlations. The work was performed a couple of years before the interest in the local component of the neutron noise became interesting, due to its role in BWRs.

Although Kostic and Seifritz used an infinite medium model in order to be able to handle a 3D model analytically, their work laid the foundations of the principles of power reactor diagnostics with gamma noise, which can easily be generalised for practical applications. Finite reactors could be easily handled in a 1D model. On the other hand, instead of the Green's function connecting one fast source neutron to the thermal neutrons at another position, one should use the neutronic transfer function between the cross-section fluctuations as the noise source when calculating the source of gamma photon fluctuations. In two-group theory, accounting for fast fission, gamma generation by fast neutrons should also be included, so all components of the neutronic Green's matrix should be included. This could be particularly necessary in fast reactors. The algorithm can also be incorporated into the numerical codes being currently developed for the two- or many-group noise simulators.

The extension of the method towards these directions was made by van Dam and Kleiss in 1985 [80]. They applied the same concept that prompt gamma photons are being generated in fission and neutron capture, but attributed the fluctuations of these to the noise sources of the neutron fluctuations, i.e. cross-section fluctuations, which makes the model more realistic. Two-group diffusion theory was used for the neutrons, and the first flight approximation for the (instantaneous) gamma transport. This latter because gamma scattering is substantially anisotropic, and would lead to significant complications to handle, therefore scattered gamma photons were neglected. With this, their treatment of gamma transport is equivalent with the attenuation kernel approach of [79]. They restricted the model to the plateau region ($\lambda \ll \omega \ll \beta/\Lambda$), where the delayed gammas can be neglected. Nevertheless, the formalism still assumes an energy threshold for the gamma detection, not because of discriminating the delayed gammas, but rather because the first flight approximation would be justified.

In their approach, in contrast to that of Kostic and Seifritz, the noise source in the fast group also contributes to the fluctuations of the gamma field, which indeed has to be considered if propagating perturbations are to be treated. A coupled system of equations was set up for the neutrons and gamma photons where the angular dependence of the gamma field was retained, since the scattering of gammas was neglected. Since the gamma generation is only by thermal neutrons, introduction of the adjoint functions was again advantageous, with ψ_1 and ψ_2 standing for the (scalar) fast and thermal adjoint for neutrons, and Ψ_γ for the (angular) gamma adjoint. The equation for the adjoints, with the source being the cross section $\Sigma_{d\gamma}$ of the gamma detector, reads (with some slight simplifications and renotations) as

$$
\begin{bmatrix}
D_1 \nabla^2 - \Sigma_1 & \Sigma_R & 0 \\
\nu\Sigma_f(1-\beta) & D_2\nabla^2 - \Sigma_2 & \int d\Omega \frac{\nu_\gamma}{4\pi}\Sigma_{a2} \\
0 & 0 & \Omega \cdot \nabla - \Sigma_\gamma
\end{bmatrix}
\begin{bmatrix}
\psi_1 \\
\psi_2 \\
\Psi_\gamma
\end{bmatrix}
=
\begin{bmatrix}
0 \\
0 \\
-\Sigma_{d\gamma}
\end{bmatrix}
\tag{2.236}
$$

where ν_γ is the average number of gamma photons per thermal neutron absorption (both fission and capture), while Σ_γ is the total cross section of γ reactions.

As van Dam and Kleiss emphasise, the neutron adjoints ψ_1 and ψ_2 here are not the same as the ones calculated in the previous sections for the neutrons, because they are the solution of a different equation, with a different source term. Unfortunately, they do not calculate these, only estimate Ψ_γ in a simplified scenario.

Using the direct equation for the neutron and gamma noise, and utilising the properties of the adjoint, the reaction rate R_γ in the γ-detector is obtained as

$$
\begin{aligned}
R_\gamma &\equiv \int d\Omega \, dV \Sigma_{d\gamma} \, \delta\Phi_\gamma \\
&= \int dV \{\delta\Sigma_R \, \phi_1(\psi_2 - \psi_1) - \delta\Sigma_{a2} \, \psi_2 - \delta\Sigma_1 \phi_1 \, \psi_1\} \\
&\quad + \int dV d\Omega \left\{\frac{\nu_\gamma}{4\pi}\delta\Sigma_{a2}\phi_2\Psi_\gamma - \delta\Sigma_\gamma \, \Phi_\gamma \, \Psi_\gamma\right\}
\end{aligned}
\tag{2.237}
$$

where Φ_γ is the angular gamma-flux.

Equation (2.237) shows that in the calculation of the response of the gamma detector, all three adjoints, i.e. ψ_1, ψ_2 and Ψ_γ play a role. This is similar to the case when in a BWR, the response of a thermal neutron detector for the perturbation of the removal cross section, both ψ_1 and ψ_2 need to be accounted for. However, in [80] the authors concentrate only on the behaviour of the angular gamma adjoint Ψ_γ, or rather the scalar gamma adjoint ψ_γ, i.e. the angular integral of Ψ_γ. It is calculated in a simple analytical model, assuming the detector as a line "source" in a 3D system with cylindrical symmetry. From the dependence of radial dependence of ψ_γ the authors draw two conclusions, namely that

- a γ-detector horizontally "sees" deeper into the fuel assembly than a corresponding neutron detector, and
- vertically, a γ-detector of a certain axial length is equivalent with an (axially) elongated neutron detector.

Both of these observations point to the direction that an in-core gamma detector is less suitable for the measurement of the transit time between two axially displaced detectors in the same instrument tube. However, this statement is solely based on the properties of the gamma adjoint, and its single spatial relaxation length. When using a gamma detector in a BWR, even the γ-noise will still largely be due to the fluctuations of the removal cross section, which are transferred by the neutron adjoints. These have not been calculated in [80], but one can expect that they still contain a local component, which counteracts the disadvantages deduced from the properties of the gamma-adjoint. Calculating the neutronic adjoints from (2.236), or the corresponding elements of the 3×3 Green's matrix, is yet to be done, but such work is outside of the scope of this book.

The questions raised in [80] were discussed further by Behringer and Nishihara [81]. The authors note that transit time measurements were made with both neutron detectors and gamma-sensitive TIP detectors, placed in the same instrument tube, in the Swiss BWR Mühleberg and the Dutch BWR Dodewaard, from which the axial void velocity profile could be determined. In the case of Mühleberg the neutron and gamma measurements gave the same velocity profile, whereas in Dodewaard the gamma-detectors yielded higher velocities in the upper part of the core than the neutron detectors. From this the authors conclude that there should be a local component also in the field of view of the gamma-detectors, but its range must be larger, such that higher void velocities, further away from the detector, are also sensed by the gamma-detectors, leading to a higher estimated velocity.

To give a qualitative explanation of the larger field of view of the gamma detectors, the authors elaborate a formal energy-dependent transport theory extension of the previous models, but without concrete calculations. Nevertheless, they identify empirically the same three components of the detector field of view as in [80], that is the two neutronic components and the single gamma component. They note that if there exists a local component, it must be in the neutronic part of the response. They argue that this component must have a wider space relaxation than in the neutron detector signals, because in the latter case the adjoint is equation has a Dirac-delta as the inhomogeneous term, whereas in the gamma case, the same function has the gamma importance as the inhomogeneous term, which itself is a solution of an equation with a Dirac-delta as the source term, hence it must be spatially more extended than a Dirac-delta. This would explain why the local component of the gamma-detector field of view is wider than for a neutron detector, which also agrees with the experimental observations.

To summarise, there is some experimental evidence and theoretical support for the fact that gamma-detectors can be used for noise measurements as a complement to, or a replacement, of neutron noise methods. Very similar information about the frequency content of the signals can be extracted from both systems. Regarding the spatial character of the transfer properties, much less is known. There are indications that local and global components may exist in the gamma signals, but their spatial properties are much less known. The space dependence of the relevant adjoints, or the components of the direct Green's matrix, have never been calculated, not even in simple models like the two-group neutronic model used in this book.

Moreover, all experimental and theoretical efforts so far have been focused on the possibilities of two-phase flow diagnostics by in-core detectors. The question of using gamma-detectors for localising perturbations other than propagating ones, has not been investigated yet, except for one early attempt [216]. Theoretically, gamma-detectors outside but close to the core, may be useful for monitoring in-core processes, at least peripheral ones. Since in some Gen-IV and SMR designs the use of in-core instrumentation may not be available, there is an interest to explore these questions. The framework is already laid out, and tackling the relevant questions appears to be straightforward, both by simple analytical models and noise simulators, if they are appended with the gamma noise option. It is anticipated that these questions will be explored in the near future.

2.3.8   Miscellaneous other applications

Our main emphasis so far has been, and will be also in the continuation, to discuss neutron noise diagnostic problems, in which the transfer properties of the core play an important role. The goal has been to study how the different material and geometrical properties, and how the ensuing different neutron physics properties of next generation systems will alter the dynamic transfer properties of the new systems as compared to the currently used reactor types, and by this also the possibilities, efficiency and accuracy, of performing noise diagnostic tasks. However, in addition, there are several diagnostic problems in which such a noise unfolding process, based on the transfer properties of the system, is not necessary, due to the nature of the disturbance/noise source and the signal processing method. Several such instances are known in the current reactors, and it is worth to have an overview whether such, or similar problems can be expected in next generation systems. Such an overview is given in the following.

2.3.8.1   Core-barrel vibrations

Flow induced vibrations of the core barrel are present in PWRs even during normal operation. This is not an operational concern, but many power utilities prefer to monitor these vibrations, in order to see whether the amplitude of the vibrations increases over a period of time, because it could be an indication of material degradation.

Core-barrel vibrations can be detected by ex-core (ex-vessel) ionisation chambers, which are usually included into the standard safety instrumentation. Several PWRs, including the Westinghouse type such as those at the Swedish power plant Ringhals, have 8 ex-core detectors at two axial level, at each level 4 detectors with 90° spacing. Some other reactors, such as the older VVER ones, have only three detector an axial level, with 120° spacing.

These detectors are installed to check that the static power distribution is not "tilted" beyond an allowed level either axially or radially. Their response is nevertheless sufficiently fast for a spectral analysis up to 50 Hz. From the point of view of core-barrel vibrations, these detectors act as displacement sensors, namely the fluctuating part of the signal at the resonant frequency of the vibrations is a mapping of the displacement of the core boundary along a radius connecting the core centre with the detector.

There are several types of vibration modes of the core barrel, the two most commonly encountered being the beam mode or pendular vibrations, and the shell mode vibrations. The beam mode is a 2D random movement of the core on a horizontal plane, whereas the shell mode which consist of periodic shape changes of the core shape from circular to elliptic, i.e. contraction along one diameter and expansion along a perpendicular diameter. The two vibrations coexist, but they can be easily separated. On the one hand, by phase relationships; for the beam mode, diagonally opposite detectors are out-of-phase, whereas for the shell mode they are in-phase, but two adjacent detectors are out of phase. On the other hand, they also occur at different frequencies, and can easily be separated in the power spectra. Typical values are 8 Hz for the beam mode, and 20 Hz for the shell mode. For more information about core motion analysis, the separation of the modes, as well as the application of the method for long-term trend analysis, we refer to the literature [17,82].

Core barrel vibrations may appear also in liquid metal cooled and gas cooled reactors, but there is no operational experience yet on these. The prerequisite of their detection is the same as with PWRs, namely a sufficient number of ex-core detectors, a minimum of three, or rather four ionisation chambers at the same axial level. Alternatively, gamma-sensitive detectors could also be used, especially if it is difficult to place ex-core detectors sufficiently close to the reactor vessel.

It would be particularly interesting to monitor core-barrel motion with ex-core detectors in sodium cooled fast reactors. During operation of the PHENIX SFR, in 1989 and 1990, one experienced four cases of sudden power drops, followed by fast oscillating transients. These were named negative reactivity transients (NRTs). A recording from a power range monitor is shown in Figure 2.30 from [83]. There is no direct evidence for the reason of these negative reactivity drops, but the most likely and generally accepted view is that they are caused by "core flowering", i.e. a sudden expansion of the core radius in all directions isotropically (see further information on this in Chapter 5). In this case the oscillations following the power drop are due to a periodic contraction and expansion of the core, the first contraction causing an overshoot, which triggered an automatic reactor trip. Obviously, core contraction, which

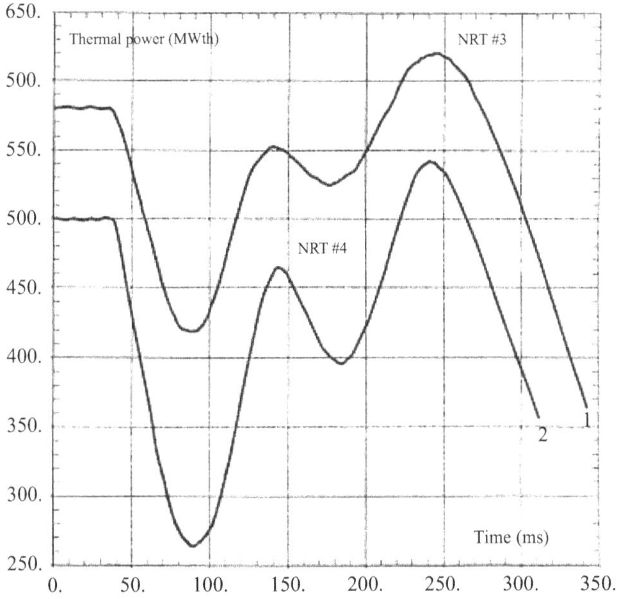

Figure 2.30   Power signal recorded during an NRT event. From Reference [83], with permission from Elsevier

induced a positive reactivity change, is a safety concern, hence the possibilities and consequences of core flowering need to be understood.

From the PHENIX end-of-life tests, performed by CEA, AREVA and EDF, it was possible to verify experimentally that core flowering indeed leads to the reactivity effects which were observed during the NRT events. This is, however, not a direct proof that the observed events were caused by core flowering. Direct proof could be obtained by a similar ex-core detector setup as the one used in PWRs for core barrel motion monitoring. The event could be identified and distinguished from beam mode and shell mode vibrations from the detector phase relationships. Signals from all four detectors in an NRT event would be in-phase, which is different from those of the beam mode and shell mode vibrations. The sensitivity of both in-core and ex-core detectors for core flowering-type periodic oscillations in an SFR has already been performed [84], showing the feasibility of detecting core barrel motion and core flowering.

By extrapolation, core flowering, leading to NRT events can be expected to occur in all Gen-IV reactors, in particular in molten metal cooled reactors, such as lead cooled reactors. There is therefore an extra incentive for monitoring liquid metal cooled reactors, but in the first place SFRs, with ex-core neutron of gamma detectors.

2.3.8.2   Reactor stability

The most common type of instability in Gen-II reactors is the coupled neutronic-thermal hydraulic oscillations occurring in BWRs. The possibility of BWR instability was predicted by Thie as early as in the late 1950s [85,86]. There are several types of BWR instabilities, such as global or core-wide, regional or out-of-phase, and local- or channel-type instabilities. BWR instability is a rather complex phenomenon, a result of an intricate interplay between core physics and thermal hydraulics. Although it is not fully understood in all details, a vast experience has been collected during the years on how to detect, quantify, as well as to predict and suppress instabilities [17,87,88].

The physical reason of the instability is the negative feedback between void fraction and core poser. A temporary increase of reactor power leads to an increase of the void fraction which, through the negative void coefficient, leads to a decrease of core power and subsequently and hence the void, fraction, until this leads to a new rise in the power. This process can lead to periodic self-sustained oscillations. However, flow oscillations in heated channels can occur even in single phase flow, without void generation, and hence without a feedback from the flow to the power generation, on a purely thermal hydraulic basis, under natural circulation. The mechanism leading to oscillations consists of that a disturbance of the flow, i.e. a temporal increase of the speed, will result in a temperature decrease, increase of density/specific weight, and slowing down the flow, which will be heated up again. Actually, this effect is encountered also in BWR stability; experience shows that BWRs are much more prone to develop instability with natural circulation or nearly natural circulation than with forced circulation.

Some of the planned Gen-IV systems, both liquid metal-cooled SMRs, as well as some MSRs, are designed to operate with natural circulation. Examples of SFRs are the GE Hitachi PRISM and the Russian BN-800 reactor design, which incorporate features that allow for natural circulation under certain operating conditions. Some designs incorporate natural circulation features to improve passive safety. The lead-cooled BREST-OD-300 in Russia is a another example, designed to utilise natural circulation for cooling. Terrestrial Energy's integral molten salt reactor (IMSR) is also designed to operate with natural circulation. Therefore, there is reason to assume that thermal hydraulic instabilities may occur even in these types of Gen-IV reactors. This possibility has already been noticed in the literature [89]. In that reference, the lead-cooled natural circulation SUPERSTAR reactor was analysed, and the conclusion was that during normal operation, there are sufficient safety margins against instability. However, unexpected changes in the operational conditions might lead to decreasing of the stability margins. For this reason, it might be advisable to monitor the stability of the core. For instance, the authors note that the core becomes less stable if the transit time between core exit and core inlet is increased, which could happen in the case of a partial blockage.

Even if the reason of the instability in natural circulation cores is of purely thermal hydraulic character, i.e. the flow instability does not affect the stability of the neutronics, the flow oscillations, through the density variations of the coolant still might lead to local flux fluctuations. This could give an opportunity to monitor the onset and evolution of a flow instability.

In an MSR, the coupling between thermal hydraulics and neutronics is two-way, hence stronger than in liquid metal-cooled reactors. An increase of the temperature in the core will lead to dilution of the salt, hence to a decrease of the amount of fissile material and delayed neutron precursors in the core, whose negative reactivity effect is amplified by a higher velocity of the circulation as well. This will lead to a decrease of the power generation, and hence to the possible oscillation of both neutronic and thermal hydraulic nature. Especially in natural circulation MSRs, both the likelihood of an instability, as well as the possibility to detect and quantify it through neutron noise measurements, is higher than in metal-cooled reactors. Here again, early warning systems with neutron noise analysis might prove useful.

2.3.8.3 Utilising spectral properties of the noise

The word "spectral" here and in this subsection refers to the energy spectrum, i.e. the division of the noise between Groups 1 and 2, in contrast to the frequency spectra, which is the main usage in the rest of the book. In most cases investigated so far, it was found that the information content in the fast and the thermal/epithermal groups was completely equivalent, except for the frequency dependence of the phase for propagating perturbations. This is largely due to the fact that in the simple bare homogeneous system used, asymptotic reactor theory holds, that is the energy spectrum is constant in space. But in addition, also because we were only investigating the space or frequency dependence of the noise for a given perturbation with constant parameters. Hence, spectral effects (with respect to energy), such as the change of the ratio of the amplitudes of the noise in Groups 1 and 2, did not occur.

The situation changes if there is a change in the perturbations, such as its amplitude. This can be best seen in the equations for the noise, expressed as a function of the noise source. These were written previously in a matrix form, but for the sake of the argument we write them out explicitly here as

$$\delta\phi_1(x,\omega) = \int G_{11}(x,x',\omega)S_1(x',\omega)\,dx' + \int G_{12}(x,x',\omega)S_2(x',\omega)\,dx' \quad (2.238)$$

and

$$\delta\phi_2(x,\omega) = \int G_{21}(x,x',\omega)S_1(x',\omega)\,dx' + \int G_{22}(x,x',\omega)S_2(x',\omega)\,dx' \quad (2.239)$$

where the noise sources are given as

$$S_1(x,\omega) = (\delta\Sigma_{a_1}(x,\omega) + \delta\Sigma_R(x,\omega) - \chi_1(\omega)\,\delta\nu\Sigma_{f1}(x,\omega))\,\phi_1(x)$$
$$- \chi_1(\omega)\,\delta\nu\Sigma_{f_2}(x,\omega)\,\phi_2(x) \quad (2.240)$$

and

$$S_2(x,\omega) = -(\delta\Sigma_R(x,\omega) + \chi_2(\omega)\,\delta\nu\Sigma_{f1}(x,\omega))\,\phi_1(x)$$
$$+ (\delta\Sigma_{a_2}(x,\omega) - \chi_2(\omega)\,\delta\nu\Sigma_{f2}(x,\omega))\,\phi_2(x) \quad (2.241)$$

It is seen that if the change of the amplitude of the perturbation affects S_1 and S_2 in a different way, this will have the effect that the ratio of the amplitudes of the noise in the two groups also change, i.e. the spectral characteristics of the noise will respond to the change of strength the perturbation. It is also seen that if the perturbation affects only one type of the cross sections, e.g. Σ_{a1} and Σ_{a2} proportionally, it will not affect the noise spectral ratio. Therefore, out of the perturbations considered so far, one can state that an increase of amplitude variations of an absorber of variable strength, and increase of the amplitude of the displacements of a fuel or an absorber pin, or the increase of the void fluctuations in a BWR will not change the spectral ratio of the noise.

On the other hand, if the perturbation affects several cross sections simultaneously, then, even if the change of the perturbation affects the fluctuations of the corresponding cross sections proportionally, this will still lead to a change in the spectral ratio of the noise, because the different cross sections appear with different weights in S_1 and S_2. This is the case of the density fluctuations of the coolant of a liquid metal-cooled reactor or that of the molten salt in an MSR (see e.g. (2.156) and (2.157)). This fact could be utilised for the detection and quantification for void in these systems.

Since the void fluctuations constitute a white band noise without resonances, the spectral ratio can be best quantified with the normalised root mean square (NRMS), NRMS of the detector signal, which is the root mean square of the normalised (with the static flux) flux fluctuations:

$$\text{NRMS} = \sqrt{\int \text{APSD}_{\delta\phi/\phi}(\omega) \, d\omega} \tag{2.242}$$

With the help of the NRMS, the spectral ratio η_{12} of the noise can be quantified as

$$\eta_{12} = \frac{\text{NRMS}_1}{\text{NRMS}_2} \tag{2.243}$$

The spectral ratio η_{12} may be used to quantify the intensity of the void fluctuations in liquid metal-cooled and molten salt reactors. For a sparse bubbly flow, there should exist a linear relationship between the average void fraction and the spectral ratio. To prove whether this is the case, it is necessary to calculate quantitatively the cross-section fluctuations that are caused by the void or the density of the coolant. This can be achieved by static ICFM codes or by the noise simulators. Such a study is beyond the scope of this book, but will have to be performed to quantify the applicability of the method.

The procedure suggested here shows similarity to, or overlapping with some methods suggested for void fraction measurements in BWRs. The use of band-limited NRMS, that is

$$\text{NRMS} = \sqrt{\int_{f_1}^{f_2} \text{APSD}_{\delta\phi/\phi}(f) \, df} \tag{2.244}$$

of thermal neutron detectors was suggested by Thie for determining the local void fraction in BWRs [5]. The suggested values for the bandwidth are 1 to 10 Hz. This is not a spectral method, since only the thermal noise is measured, but it demonstrates that due to the wide-band character of the boiling induced noise, or in general neuron noise induced by void fluctuations, the information content of the NRMS is intrinsically related to the average void fraction.

Spectral methods were suggested for BWRs by Loberg *et al.* [54], although not for neutron noise measurements, rather to the dependence of the spectral ratio of the static flux. They showed that an increasing global (core-wide) void fraction lead to a monotonically hardening neutron spectrum, and they found a linear relationship between the void fraction and the spectral ratio between the fast and thermal flux.

The physical mechanisms behind the method suggested in [54] are though different from what we suggest here and, correspondingly, the areas of applicability are also different. The method behind the static spectral ratio is that the presence of the void changes the properties of the system (the core) itself on a global scale. For the noise-based method, this would be equivalent with the change of the transfer properties of the system, i.e. a change of the elements of the Green's matrix (or actually to the breakdown of linear system theory where the effect of the perturbation is transmitted by the properties of the unperturbed system). This would happen only at relatively high void fractions existing everywhere in the core, above 20–30%.

From physical intuition it can expected that the static spectral method is not suitable for detecting incipient boiling, or the occurrence of a low void fraction. The noise-based method, geared for treating small perturbations, is expected to be much more sensitive to detect and quantify low void fraction, such as sparse bubbles, which is the realistic scenario for Gen-IV systems. On the other hand, the noise-based method is not suitable to handle large void fractions. Not only because of the non-linear character of the system such that the transfer functions themselves would be affected, but also with a void fraction above 50%, the identification of the void fraction fluctuations with the perturbation is not straightforward. It has to be added that noise methods for measuring void fractions above 50% also suggested [56], but those are not based on spectral methods.

2.3.8.4   Determination of reactivity coefficients with noise methods

A hitherto unexplored possible area of noise applications in Gen-IV systems and SMRs is the determination of reactivity coefficients. Reactivity coefficients are properties of the neutronic transfer characteristics of the system, therefore their determination follows a different strategy than the determination of the parameters of a perturbation. To determine the parameters of the transfer function, one needs to measure both the input and the output.

The main application of noise methods in current reactors is the online monitoring of the MTC in PWRs from the correlation between in-core neutron and temperature noise [90,91]. In this case the temperature fluctuations are the input, and the neutron noise is the system response. In fast reactors, the MTC will not play a role. In SMRs of the PWR type, if they will use boron, monitoring the MTC during the cycle could be interesting. However, most designs, such as the Rolls-Royce SMR, are planned to be boron-free.

In addition, previous work shows that due to the very weak radial correlation of the inlet temperature fluctuations in PWRs (which are the driving force of the neutron noise), the temperature fluctuations should be measured at a relatively large number of radial positions in the core (≈ 10). This is unlikely to be included into the design of future SMRs.

The important reactivity coefficient in sodium fast reactors is the void reactivity coefficient. However, for this one should measure both the void fraction fluctuations, and the induced neutron noise. There are no methods available to measure directly the void fraction, or its local fluctuations. Hence it appears that with the assessment of reactivity coefficients, one has to resort to numerical estimates.

Chapter 3

A basic study of the structure of the core response in the basic planned Gen-IV types (metal-cooled, gas-cooled and molten salt reactors)

Imre Pázsit[1], Zsolt Elter[2] and Hoai-Nam Tran[3]

3.1 Introduction

In this and the next chapter, analysis of various Generation-IV (Gen-IV) and small modular reactor (SMR) systems is performed for power reactor diagnostics, whereafter in Chapter 6, the extensions judged to be necessary for applications of zero-power noise analysis in next generation systems will be described. Power reactor applications will take a larger extent; zero-power noise applications will be described in a smaller scale. In addition, we treat first the power reactor applications in three chapters, and the zero-power applications follow thereafter in only one chapter. This is in reverse order as compared to the chronology of how these two methods were developed.

This is for two reasons. The first is that the large variety of next generation reactors has a much smaller effect on the extension of the existing methods for zero-power noise than for power reactor noise. The second is that unlike for power reactor noise, the algorithms of zero power noise are generic in character, they do not require the development of system-specific system codes, they can be implemented directly from the formalism presented in this book. This is largely the consequence of the fact that the origin of zero-power noise itself is generic: it is the branching process, which is universal, and in the space-independent model used for its description, the reactor properties (small vs large system, frequency of the perturbation, etc.) do not play a role. This is not the case for power reactor noise, where there exist a multitude of different types of perturbations as the reason of the neutron noise, and the material and geometrical properties of the system play a significant role in how the noise propagates from the source to the neutron detectors.

The purpose of the present chapter is thus to explore how the neutronic transfer properties of the next generation reactors, Gen-IV systems will be different from the current light water ones due to the different type of fuel, coolant, enrichment,

[1]Division of Subatomic, High Energy and Plasma Physics, Department of Physics, Chalmers University of Technology, Sweden
[2]Department of Physics and Astronomy; Division of Applied Nuclear Physics, Uppsala University, Sweden
[3]Phenikaa Institute for Advanced Study, Phenikaa University, Hanoi, Vietnam

delayed neutron fraction and neutron spectrum. By properties we mean the space and frequency dependence of the transfer function (or the propagation adjoint) between the perturbation and the induced noise in the various groups when using a two-group description. This is because these properties determine the possibility or efficiency of performing the diagnostic tasks, such as locating vibrating components or local channel instabilities, local boiling, or determining flow velocity or void fraction in the coolant, to name a few. A strong point-kinetic behaviour (response) makes it hard to locate localised perturbations; a space-dependent component with a fast spatial relaxation/attenuation means that only detectors close to the noise source will detect the effect of the noise source (and hence detect and locate it); and a local component with a low amplitude means that thermal hydraulic parameters of the propagating coolant will be difficult to measure.

For the quantitative investigations with the two-group one-dimensional (1D) models developed in the previous chapters, in the first-step two-group cross sections, group velocities, and β and λ values need to be generated from the original full 3D geometry and continuous energy model of these systems. Homogenised and collapsed two-group cross sections were generated with the Serpent2 continuous energy Monte Carlo code [92]. The simulations model the active core and reflector regions of the reactors both radially and axially, and in some cases even the surroundings of the reactors, such as the shielding and the pressure vessel, are included. The reactor cores were always considered to be in fresh, hot conditions. The region of interest for the group constant homogenisation was set to the active core. The definition of the active core can be ambiguous in cases where the periphery of the core is filled with a blanket for breeding or in hexagonal lattices where Serpent2 does not allow for simple definition of the outer surface. Nevertheless, for our demonstration purposes, the inconsistency due to these small ambiguities is considered here insignificant.

In the simulations, the JEFF 3.1 cross-section library was used, whereas the thermal scattering data were adopted from JEFF 3.3. The intermediate multi-group structure was set to the Scale-252 structure, and the final two-group constants were collapsed without critical spectrum correction. The leakage spectrum correction was performed outside of the Serpent calculation. The energy boundary between Groups 2 and 1, i.e. the thermal/epithermal and fast energy groups, was set to 0.625 eV for thermal systems and to 1.35 MeV for fast systems.

In the second step, these data, which still belong to a 3D system, had to be modified for radial and/or axial leakage, in order to emulate the behaviour of the system in the relevant selected dimension. This latter depends on the type of perturbation that is to be diagnosed. The space dependence of the noise induced by a variable strength absorber and a vibrating fuel or absorber rod, as well as the localisation of these perturbations, is relevant in the horizontal direction. For these perturbations, in the simulations the width of the 1D system (slab) was chosen to be equal to the diameter of the original 3D system. In order to compensate for the leakage in the axial direction and for the difference between the leakage of a cylinder and a slab, respectively, the absorption cross sections had to be adjusted such that the 1D system remained critical.

With regard to the propagating perturbations, these take place in the axial direction, so in the 1D model, the width of the slab was chosen to be equal to the height of the core, and a similar compensation for the radial leakage had to be performed. Since in general the height of the core is different from its diameter, this leads to slightly different cross sections than in the previous case. The different radial and axial dimensions also incur different dynamic behaviour in the two directions, which was this way captured by the 1D model.

In line with the conclusions drawn in Chapter 2, Section 2.3.6.2, molten salt reactors (MSRs) were modelled with static fuel, but with the parameter β rescaled by H/T, H being the height of the core and T the total length of the recirculation loop. This approximation makes it possible to model also fast spectrum MSRs, since for these the two-group theory is not elaborated yet, not even in the prompt recirculations approximation.

In this chapter, three Gen-IV cores are analysed: a sodium-cooled fast reactor (SFR), a gas-cooled high-temperature reactor (Allegro), and a thermal MSR. Several SMR types will be discussed in Chapter 4.

As was mentioned in the Introduction, all numerical work in Chapters 2–4, what regards the quantitative evaluation of the analytical formulae developed in the previous chapter, was performed in the symbolic manipulation language Wolfram Mathematica [9]. The Wolfram Mathematica notebook [10] is made public by having uploaded it to the Wolfram Notebook Archive [11]. The link to the notebook is in [10]. By downloading the notebook or running it on the Wolfram cloud, the reader can retrieve all results/figures in Chapters 3 and 4 of this book, or make new runs with new input data. Instructions of accessing and using the Notebook, with a user manual, are found in Appendix A.

3.2   Sodium-cooled fast reactor

SFR were built and used already in the past, in particular in France (Masurca, Phenix, Super-Phenix). These reactors are now all shut down. There were also plans on a new demonstration reactor with the name ASTRID, but it was also abandoned.

Currently the OECD Nuclear Energy Agency (NEA) has conducted benchmarks to analyse the performance and safety of these reactors. The analysis next is performed on an SFR core based on the 3600 MWth oxide core description of the OECD NEA benchmark [93]. The reactor consists of an inner core, an outer core and a reflector region surrounding it. The fuel assemblies are axially heterogeneous: the active part of the core consists of five 20 cm height zones and is surrounded axially by steel reflectors. The active core has a radius of 256 cm and a height of 100 cm. The control assemblies are omitted from the model. i.e. it is a rather flat design, which is customary with SFRs. The layout of the core is shown in Figure 3.1.

The two-group nuclear and kinetic parameters of the core are shown in Table 3.1. Since the SFR is a fast spectrum core, the separation energy between the groups was taken to be 1.35 MeV when generating these data from the Serpent input file. Hence instead of fast and thermal groups, we refer to these as Groups 1 and 2, or the fast and the epithermal groups, respectively.

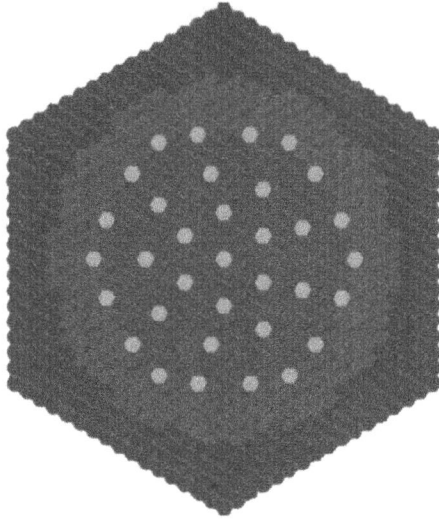

Figure 3.1   The geometry of the SFR core with reflectors

Table 3.1   Group constants and kinetic parameters of the SFR core

Group	D	$\nu\Sigma_f$	Σ_a	Σ_R	χ_p	χ_d
1	2.66184	0.021753	0.008093	0.040346	0.5978	0.04177
2	1.24986	0.005828	0.005853	0	0.40218	0.9582

β	λ	v_1 [cm/s]	v_2 [cm/s]	R [cm]	H [cm]
0.00448	0.60072	$2.1071 \cdot 10^9$	$2.3861 \cdot 10^8$	256	100

In the 1D simulations of the horizontal direction, the core radius was used as the half-width of the equivalent slab. The absorption cross sections were modified from their value in the Table, to account for leakage, due to the difference between the material buckling and the geometrical buckling of the slab arrangement.

The static fluxes in Groups 1 and 2 are shown in Figure 3.2. Like with other fast systems (seen in the previous chapter and further confirmed in subsequent cores), with the threshold energy of 1.35 MeV, the flux in Group 2 becomes much higher than in Group 1. This is in sharp contrast to the thermal LWR systems, seen in the previous chapter, where the fast flux is higher than the thermal one. This is due to the different definition of the groups which, among others, is reflected in the group neutron velocities. The difference in the neutron velocities between Groups 1 and 2 is much smaller than in the thermal LWR cores.

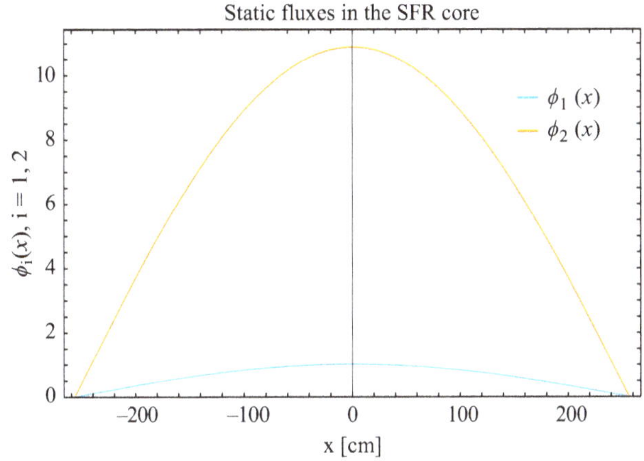

Figure 3.2   *The static fluxes in the SFR core*

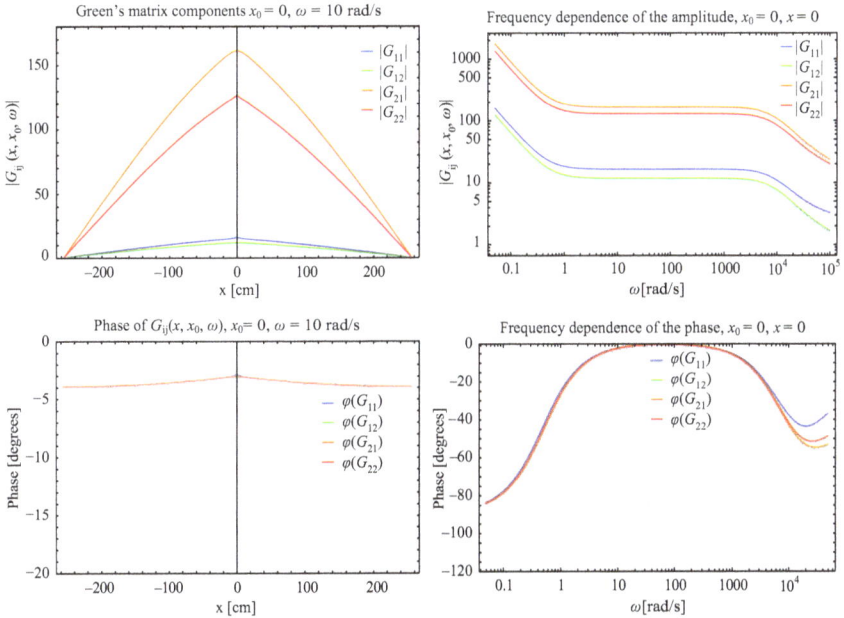

Figure 3.3   *Space dependence of the amplitude and the phase of the components of the Green's function for $\omega = 10$ rad/s (left figures), and frequency dependence of the same for $x = x_0 = 0$ (right figures) in the SFR core*

3.2.1   The Green's functions

The space and frequency dependence of the amplitude and the phase of the Green's functions of the SFR core along a horizontal cross section are shown in Figure 3.3.

Due the specific flat geometry of the core, i.e. very large difference between the radial and axial positions, it is expected that the dynamic characteristics of the core will be rather different into the radial and the axial directions, respectively. Due to the large diameter of the system, the core is expected to behave in a relatively strong space-dependent manner. Figure 3.3 indeed confirms this expectation, although the space dependence is not as strong as in the case of the commercial boiling water reactor (BWR) treated in the previous chapter. This is despite that the diameter of the SFR is considerably larger than the height of the BWR. The main reasons for this are the stronger neutronic coupling in fast spectrum cores, as well as the smaller β. As was discussed previously, at the plateau frequency, a smaller delayed neutron fraction β increases the relative weight of the point kinetic component.

Similarly to the static flux, the amplitudes of the components of the Green's matrix in Group 2 (i.e. $G_{2i}, i = 1, 2$) are larger than in Group 1 (G_{1i}). At the same time, the local component appears in Group 1, i.e. the fast group. As it will be seen, these are also generic properties of the two-group description of cores with a fast spectrum.

The frequency dependence of the amplitude and the phase shows that, due to the high neutron velocities (see Table 3.1), the plateau region is exceptionally wide, stretching from about 0.5 rad/s to about $5 \cdot 10^4$ rad/s. This means that perturbations up to quite high frequencies will be possible to monitor in the SFR. It is also seen that over a large part of the wide plateau region, the phase is very close to zero. This means that for perturbations within the plateau region, the frequency and temporal behaviour can be treated interchangeably. To put is in a different way, the temporal behaviour of the perturbation will be reflected in the temporal behaviour of the neutron flux fluctuations, without time delay effects.

A clarification has to be added at this point. When plotting the space and frequency dependence of the phase, it is the phase of $-G_{ij}$ which is plotted. This means that $180°$ was added to the (negative) phase of G_{ij}. This is simply for reasons of convenience. Namely, from the formalism used in both one-group and two-group theory, it is seen that the elements of the Greens' function transfer the changes of the absorption cross section with a positive sign, whereas the response needs to be negative (increase of the absorption leads to the decrease of the flux). Hence such an "out-of-phase" property is built in to the Green's function. To avoid showing phase values which are permanently below $180°$, we introduced this phase shift into the plots of the Green's functions.

This convention also has the advantage that the results, in particular the frequency dependence of the phase, will be easier to compare with the phase of the zero-power transfer function $G_0(\omega)$. This latter does not have the built-in phase shift property, since it gives the response of the system to the change of the reactivity. Hopefully this plotting convention will not lead to any confusion. It is only applied to the Green's functions, but not to the phase of the noise, shown in the subsections which will follow.

3.2.2   *Absorber of variable strength*

The noise induced by a variable strength absorber in both groups is a linear combination of two Green's matrix elements, weighted with the relative weights of the absorber cross sections in the two energy groups. Hence the space dependence of the noise in the two groups looks rather similar to that of the individual components of the Green's matrix. In the present case this means that the noise induced by an absorber of variable strength shows a not too strong, but definitely not negligible spatial dependence, i.e. a deviation from the point kinetic behaviour. This is confirmed by the left side of Figure 3.4. The phase and the amplitude are shown only for a central absorber, since there is no new information in the noise with a different position of the absorber. Here, when plotting the phase, the convention regarding adding an extra phase shift was not used, to show the real phase delay of the noise.

The amplitude and phase localisation curves $\Delta(x_p)$ and $\theta(x_p)$, defined in (2.138) and (2.139), respectively, are shown in the right column of Figure 3.4. Corresponding to the appreciable space dependence of the induced noise, the ratio of the APSDs of the two peripherally placed detectors shows a strong space dependence. Interestingly, the phase is rather insensitive to the position of the perturbation, but this is a consequence that the space dependence of the phase is rather weak in the Green's functions as well, as it is seen in Figure 3.3. The conclusion is that the chances are good that the radial position of a local perturbation of the variable strength absorber type can be extracted from the amplitude of the APSDs of suitably placed detectors.

3.2.3   Vibrating fuel rod

Figure 3.5 shows the amplitude and phase of the neutron noise at two different equilibrium fuel pin positions: a central one and one 45 centimetres off the core centre. In

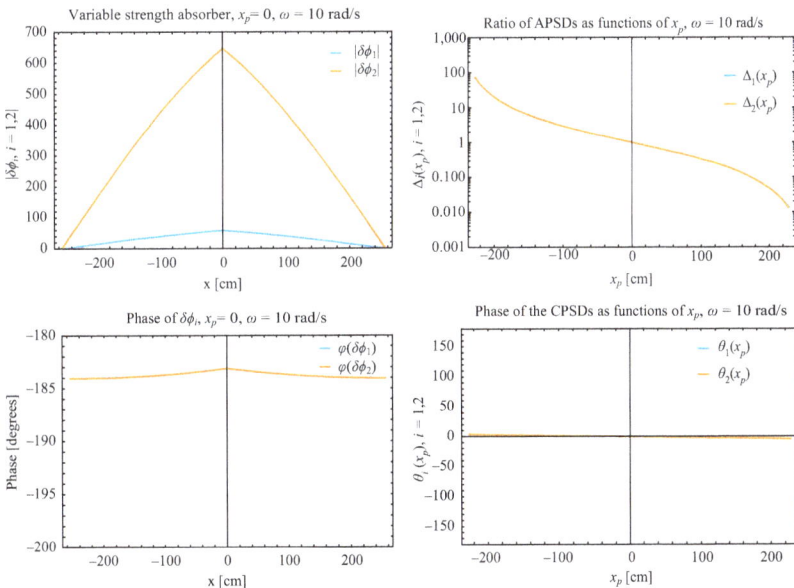

Figure 3.4   Space dependence of the amplitude and the phase of the noise in the two energy groups, induced by an absorber of variable strength for $x_p = 0$ (left column) and the amplitude and phase localisation curves for the absorber of variable strength for the SFR core (right column)

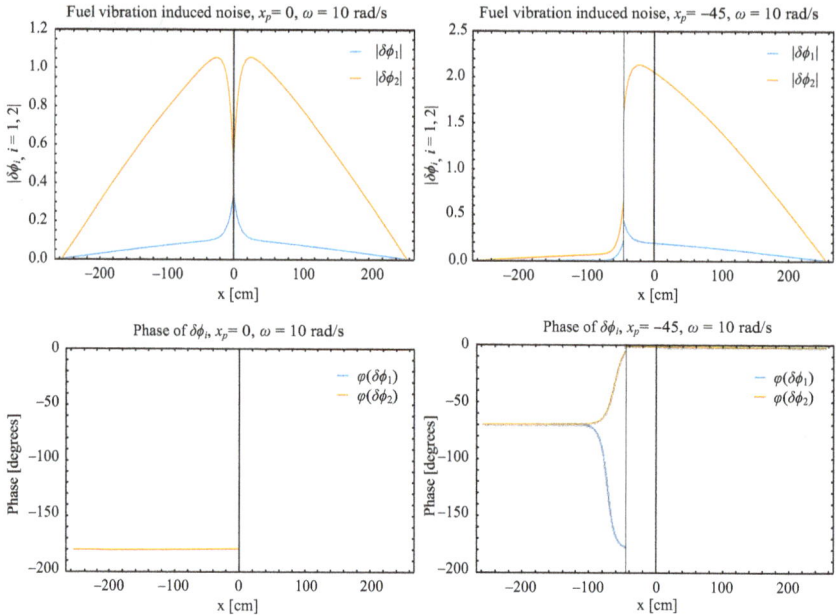

Figure 3.5   Space dependence of the amplitude and the phase of the noise in the two energy groups, induced by a vibrating fuel rod, for $x_p = 0$ (left column) and $x_p = -45$ cm (right column), respectively, in the SFR core

general, due to the fact that the relative weight of the reactivity- and space-dependent components is also dependent on the position of a vibrating fuel pin, the dependence of the induced noise in detectors at different positions shows a stronger dependence on the position of the perturbation for a vibrating fuel pin than for an absorber of variable strength, which is confirmed by the figure.

The space dependence of the amplitude of the noise due to a central vibrating fuel pin (left side of 3.5) shows a relatively large local peak in the noise in Group 1. Its relaxation length is between 10 to 20 cm, larger than that in LWRs. This is also a generic feature of fast reactors. The phase is zero on the r.h.s. and $-180°$ on the l.h.s., which is expected, since the reactivity effect is zero for a central vibrating component, and the space-dependent and local components are out of phase on the two sides.

When the equilibrium position of the vibrating fuel pin is moved out from the central position, the reactivity component appears, and its relative weight increases with increasing distance from the centre. This is simply due to the increase of the magnitude of the spatial derivative of the static flux. There is a point when the absolute values of the point kinetic and space-dependent parts are roughly equal in amplitude, and on the l.h.s. of the vibrating pin, where the two components are out of phase, the amplitude of the noise left to the pin is minimum. For this core this happens at around $x_p = 40$–50 cm. The right column of Figure 3.5 shows that with $x_p = -45$ cm, apart from the local component, the amplitude of the noise nearly vanishes on the left side of the pin, and the phase takes the intermediate value of about $-70°$. Moving even farther from the core centre, the point kinetic part of the noise gradually takes over. The gap in the amplitude at the two sides of the pin disappears, and the phase is close to zero in the whole core.

Based on these observations, it is easy to interpret the behaviour of the localisation curves for the ratio of the APSDs and the phase of the CPSD, which is shown in Figure 3.6. These localisation curves are typical to medium-to-large size reactors. There is a considerable dependence of the ratios of the APSDs on the position of the vibrating pin, but the inverse of the curve is many-valued, so the position of the pin cannot be determined uniquely from the APSD ratios. Some help can be obtained from the phase, which also shows large variations, with a single-valued inverse. However, this only concerns pin positions between about −80 to 80 cm. On the other hand, the ambiguity and possible multiple guessed pin positions can be helped in practice by using multiple detectors, which will be necessary anyway in the real case where the problem is to find the location of the vibrating component on the two-dimensional (2D) horizontal plane. The strong dependence of the APSD ratios on the position of the pin is a more important indicator of the possibility of the localisation than the possible ambiguity of the predicted location.

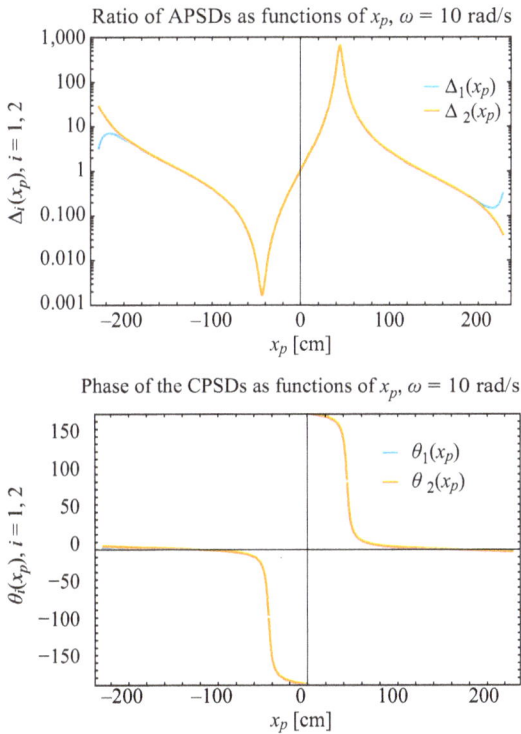

Figure 3.6   Dependence of the amplitude and phase localisation functions on the position of the vibrating fuel rod in the SFR core

3.2.4   Propagating perturbations

For the investigation of the neutron noise induced by propagating perturbations, the core will be modelled in its axial direction. Here we encounter a very special circumstance, which is typical for sodium fast reactors, namely that the height of the core is substantially smaller than its diameter. The SFR is a very flat "pancake core", with a diameter more than 5 times larger than the height of the core.

Since the propagation of the coolant is along the axial direction, in the 1D model the thickness of the slab has to be chosen equal to the height of the core. To have the 1D model critical, the cross sections need to be readjusted for axial leakage, which will, among others, lead to a substantially larger static buckling than in the case when the radial dimensions were modelled. Whereas in the radial direction the behaviour deviated appreciably from the point kinetic, showing a relatively emphasised space-dependence (see Figure 3.3), in the axial direction the behaviour is expected to be strongly point kinetic.

We do not show the recalculated values of the components of the Green's function matrix with the updated cross sections and static buckling, the behaviour will be seen on the shape of the propagation adjoints. As described in the previous chapter, these consist of combinations of the corresponding individual components of the Green's matrix, with weighting factors α_1 and α_2, determined by the type of the perturbation.

As is known [80], and was mentioned before, in BWRs the removal adjoints/Green's functions are the difference of a peaked and a non-peaked function, which enhances the weight of the local component, and which is beneficial for the diagnostics of propagating perturbations. For the sodium-cooled fast core, it is reasonable to assume that the density fluctuations of the coolant affect only the absorption and removal cross sections, with the same relative weight as the corresponding static cross sections (see (2.159). With this assumption one obtains

$$\alpha_1 = 0.894534$$

and

$$\alpha_2 = 0.394153$$

Since in calculating these coefficients, the absorption cross sections dominate over the removal cross section, the two coefficients have the same sign. Hence no difference between the individual Green's function components appears, which means that, unlike in the removal adjoints of the BWRs, the local component will not be amplified in the propagation adjoints (transfer functions) of the SFR.

In addition, as surmised, the axial behaviour of the core is dominated by the point kinetic component, and these two facts together result in a complete suppression of the local component. This is confirmed by Figure 3.7, which shows the spatial dependence of the amplitude and the phase of the propagation adjoints.

Unlike the radial Green's function components, these show no visible space dependence, and no presence of the local component either. Still an attempt was made to see if any linear dependence of the phase could be seen. For the calculations, the velocity of the flow was chosen to be 150 cm/s. No data on the velocity (or transit

Propagation adjoints, $z_0 = H/2$, $\omega = 10$ rad/s

Phase of ψ_i, $z_0 = H/2$, $\omega = 10$ rad/s

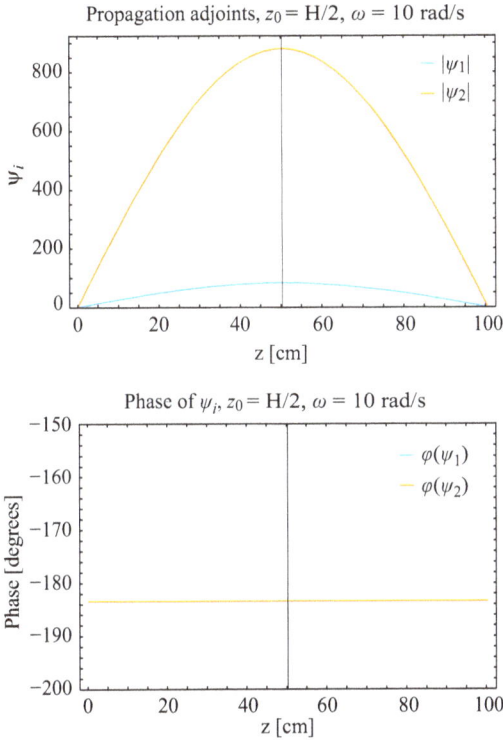

Figure 3.7   *Space dependence of the amplitude and the phase of the fast and thermal propagation adjoints for $z_0 = H/2$ in the SFR*

time) of the sodium in the core is given in [93], but according to [94], the flow speed in SFRs is between 1 to 2 m/s.

The results are shown in Figure 3.8. Not surprisingly, the phase of the cross-spectrum shows no linear dependence on the frequency, rather it shows large periodic oscillations. Somewhat surprisingly, these oscillations still take place around the theoretical linear slope of the phase, which one would get with a large local component with a very short spatial relaxation. As discussed in [47], the large oscillations are due to the interference between the point kinetic and the local components. Although globally, the oscillations follow the proper slope of the straight line, it is questionable whether the proper transit time could be obtained from a measurement. There is also an added complication, noted already in [52], that the phase angle is computed in the principal value of $(-\pi/s, \pi/s)$, which would make the plot of the phase discontinuous, and turning it into a continuous function by eliminating the jumps from $-\pi/s$ to π/s is somewhat ad-hoc and involves a guess work. One has to add that still the phase shown in Figure 3.8 is made on "clean" data, i.e. computed signals without any extra noise in the process and the data collection.

In summary, in an SFR core with these characteristics, measurements of transit time of the flow with in-core neutron noise measurements is hardly possible. This

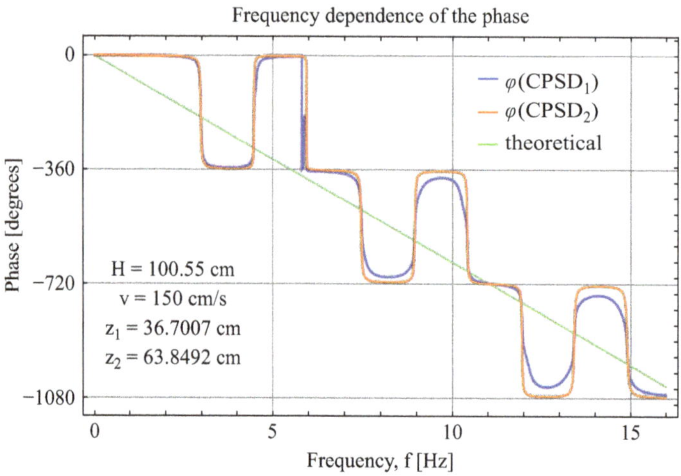

Figure 3.8   Dependence of the phase of the cross-correlations on the frequency between two axially displaced detectors for propagating perturbations in the SFR

was expected to be the case in a tightly coupled core with an extremely small height. However, flow velocity measurements in SFRs are possible with other means, such as with an eddy current flow meter (ECFM), electromagnetic (EM) flow meters and ultrasound Doppler velocimetry (UDV). These will be treated in Chapter 8. On the other hand, the appearance of void and a possible quantification of the local void fraction might be still possible with in-core neutron detectors. As discussed in Section 2.3.8.3, this does not require the existence of a strong local component.

3.2.5   Detection of core-barrel vibrations and core flowering

To monitor core-barrel vibrations, and in particular the so-called core-flowering event, ex-core detectors could be used the same way as with the pressurised water reactors (PWRs) presently in operation. This possibility was discussed in the previous chapter, Section 2.3.8.1. The sensitivity of ex-core detectors for detecting these phenomena has not yet been performed. Previous studies aimed at either confirming that core flowering does lead to reactivity changes [83], or calculating the in-core neutron noise with a realistic model of the core deformation and using the noise simulator CORE SIM [84]. Studies of ex-core neutron noise due to core deformation should be performed along the lines of this latter publication.

3.3   Gas-cooled fast reactor

The gas-cooled fast reactor (GFR) is represented here by the Allegro 75 MWth demonstrator reactor concept developed in Europe to prove the feasibility of gas cooled reactor technology [95]. The Serpent2 input files were developed within the

SafeG/TREASURE European project. The core studied contains ceramic, carbide fuel cooled with Helium at 70 bar pressure. The active core is relatively small with its height of 87 cm and radius of 61 cm. The layout of the core is seen in Figure 3.9.

Despite the small size and low power of the currently studied design, it is not classified as an SMR, rather as a Gen-IV demonstrator, this is why it is included in this chapter on Gen-IV reactors. On the other hand, the conclusions drawn from the study of the current design state may be changed when the reactor gets scaled up to the range of power reactors.

The hexagonal core was approximated by a cylinder with radius $R = 61$ cm, and in the 1D simulation this was used as the half-width of the equivalent slab. The material data of the core are shown in Table 3.2. These were calculated with $E = 1.35$ MeV as the separation energy between the two groups.

For the modelling of the core in the horizontal cross section with an 1D model with a slab thickness equal to the diameter, the absorption cross sections were modified to account for axial leakage. The static fluxes in Groups 1 and 2 are shown in Figure 3.10. Like with other fast systems, the flux in Group 1 is lower than in Group 2.

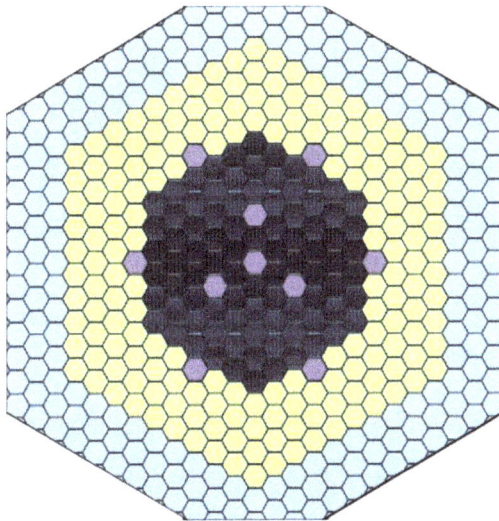

Figure 3.9   The geometry of the ALLEGRO core with reflectors

Table 3.2   Group constants and kinetic parameters for the Allegro core

Group	D	$\nu\Sigma_f$	Σ_a	Σ_R	χ_p	χ_d
1	3.81018	0.020742	0.007183	0.03359	0.5989	0.0412
2	1.92671	0.009046	0.006529	0	0.4011	0.9588

β	λ	v_1 [cm/s]	v_2 [cm/s]	R [cm]	H [cm]
0.003903	0.565632	$2.1166 \cdot 10^9$	$2.1713 \cdot 10^8$	61	87

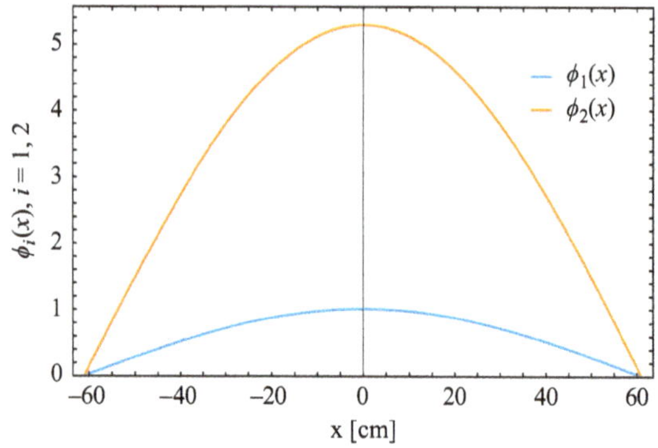

Figure 3.10   The static fluxes in the ALLEGRO core

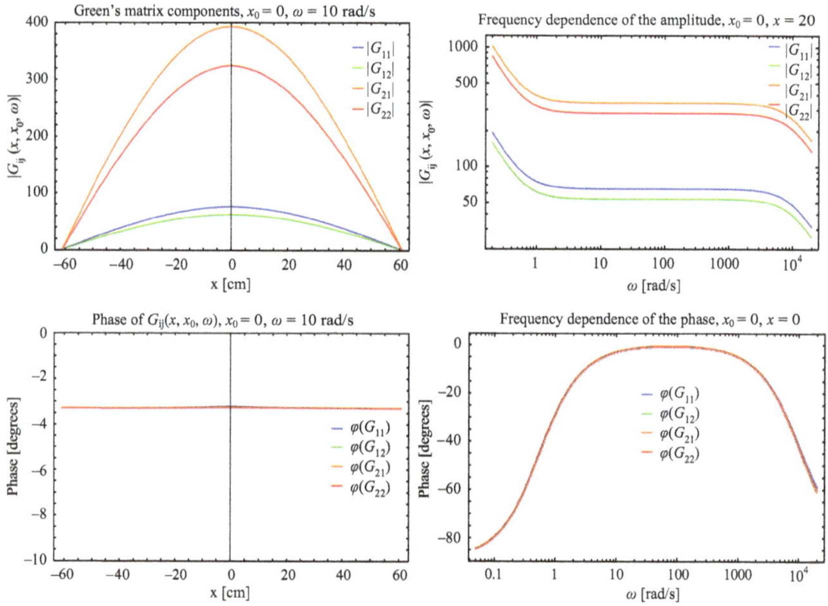

Figure 3.11   Space dependence of the amplitude and the phase of the components of the Green's function for $\omega = 10$ rad/s (left figures), and frequency dependence of the same for $x = x_0 = 0$ (right figures) in the Allegro core

3.3.1   The Green's functions

The space and frequency dependences of the amplitude and the phase of the Green's function of the Allegro core along a horizontal cross section are shown in Figure 3.11.

As it was already seen with the SFR, the amplitude of the components in Group 2 are larger than in Group 1. The figure shows that, as it could be expected, the core behaves in a rather point kinetic way. This is not surprising in view of the small

size and the generally stronger neutronic coupling in fast systems. In agreement
with this, the local component is nearly invisible. This has though not much practical importance, since in a gas-cooled reactor one cannot expect strong propagating
perturbations that would induce measurable neutron noise, whose diagnostics would
be the main use of the local component. Due to the high neutron velocities (see
Table 3.2), the plateau region is exceptionally wide, stretching from about 0.5 rad/s
to about $5 \cdot 10^4$ rad/s. This means that perturbations up to quite high frequencies will
be possible to monitor, similarly as in the SFR.

Due to the smallness and limited value of the local component, what regards
the possibilities for diagnostics, only the localisation of a reactor oscillator and a
vibrating component will be investigated. The effect of propagating perturbations
will not be considered for this core.

3.3.2 Absorber of variable strength

Based on the small, tightly coupled core, which is also seen on the space dependence
of the Green's functions, the space dependence of the noise by a variable strength
absorber is expected to follow that of the static flux, with an essentially constant phase
delay. This is confirmed by the left side of Figure 3.12. The phase and the amplitude

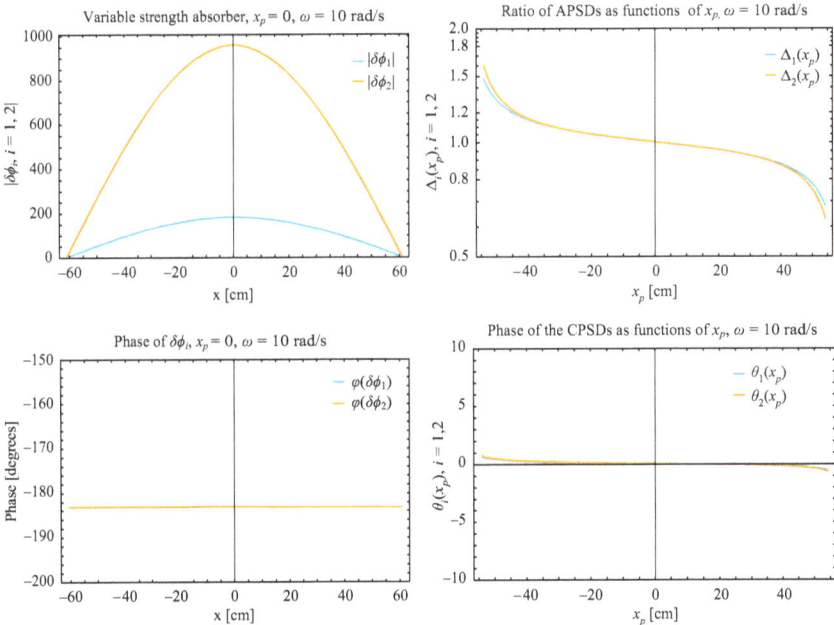

Figure 3.12   *Space dependence of the amplitude and the phase of the noise in the
two energy groups, induced by an absorber of variable strength for
$x_p = 0$ (left column) and the amplitude and phase localisation curves
for the absorber of variable strength (right column) for the Allegro
core*

are shown only for a central absorber, since there is no new information in the noise with a different position of the absorber.

The amplitude and phase localisation curves are shown in the right column of Figure 3.12. They simply confirm that in such a small, tightly coupled system finding the position of a localised perturbation of variable strength is not possible. It is interesting to notice that compared to the PWR SMR (see Chapter 4), in the vicinity of the detectors, the amplitude localisation curves differ over a larger distance. In other words, the field of view of the local component is larger than that of a PWR of the same size. This is also confirmed by the fact that the root ν of the characteristic equation* is about 0.1 cm^{-1}, a factor 4–5 times smaller than the corresponding value of ν of the PWR SMR, or any other LWR. The 4–5 times larger relaxation length may be an additional reason, besides of its small numerical weight, why the local component is not visible in the Green's functions.

3.3.3   Vibrating fuel rod

Theoretically, there is a somewhat better chance to observe some space dependence (dependence on the position of the perturbation) for a vibrating fuel pin. Figure 3.13

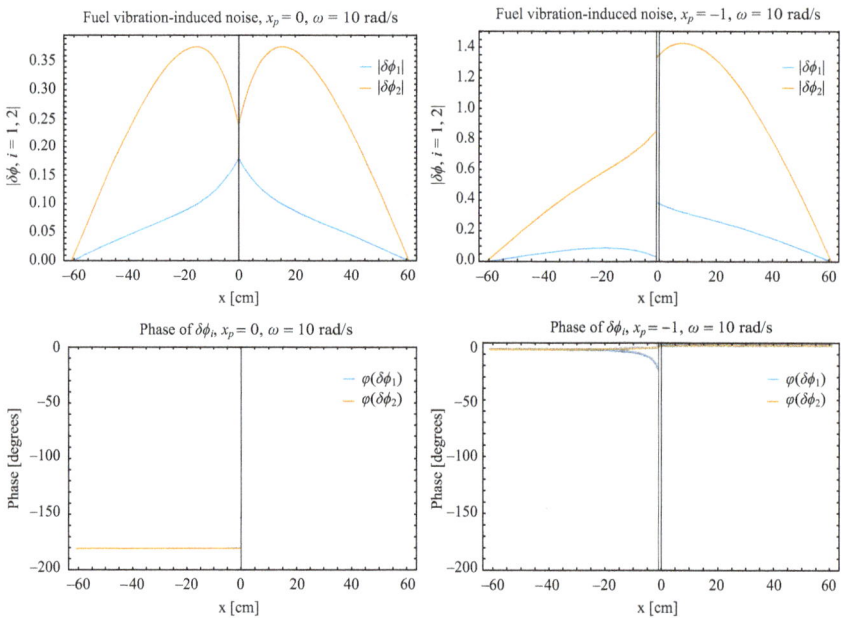

Figure 3.13   *Space dependence of the amplitude and the phase of the noise in the two energy groups, induced by a vibrating fuel rod, for $x_p = 0$ (left column) and $x_p = -1$ cm (right column), respectively, in the Allegro core*

---

*The parameter ν here stands for the local root of the characteristic equation of (2.111), and should not be mixed up with the mean number ν of neutrons per fission.

shows the amplitude and phase of the neutron noise by two fuel pin positions: a central one and one a few centimetres off the core centre.

The space dependence of the amplitude of the noise due to a central vibrating fuel pin (left side of Figure 3.13) shows a relatively large local peak in the noise in Group 1. It is also seen that its relaxation length is about 10 cm, larger than that in LWRs. The large amplitude of the local component might be somewhat surprising, in view of the fact that the local component is not seen either in the Green's functions, or in the noise of an absorber of variable strength. Moreover, in the case of the PWR SMR, although the local peak is clearly visible in the thermal Green's function G_{22}, the local peak in the noise in Group 1 is much smaller than for the present case.

The solution of this apparent contradiction lies in the fact that we are considering a vibrating fuel pin. In a thermal reactor, the local peak is in the Green's function component G_{22}, which connects the perturbation in Group 2 with the induced noise in Group 2. However, in a thermal system, a perturbation of the fission cross sections generates noise in the thermal group only via G_{21}, since $\chi_2 = 0$. This explains the relatively small peak in an LWR in the noise in the fast group, since there is no local peak in G_{21}. Hence, only a large sink is seen in the noise in Group 2.

In a fast system, where $\chi_2 > 0$, the vibrations of the fission cross section will generate a noise directly in Group 2 via the derivative of G_{22}, which will amplify the relative weight in the induced noise much more than G_{12} or G_{21} does. This explains the huge difference in the spatial structure of the fast noise for fuel pin vibrations between a fast and a thermal system. One can also add that if one considered the vibrations of a thermal absorber instead of a fuel pin, then one would have a large local peak in the thermal noise instead of the small sink that the vibrating fuel rod causes.

It is also seen in the right column of Figure 3.13, that moving the fuel pin only 1 cm off the centre, the reactivity term gets already significant enough such that the phase gets close to zero at both sides of the fuel pin, except at positions close to the fuel pin on the l.h.s. This shows that the reactivity component starts to dominate already for slightly of-centre fuel pins.

These characteristics are also reflected in the localisation curves, shown in Figure 3.14. Apart from a fuel pin position slightly off the centre, the values of the amplitude and phase localisation functions are nearly constant. The slight dependence on the position of the fuel pin will be completely hidden behind the background noise in a realistic case.

3.3.4   *Propagating perturbations*

In view of the extremely small magnitude of the local component, its relatively large spatial relaxation, and the presumably very small perturbations in the gas coolant, there is no possibility to use neutron noise signals for gas flow velocity measurement in Allegro. In general, in-core neutron noise measurements cannot be used for anything else than reactivity measurements, or to detect the appearance of local perturbations, but it will not be possible to locate them from neutron noise measurements.

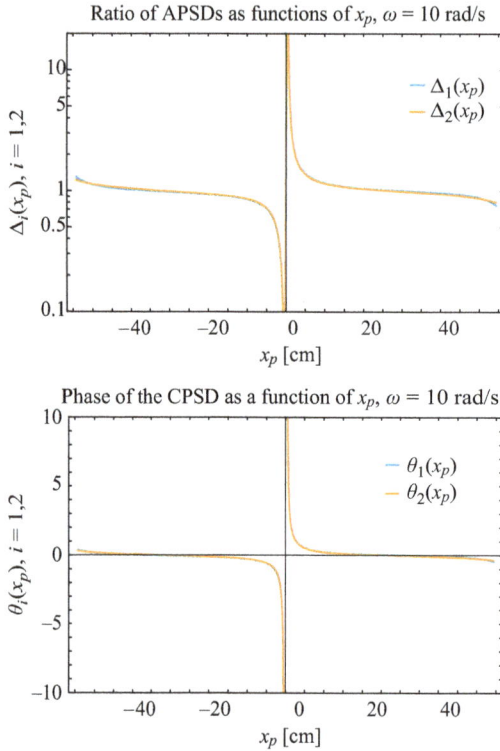

Figure 3.14   Dependence of the amplitude and phase localisation functions on the position of the vibrating fuel rod in the Allegro core

3.4   Molten salt demonstration reactor (MSDR)

The MSDR is a 750 MWth design, with graphite moderation and LiF-BeF2-ThF4-UF4 fuel salt. The active core radius is 304 cm and the core height is 640 cm. The MSDR input files were modified based on open source information. The layout of the system is shown in Figure 3.15.

The group constants and the kinetic parameters used in the calculation are shown in Table 3.3. Given that the MSDR is a thermal system, Groups 1 and 2 are the conventional fast and thermal groups, respectively. As with all systems treated in this book, these are the original group constants derived from the Serpent input file, which then will be modified for leakage when the core is modelled in the horizontal or vertical dimensions.

The static fluxes in Groups 1 and 2 in the horizontal cross section of the core, along a diameter, are shown in Figure 3.16. Interestingly, the thermal flux is larger in the system than the fast one, which is different from that of the thermal light water systems.

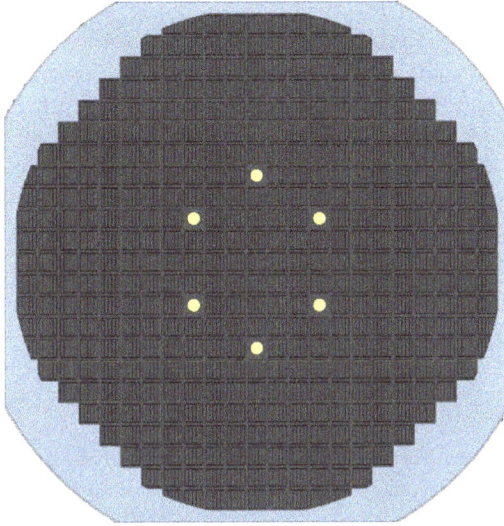

Figure 3.15   The geometry of the MSDR core

Table 3.3   Group constants and kinetic parameters for the MSDR core

Group	D	$\nu\Sigma_f$	Σ_a	Σ_R	χ_p	χ_d
1	1.061	0.000104	0.000453	0.0071	1	1
2	0.841634	0.001505	0.001045	0	0	0

β	λ	v_1 [cm/s]	v_2 [cm/s]	R [cm]	H [cm]
0.005476	0.40735	$7.666 \cdot 10^6$	461070	304	640

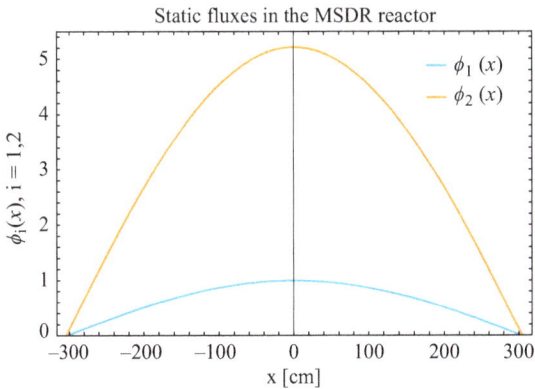

Figure 3.16   The static fluxes in the MSDR core

3.4.1   The Green's functions

According to the conclusions of Chapter 2, Sections 2.3.6.1 and 2.3.6.2, the dynamic properties of the MSRs will be calculated as they were a traditional core, i.e. with fuel velocity $u = 0$, but with a reduced effective delayed neutron fraction β_{eff}, (2.202). The reduction is by the factor H/T, that is the ratio of the core height to the total length of the recirculation loop, or rather the time spent in the core vs the total recirculation time. There are no concrete data available on this factor, hence we take an estimate that the external loop is 1.5 times longer than the core height (the transit time outside the core is 1.5 times of the transit time), which yields a factor $1/2.5 = 0.4$ for the reduction of the delayed neutron fraction. All dynamic calculations were made with this reduction factor.

The interpretation of the dynamic properties of the MSDR is somewhat complex. Although the investigated MSDR design is relatively large by geometrical size (6×6 m), the thermal diffusion length is also large in a graphite moderated system, so in terms of the mean free path, although the core is still large, it is not excessively large. Besides, the reduction of the delayed neutron fraction, which emulates the dynamic properties of MSRs as compared to traditional reactors, increases the weight of the point kinetic component further.

On the other hand, both the low neutron velocities in both groups, as well as the low value of β_{eff} act into the direction that the upper break frequency of the neutronic transfer functions (shown soon) is considerably lower than in both LWRs and fast reactors. This means that at the plateau frequency of the LWRs, which coincides with the frequency of the band-limited perturbations, such as vibrations, the amplifying effect of the low β_{eff} on the point kinetic component is already counteracted.

This is illustrated in Figure 3.17, which shows the space dependence of the amplitude and the phase of the Green's functions for the MSDR core along a horizontal cross section of the reactor. Unlike in most of the similar figures for the other designs, these functions are shown here not only for two different positions x_0, but also for two different frequencies. In all other figures, a frequency of 10 rad/s was used, corresponding to about 1.6 Hz, which is well within the plateau region for all other reactor designs, and also corresponding to a characteristic eigenfrequency of mechanical vibrations of control and fuel assemblies. However, as Figure 3.18 will show, in this particular core 10 Hz is already beyond the upper break frequency. Hence the space dependence of the Green's functions is shown here for both 1 and 10 rad/s.

The Greens functions show a noticeable departure from point kinetic behaviour at 10 rad/s but this deviation is relatively moderate, smaller than with the commercial BWR shown in Chapter 2. The large size of the core affects much more the space dependence of the phase which, at 10 rad/s, varies about 30° from the centre of the core to the core boundary. This is the largest variation of the phase among all the designs considered, which indicates that in this reactor, also the phase can be helpful for locating localised perturbations.

The right column of the figure shows that at $\omega = 1$ rad/s, the neutronic response is significantly more point kinetic, which is due to the amplified weight of the point

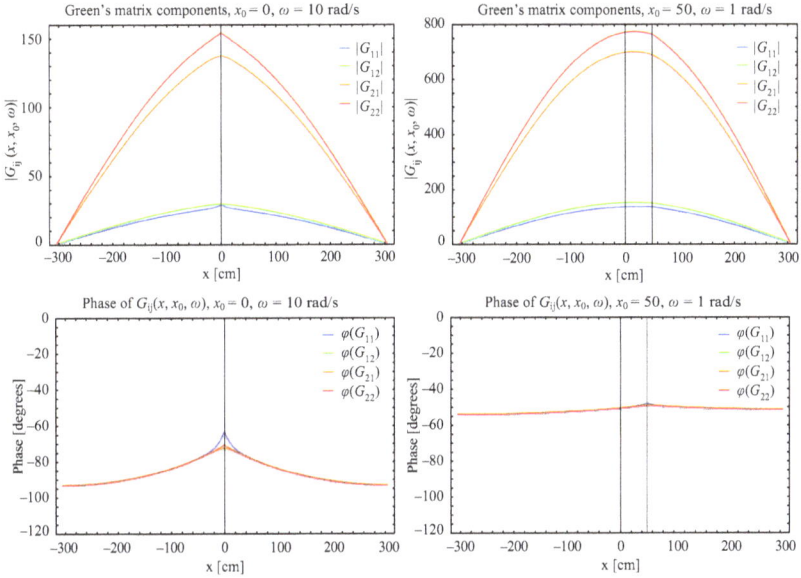

Figure 3.17   Space dependence of the amplitude and the phase of the components of the Green's function for $x_0 = 0$ and $\omega = 10$ rad/s (left figures), and $x_0 = 50$ cm and $\omega = 1$ rad/s (right figures) in the MSDR core

kinetic component in an MSR. The variation of the phase across the core is also much more moderate than at 10 rad/s.

One peculiar characteristic of the Green's functions, which is in line with the behaviour of the static flux, is that in this nearly perfectly thermalised core, the amplitudes of the fast components G_{11} and G_{12} are smaller than that of the thermal ones G_{21} and G_{22}. Also, the small local peak is only visible in the fast component G_{11}. Such amplitude relationships between Groups 1 and 2 are seen in cores with a fast spectrum, but not in the thermal spectrum LWRs. It is therefore somewhat of a contradiction that in an even better moderated core than the light water cores, one observes characteristics of fast systems. On the other hand, we need to keep in mind that Groups 1 and 2 have a different meaning in the fast cores and in the thermal cores, which might be part of the explanation of this apparent contradiction.

Because of the very low upper break frequency of the MSDR core, all further figures with a fixed frequency will be shown for both 1 and 10 rad/s. This is because, although in a general case the plateau frequency behaviour is the most interesting, on the other hand the majority of the interesting perturbations related to mechanical vibrations are in the range of 1 Hz or above, therefore the response of the MSDR at those frequencies is also interesting.

The frequency dependence of the amplitude and phase of the Green's functions is shown in Figure 3.18. It is seen that the plateau region is so narrow, that there is no frequency region where the amplitude is even approximately constant. The upper break frequency is around 2 rad/s, which is extremely low. Also, the phase behaviour

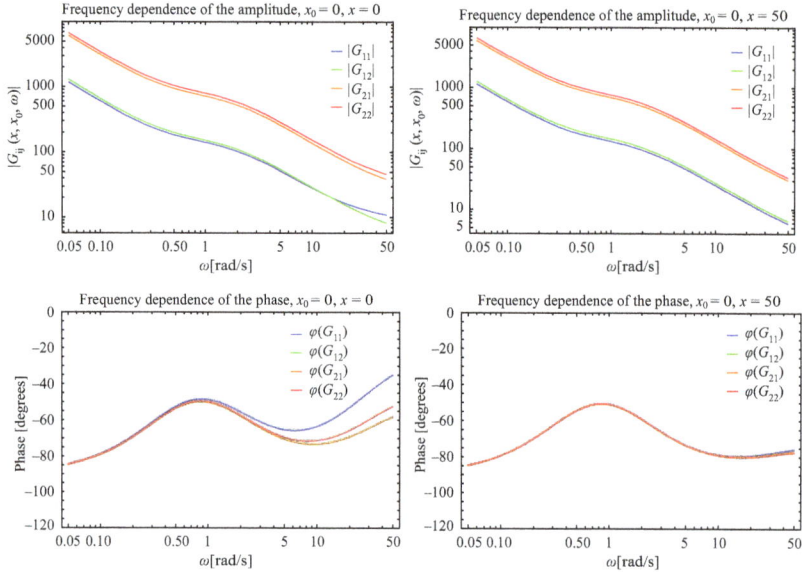

Figure 3.18   Frequency dependence of the amplitude and the phase of the components of the Green's function for $x_0 = 0$ and $x = 0$ (left figures), and $x = 50$ cm (right figures) in the MSDR core

is markedly different from the other reactor types in that the minimum phase delay, i.e. the top of the phase curve at the plateau region, is around $40°$, which is much larger than with the other reactor types.

There is not much visible difference in the frequency dependence of the amplitudes for $x = 0$ (i.e. at the point of the perturbation) and $x = 50$ cm. However, the behaviour of the phase above 2 rad/s, i.e. above the upper break frequency, is markedly different. At the point of the perturbation (bottom left figure), the effect of the local component leads to the decrease of the phase delay (the upward trend in the phases) in all components, but primarily in the fast component G_{11}. At a point 50 cm away from the perturbation (bottom right plot), the effect of the local component diminishes. At points even further away from the perturbation (not shown here), the phase delay increases monotonically (the phase decreases monotonically) as a function of the frequency.

The behaviour of the amplitude and the phase as functions of the frequency shows a very distinct difference from both the thermal spectrum PWRs and BWRs, as well as from the fast reactors. In general, in the noise diagnostics literature, there have been no cases of graphite moderated reactors investigated for their dynamic transfer properties, so the study presented here will therefore be useful not only for the MSDR.

3.4.2   Absorber of variable strength

Because of the low value of the upper break frequency of the neutronic transfer of the MSDR, we investigate the noise of the absorber of variable strength both at 1 and 10 rad/s. The first case is shown in Figure 3.19.

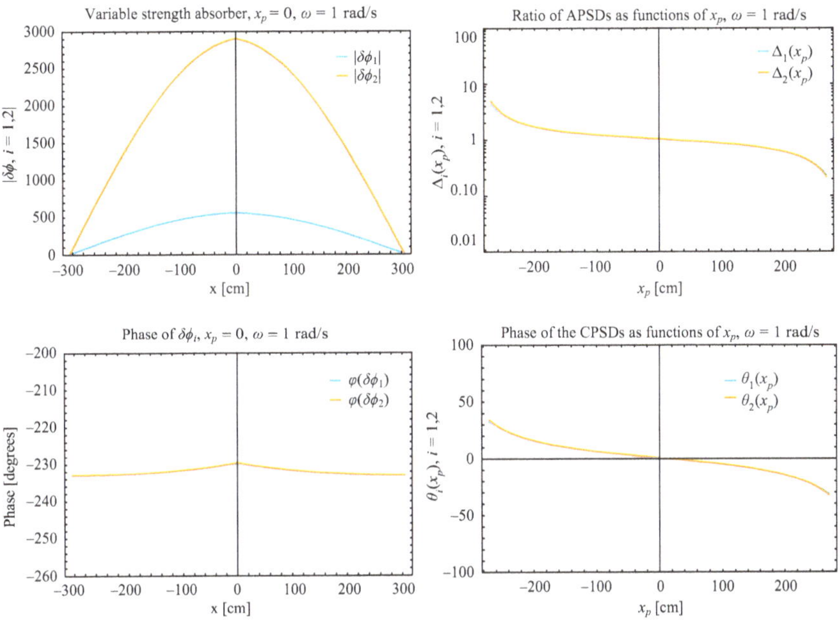

Figure 3.19   Space dependence of the amplitude and the phase of the noise in the two energy groups, induced by an absorber of variable strength for $x_p = 0$ (left column) and the amplitude and phase localisation curves for the absorber of variable strength (right column) for the MSDR core at the frequency of 1 rad/s

The space dependence of the amplitude and the phase at this frequency (plots on the l.h.s. of the figure), which is within the plateau region, shows a relatively moderate deviation from point kinetics. The deviation from point kinetics is more visible on the space dependence of the phase. The space dependence manifests itself much stronger on the localisation curves (r.h.s. plots). Both the ratio of the APSDs, as well as the phase of the CPSDs of the two detectors vary moderately, but sufficiently much that both can be used to find the location of a source of the type of a variable strength absorber, even at this low frequency.

The induced neutron noise and the localisation curves at 10 rad/s are shown in Figure 3.20. The situation is markedly different from that of 1 rad/s. Interestingly, the space dependence of the amplitude of the noise itself is still less pronounced than for the commercial BWR or for the PWR-type SMRs (see the next chapter), but the space dependence of the phase shows a large deviation from point kinetics. This is mirrored in the dependence of the localisation curves (right column). The phase of the CPSD of the two peripheral detectors show a much larger variation than in any of the other reactor types considered.

From what has been seen so far, one can draw the conclusion that in the MSDR with a large geometrical size, the deviation from point kinetics can be noticed primarily on the space dependence of the phase. Compared with the PWR SMRs, the space dependence of the amplitude of the noise in the MSDR is more point kinetic, whereas

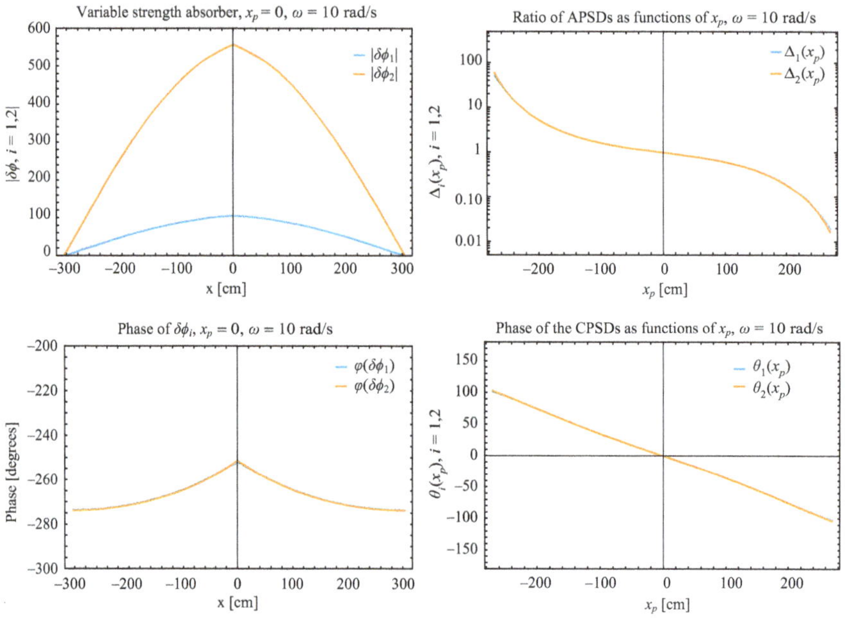

Figure 3.20   Space dependence of the amplitude and the phase of the noise in the two energy groups, induced by an absorber of variable strength for $x_p = 0$ (left column) and the amplitude and phase localisation curves for the absorber of variable strength (right column) for the MSDR core at the frequency of 10 rad/s

that of the phase is less point kinetic in the MSDR. This is a new observation in the sense that in the traditional systems the shift from point kinetic to space-dependent, as a function of the frequency or the system size, takes place simultaneously for the amplitude and the phase, or even faster for the amplitude than for the phase. Now it is seen that for an MSDR, the transition from point kinetic to space-dependent behaviour goes in a different way for the amplitude and the phase, the latter changing faster than the amplitude.

3.4.3   Vibrating absorber rod

Although for the solid fuel reactors we considered the neutron noise induced by the vibrations of a fuel rod, this is not practical for the MSRs, since there are no solid fuel elements which could vibrate laterally. On the other hand, in thermal MSRs with a solid moderator such as a graphite block of the MSDR, control rods may be used and those might be exposed to mechanical perturbations which can make them to vibrate.

The spatial structure of the neutron noise and the localisation possibilities will be shown for both $\omega = 1$ and $\omega = 10$ rad/s. The space dependence of the amplitude and the phase of the induced noise for $\omega = 1$ rad/s is shown in Figure 3.21. Corresponding to the fact that both the thermal absorption cross section, as well as the thermal flux

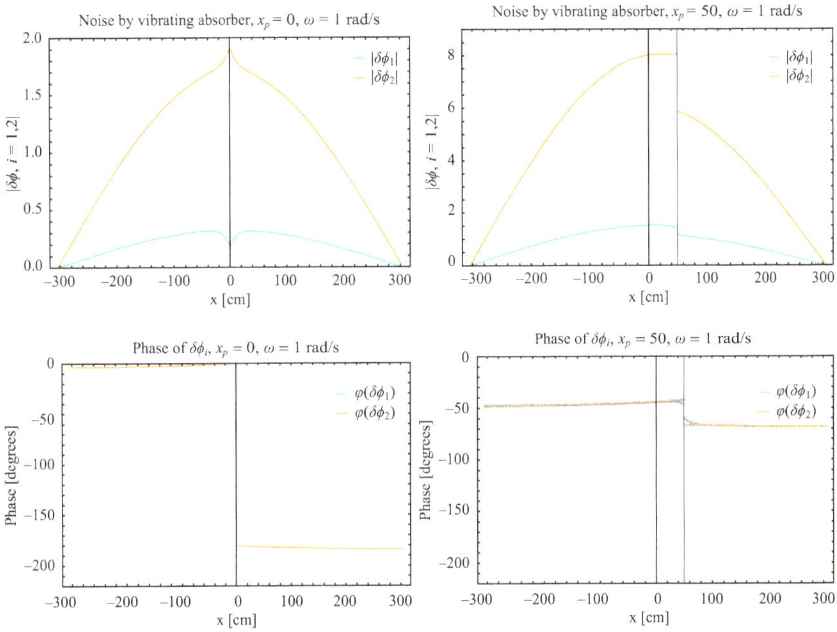

Figure 3.21   *Space dependence of the amplitude and the phase of the noise in the two energy groups, induced by a vibrating absorber, for $x_p = 0$ (left column) and $x_p = 50$ cm (right column), respectively, at $\omega = 1$ rad/s in the MSDR core*

are larger than their fast counterparts, the thermal noise is larger, and the local peak is in the thermal noise. For the central absorber, the phase is opposite at the two sides of the absorber, similarly to the noise induced by a vibrating fuel pin, but with the trivial difference that the noise is in-phase on the l.h.s. and out-of-phase on the r.h.s. of the absorber, which is opposite to the noise by the vibrating fuel rod. At the equilibrium absorber position at $x_p = 50$ cm, which is only moderately off the centre, the contribution from the reactivity term is already large enough such that the values of the phase at the two sides of the absorber are already quite close to each other.

A similar plot with a central and off-central vibrating absorber rod, but at the frequency of $\omega = 10$ rad/s is shown in Figure 3.22. Regarding the space dependence, the same differences can be seen as compared to the $\omega = 1$ rad/s case as with the absorber of variable strength, namely there is very little change in the amplitude, but a more significant change in the phase, which is now decaying must faster as the distance to the perturbation increases. Another difference is the dependence of the noise on the equilibrium position of the absorber. At this frequency the contribution of the reactivity term is noticeably smaller. This is seen on the fact that at a distance from the centre twice as large as for $\omega = 1$ rad/s, the gap in the phase at the two sides of the absorber is still larger than in the previous case.

The localisation curves for $\omega = 1$ and 10 rad/s, respectively, are shown in Figure 3.23. These once again show a structure which is different from all the other

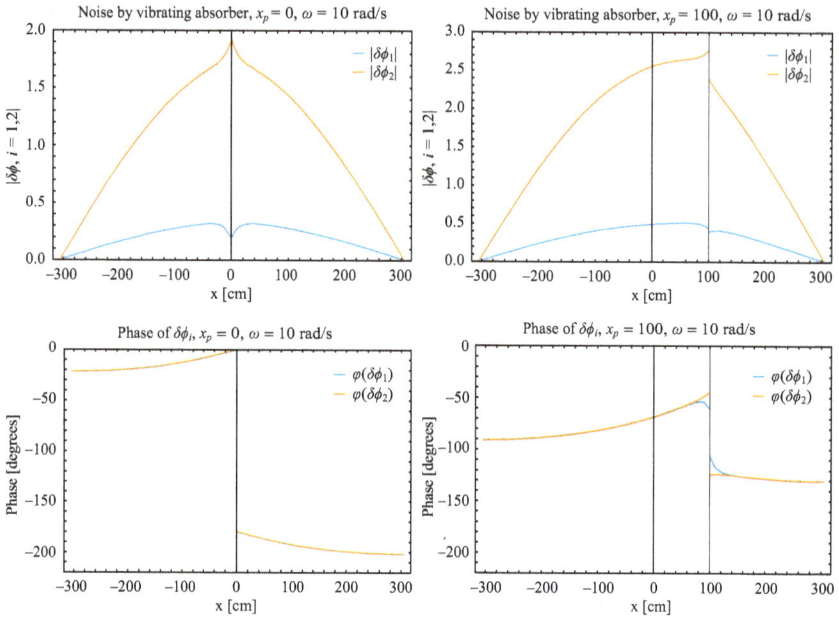

Figure 3.22   Space dependence of the amplitude and the phase of the noise in the two energy groups, induced by a vibrating absorber, for $x_p = 0$ (left column) and $x_p = 100$ cm (right column), respectively, at $\omega = 10$ rad/s in the MSDR core

reactor types, but conform with the tendencies which were observed before on the different behaviour of the amplitude and the phase. At both frequencies, the inverse of all curves is two- or many-valued, hence the localisation is not unambiguous. Interestingly, at $\omega = 1$ rad/s, the space dependence of the amplitude localisation is more marked than at $\omega = 10$ rad/s. The phase localisation curves, even if their inverse is also two-valued, change monotonically in both halves of the core. The ambiguity of identifying the location of the vibrating absorber is larger at 1 rad/s; in the case of 10 rad/s, the amplitude and phase localisation curves together are suitable to locate the position of the vibrating absorber unambiguously. The ambiguity at $\omega = 1$ rad/s means that for a given value of the amplitude and phase localisation curves can be attributed to two distinct positions; on the other hand these positions can be estimated with good accuracy. In a real practical case the problem is 2D and at least three detectors need be used, which will eliminate the ambiguity which is seen in the present 1D calculations.

3.4.4   Propagating perturbations

As usual, for the calculations of the propagating properties, the height of the core was selected as the width of the 1D model, and the absorption cross sections were readjusted for the changed leakage.

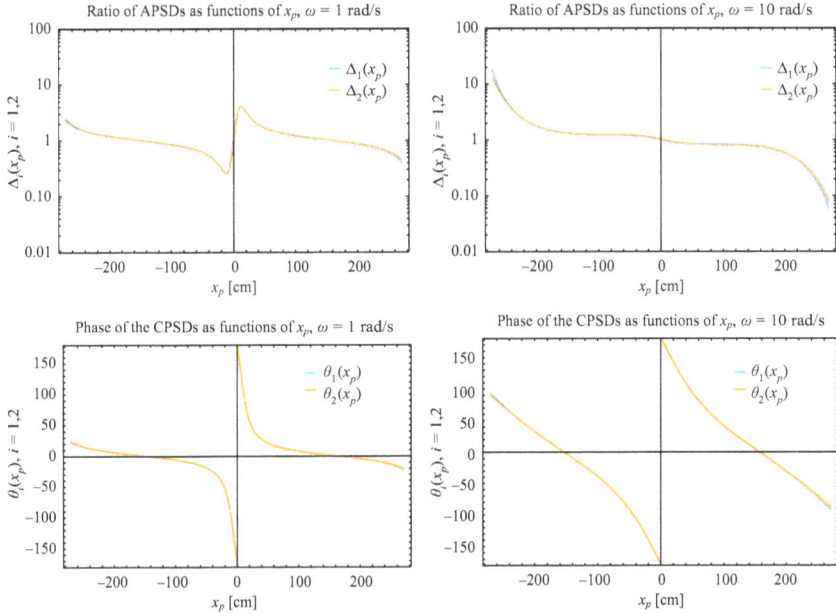

Figure 3.23 *Space dependence of the amplitude and the phase of the localisation curves for a vibrating absorber for the MSDR core at the frequency $\omega = 1$ rad/s (left column) and for $\omega = 10$ rad/s (right column)*

Similarly to other non-light water systems, one needs to specify the effect of the density perturbations of the coolant/fuel on the various cross sections. In an MSR, the fuel is dissolved homogeneously in the coolant, and a density fluctuation will perturb both the absorption and the fission cross sections. Because the thermalisation takes place in the graphite moderator, it is reasonable to assume that the removal cross section is not perturbed by the density fluctuations of the fluoride salt. With regard to the internal magnitude relationships in the fluctuations of the absorption and fission cross sections, we use the same assumption as with the SFR, namely that the magnitudes of these cross-section fluctuations are proportional to their static values. Hence we define the (arbitrarily normalised) coefficients α_1 and α_2 as follows:

$$\Sigma_{tot} = \Sigma_{a1} + \Sigma_{a2} + \nu\Sigma_{f1} + \nu\Sigma_{f2} \tag{3.1}$$

Then, accounting for the fact that in (2.120) and (2.121) one can use the relationship $\phi_2(z) = c_\mu \phi_2(z)$, we define

$$\alpha_1 = \frac{\Sigma_{a1} - \nu\Sigma_{f1} - c_\mu \nu\Sigma_{f2}}{\Sigma_{tot}} \tag{3.2}$$

and

$$\alpha_2 = \frac{c_\mu \Sigma_{a2}}{\Sigma_{tot}} \tag{3.3}$$

In (3.2) the factor

$$\left(1 - \frac{i\omega\beta}{i\omega + \lambda}\right) \tag{3.4}$$

which should multiply the fission cross sections $\nu\Sigma_{f1}$ and $\nu\Sigma_{f2}$ (see (2.79)) was neglected. This is because the very week frequency dependence of (3.4), which is corroborated by the very small value of β_{eff}, is completely negligible besides the fast oscillating frequency dependence of the exponential factor in (2.155), which expresses the propagating character of the perturbation.

With the definitions (3.2) and (3.3), for the MSDR core one has

$$\alpha_1 = -1.88868 \tag{3.5}$$

$$\alpha_2 = +1.85388 \tag{3.6}$$

These coefficients need to be used in the calculation of the fast and thermal propagation adjoints Ψ_1 and Ψ_2 in (2.162) and (2.163).

Equations (3.5) and (3.6) show a remarkably interesting result. Despite the fact that the fluctuations of the removal cross sections do not enter the formula, one still has $\alpha_1 \approx -\alpha_2$, similarly to the BWR case when only the removal cross section is perturbed. The reason for this is that the absorption cross sections appear in the noise source with an opposite sign as the fission cross sections. Since the MSDR is thermal system, these latter appear only in α_1 and are the dominating part, hence α_1 is negative. In α_2 only the absorption cross section appears, therefore it is positive. The fact that they are quantitatively close to each other is due to the specific nuclear parameters of the core, and cannot be regarded as universal, but in thermal MSRs at least α_1 and α_2 will have opposite sign. As long as their magnitude is close to each other, the propagating adjoints will amplify the presence of the local component, similarly to the removal adjoint of BWRs.

This is illustrated in Figure 3.24, which shows the spatial dependence of the amplitude and the phase of the propagation adjoints for $\omega = 1$ and 10 rad/s, respectively.

The local component is much more visible than in the pure Green's functions, even at $\omega = 1$ rad/s, although it appears in the fast propagation adjoints as a dip. At $\omega = 10$ rad/s, a large and quite sharp local component is visible in both the fast and the thermal propagation adjoints. These figures also demonstrate that with increasing frequency, the deviation of the shape from the point kinetic one, i.e. assuming a more "peaked" form, becomes more obvious.

The frequency dependence of the phase is shown in Figure 3.25 for a propagation velocity $v = 400$ cm/s. As could be expected from the peaked character of the propagation adjoints, the phase curves show a smooth linear dependence on frequency. After the initial fluctuations, they both decay with the same slope. The phase of the fast CPSD has a somewhat peculiar behaviour, namely it has a 180° jump around 3 Hz, although even in the continuation it keeps the same slope as the theoretical one. This is consistent with the fact that the local component of the fast propagation Green's function also has a similar shift between $\omega = 1$ and 10 rad/s, as is seen from Figure 3.24, namely that the sink at 1 rad/s turns to a peak at 10 rad/s.

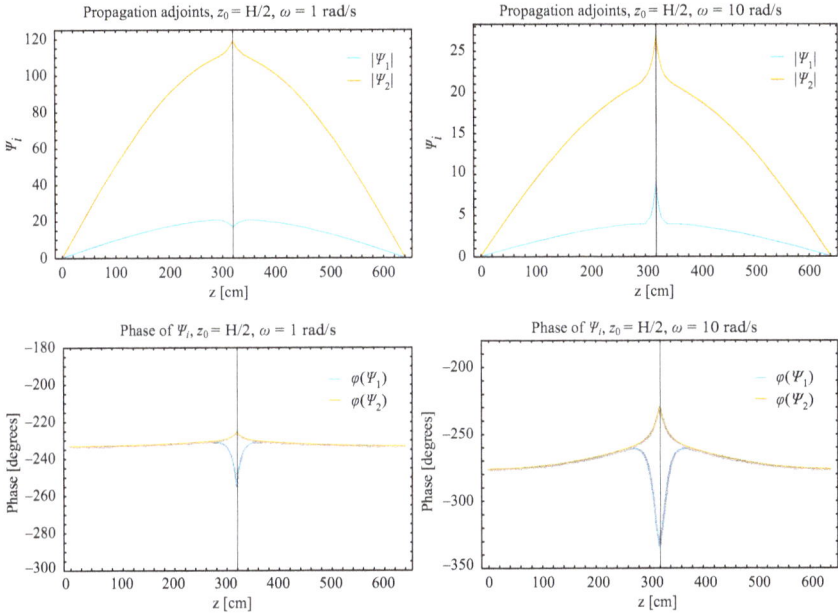

Figure 3.24 Spatial dependence of the amplitude and the phase of the propagation adjoints for the MSDR core at the frequency $\omega = 1$ rad/s (left column) and for $\omega = 10$ rad/s (right column)

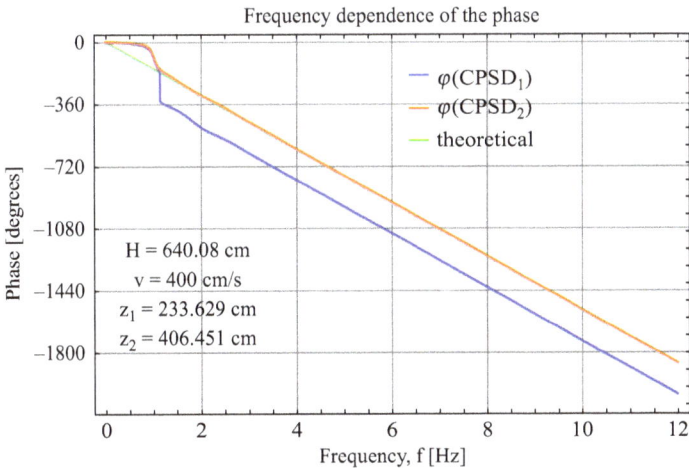

Figure 3.25 Dependence of the phase of the cross-correlations on the frequency between two axially displaced detectors for propagating perturbations in the MSDR

The conclusion is that despite the weak space dependence of the transfer properties of the MSDR core, what regards the transfer functions of the propagating perturbation, the same advantageous situation prevails as with BWRs, namely that due to the characteristics of the noise source, the local component is amplified in the propagation adjoints. This expedites the determination of the transit time between axially displaced detectors, and consequently the surveillance and diagnostic of void generation and transport in the core as well. The fortuitous quantitative values of the weight parameters α_1 and α_1 depend though on partly the core parameters, and partly on the assumptions made on the relative weight of the absorption and fission cross-section fluctuations due to density fluctuations of the molten salt.

A more reliable estimate of the relative weights of the cross-section fluctuations for a change in the temperature or density of the fluoride salt can be calculated by using in-core fuel management neutronic system codes, such as CASMO and SIMULATE-3 [96]. Such an approach was used in [33], where such calculations were made for the determination of the variation of the group constants in light water and HWRs. Similar calculations should be also made for the MSDR, but these are outside the scope of this book.

Chapter 4

A basic study of the structure of the core response in the basic planned SMRs

Imre Pázsit[1], Zsolt Elter[2] and Hoai-Nam Tran[3]

4.1 Introduction

In this chapter, the dynamic properties of some planned small modular reactor (SMR) types are investigated with the analytical model, in a similar way as in the previous chapter. This is made in the same format: first the space and frequency dependence of the amplitude and the phase of the components of the Green's function matrix are calculated. Then the system response, i.e. the neutron noise induced by three different perturbations, is investigated, namely the noise induced by an absorber of variable strength, by a vibrating absorber, and by propagating perturbations. The possibilities of diagnostics, i.e. locating the position of the first two perturbations, and determining the transit time of the latter, are assessed.

The following SMR types are considered:

1. A small 200 MWth core, which is constructed from shortened fuel elements of a typical pressurised water reactor (PWR) (AP1000);
2. An advanced LWR SMR
3. A fast spectrum molten salt reactor (MSFR)
4. A lead-cooled fast SMR (LFRSMR)
5. A fluoride salt-cooled high-temperature reactor (FHR)

4.2 A small 200 MWth AP-1000 core (PWR SMR)

The first SMR to be investigated is a small 200 MWth PWR core with a thermal spectrum, which is constructed from fuel elements of a typical PWR (AP1000), but with a shortened core height of 220 cm. The core radius is 73.5 cm. The layout of the system is shown in Figure 4.1. The data used in the calculations are listed in Table 4.1.

The static fluxes in Groups 1 and 2 are shown in Figure 4.2. These show the usual characteristics of light water-moderated thermal systems, with a fast flux higher than

[1]Division of Subatomic, High Energy and Plasma Physics, Department of Physics, Chalmers University of Technology, Sweden
[2]Department of Physics and Astronomy; Division of Applied Nuclear Physics, Uppsala University, Sweden
[3]Phenikaa Institute for Advanced Study, Phenikaa University, Hanoi, Vietnam

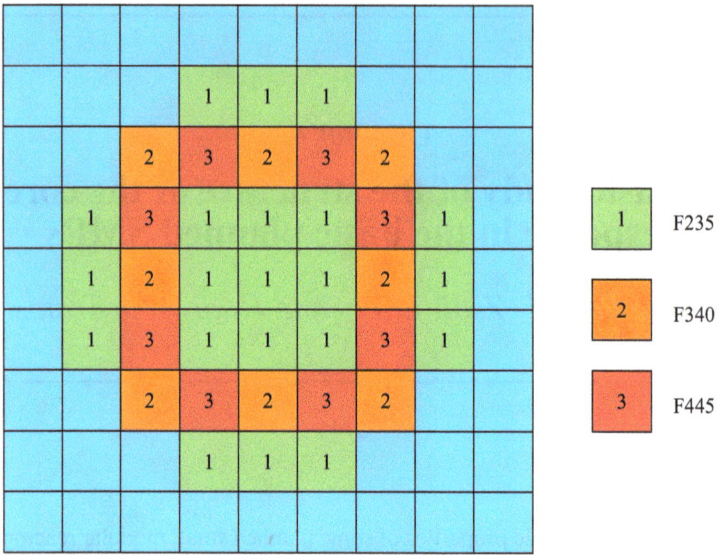

Figure 4.1   The geometry of the PWR SMR core

Table 4.1   Group constants and kinetic parameters for the PWR SMR core

Group	D	$\nu\Sigma_f$	Σ_a	Σ_R	χ_p	χ_d
1	1.43756	0.00566	0.00733	0.015144	1	1
2	0.37226	0.14249	0.10167	0	0	0

β	λ	v_1 [cm/s]	v_2 [cm/s]	R [cm]	H [cm]
0.00580	0.0848	$1.82020 \cdot 10^7$	$4.1285 \cdot 10^5$	73.5	220

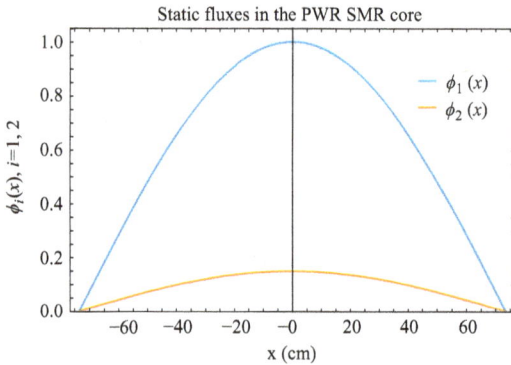

Figure 4.2   The static fluxes in the PWR SMR core

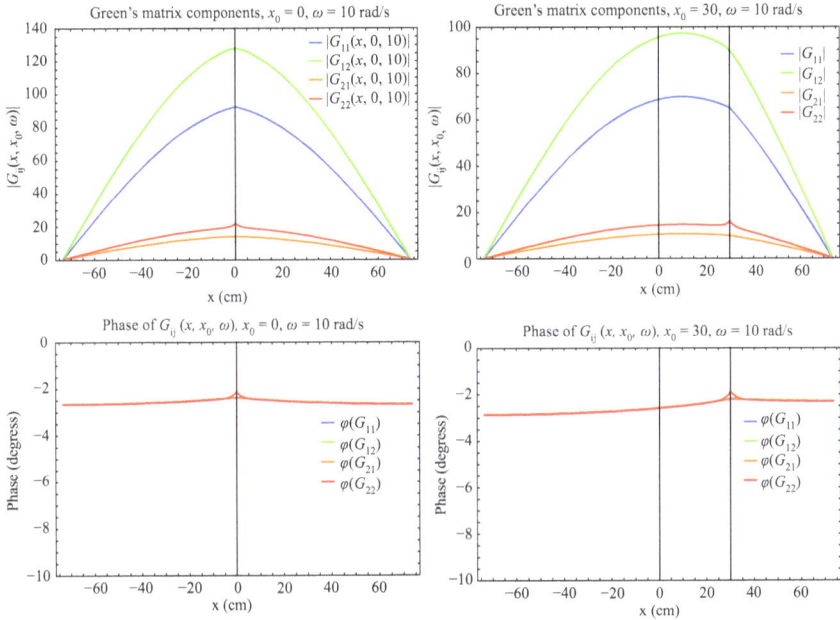

*Figure 4.3   Space dependence of the components of the Green's function (left
column) and the thermal and fast removal adjoints (right column) in a
PWR SMR core*

the thermal one. The spectrum is somewhat harder than the average, with a factor
approximately 5 between the flux amplitudes in the two groups.

4.2.1   The Green's functions

The space dependence of the amplitude and the phase of the components of the
Green's matrix is shown in Figure 4.3. Here the x-axis corresponds to the radial
dimension of the core, with $R = 73.5$ cm. Not surprisingly, they are similar to that of
the small boiling water reactor (BWR) core, discussed in Chapter 2, due to the small
size, but show somewhat more pronounced space dependence of the global component. Likewise, the local component is somewhat more visible in both the amplitude
and in the phase.

The frequency dependence of the amplitude and the phase of the four components of the Green's function for the PWR SMR are shown in Figure 4.4 for two
different detector positions: one at the position of the perturbation (left column), and
one at 30 cm. A behaviour very typical for a small LWR is seen, with a relatively
wide plateau stretching from 0.1 to over 100 rad/s.

The dependence of the phase on frequency shows that in the plateau region, the
maximum of the phase is about $-5°$. The value of the phase in the plateau region is
an interesting parameter, which can be used to classify the different systems. In fast

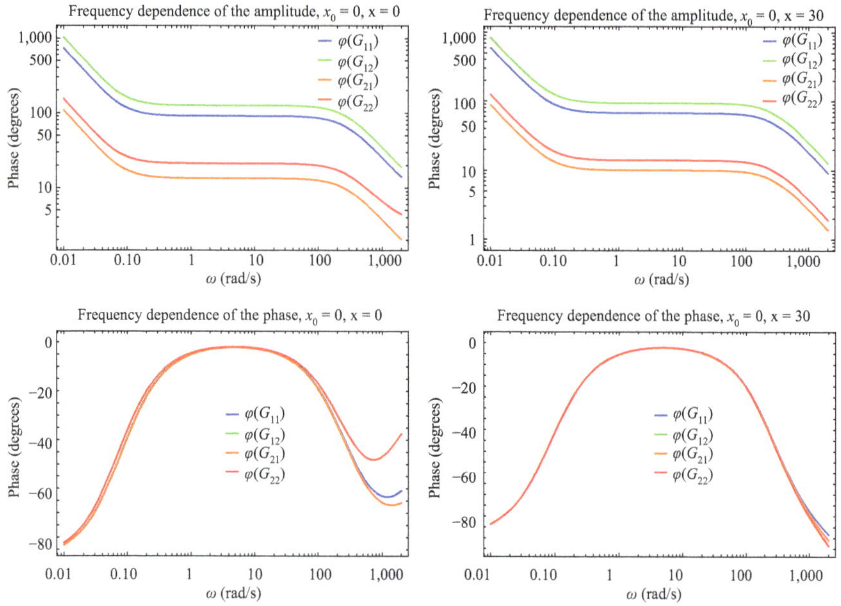

Figure 4.4   Frequency dependence of the amplitude and the phase of components of the Green's function for $x_0 = 0$, with $x = 0$ (left column) and $x = 15$ cm (right column), respectively, in a PWR SMR

reactors, the phase is very close to zero, as one could see it in the previous chapter in connection to the SFR and Allegro systems, which are both fast spectrum reactors. For extremely well-thermalised systems, such as the graphite moderated MSDR (and, as will be seen later, in the fluoride salt high temperature pebble bed reactor, where the moderation takes place in the graphite of the pebbles), the phase delay is between $20°$ and $40°$.

4.2.2   Absorber of variable strength

The amplitude and the phase of the neutron noise induced by a variable strength absorber in the PWR SMR are shown in Figure 4.5 for two different absorber positions x_p. As in the previous cases, the space dependence of both the amplitude and the phase are similar to those of the components of the Green's function. With regard to the phase, it lies moderately below $180°$, because the induced fluctuations are out of phase for variations of the absorbing cross sections.

As in the previous chapter, the possibilities of localising the absorber are explored by placing two detectors, one at $x = -0.9a$, and one at $x = 0.9a$, and taking the ratio of their APDSs, and the phase of the CPSDs. This is performed by the amplitude and phase localisation functions $\Delta_i(x_p)$ and $\theta_i(x_p)$, i = 1.2. The results are shown in Figure 4.6.

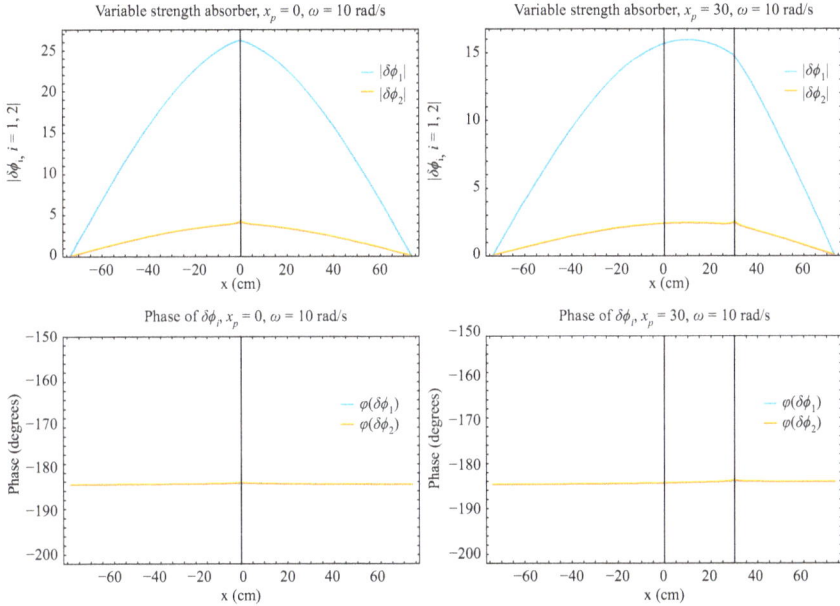

Figure 4.5   Space dependence of the amplitude and the phase of the noise in the two energy groups, induced by an absorber of variable strength for $x_p = 0$ (left column) and $x_p = 30$ cm (right column), respectively, in a PWR SMR core

Like in the case of the small BWR discussed in Chapter 2, both the fast and the thermal noise are suitable for the localisation of the absorber through the ratio of the detector APSDs. The variations of the phase are very minor and are of no help in this task.

4.2.3   Vibrating fuel pin

The amplitude and the phase of the neutron noise (divided by the vibration amplitude $\varepsilon(\omega)$), induced by the vibrations of a fuel rod at two different equilibrium rod position are shown in Figure 4.7.

The amplitude and phase for a central vibrating fuel rod always show similarities between any reactor; the phase is zero on the r.h.s. and is $-180°$ left from the absorber. The amplitude is continuous, but its derivative is discontinuous. Since this is a small reactor, when the equilibrium position of the fuel rod is moved off the centre to -10 cm, the reactivity component already starts dominating. The amplitude has a discontinuity in the fast flux, and the amplitude of the global component of the thermal flux has a discontinuity, but together with the local component, the thermal noise is continuous at the fuel pin position x_p.

Because of the rapidly increasing dominance of the point kinetic component for fuel pin positions at increasing distances from the centre of the core, locating the

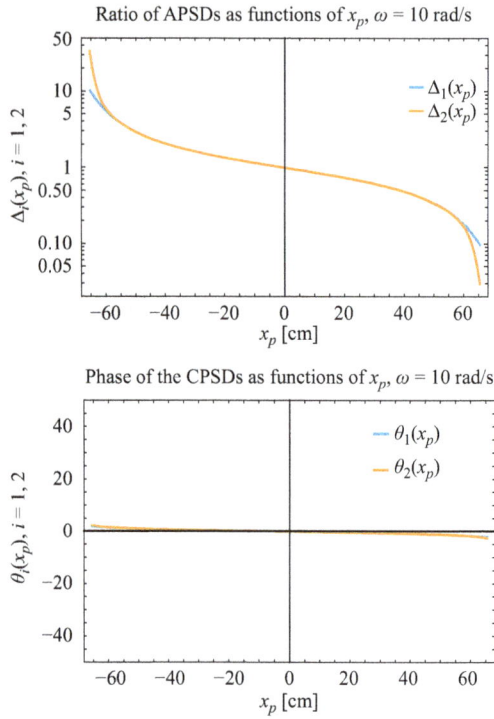

Figure 4.6   *Dependence of the amplitude and phase localisation functions on the position of the variable strength absorber in a PWR SMR*

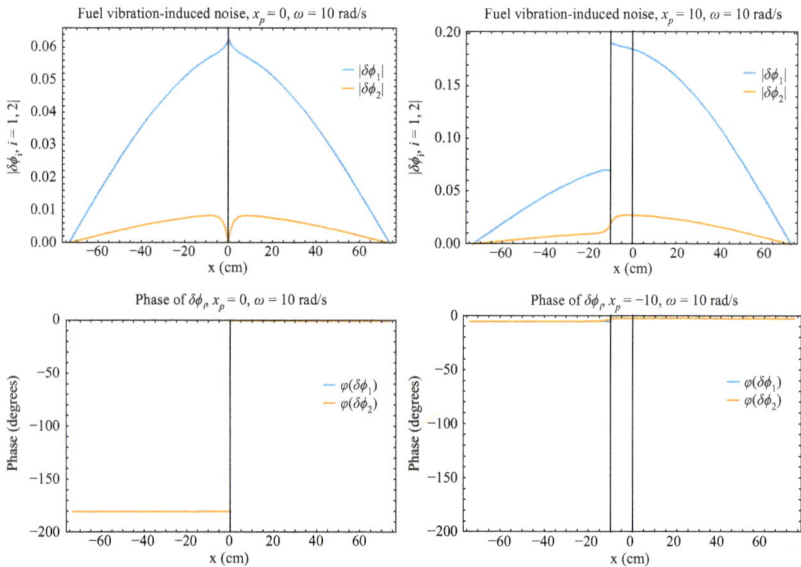

Figure 4.7   *Space dependence of the amplitude and the phase of the noise in the two energy groups, induced by a vibrating fuel rod, for $x_p = 0$ (left column) and $x_p = -10$ cm (right column), respectively, in a PWR SMR*

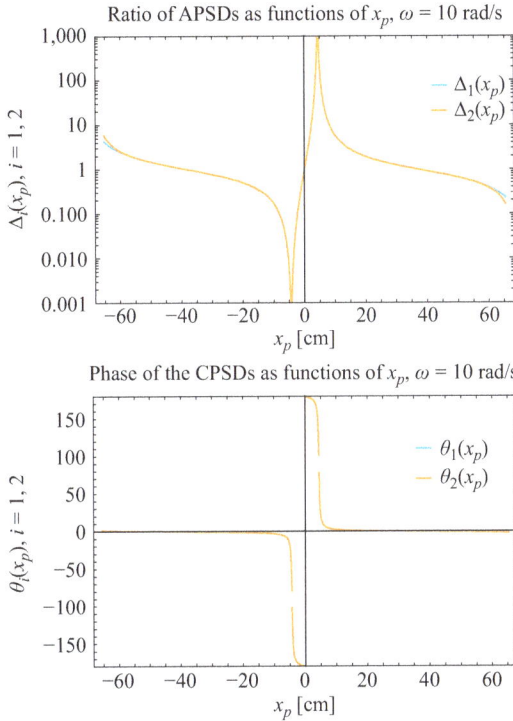

Figure 4.8   Dependence of the amplitude and phase localisation functions on the position of the vibrating fuel rod in a PWR SMR

vibrating fuel pin is significantly more complicated in this small system than in the large commercial cores. This is seen on the localisation curves, Figure 4.8.

A vibrating rod close to the core centre can be easily localised. Even if the amplitude localisation function is not monotonic, i.e. the result of the localisation is not single-valued, they give two possible positions; these are close to each other. Further away from the centre of the core than 10 cm in either direction, the rate of change of the amplitude localisation function is comparable with that of the absorber of variable strength. This also means that unlike in large systems, the spatial dependence of the measure of the interference between the reactivity and the space-dependent terms is only helpful for central vibrating fuel pins.

4.2.4   Propagating perturbations

The coordinate system is changed from the x to the z scale, and also the system size is changed to match the height of the core. The thickness of the one-dimensional (1D) slab modelling the core in the z-direction is changed from $2r = 147$ cm to $H = 220$ cm. This requires to change also the absorption cross sections, because the compensation for leakage is different from that used previously.

The change of the absorption cross sections will also change the spatial and frequency response of the system, but we do not show the full space- and frequency dependence of the modified Green's functions. Instead, we only show here the so-called fast and thermal removal adjoints. Since this SMR is a thermal system, the propagation adjoints are equal to the removal adjoints. The relative amplitude of the local component in these functions is what determines the possibilities of diagnosing propagating perturbations.

The spatial dependence of the amplitude and the phase of the removal adjoints are shown in Figure 4.9.

Again, the same similarities and differences can be seen as with the space dependence of the amplitude and phase of the Green's function components. In the thermal removal adjoint the peak is somewhat more pronounced than for the small BWR of Chapter 2. Since this SMR is a PWR, propagating perturbations are not a concern, so the local component plays a smaller role.

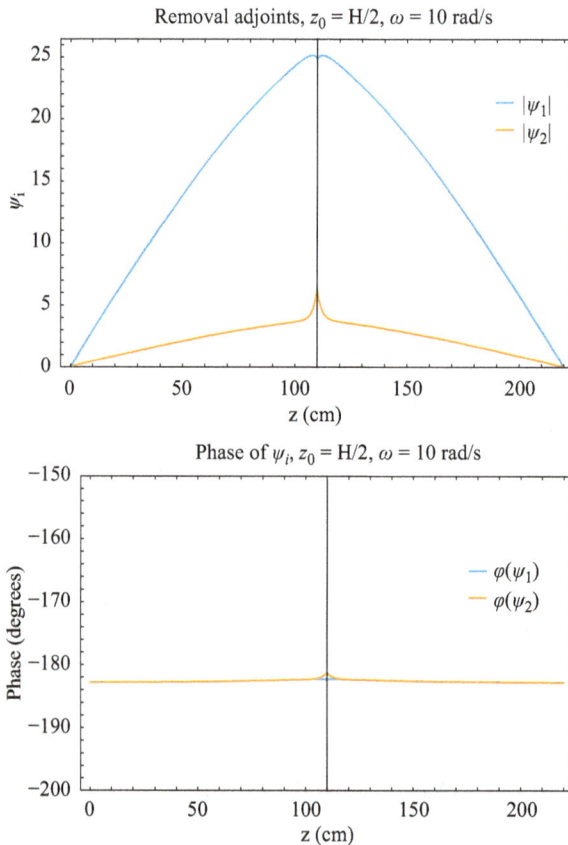

Figure 4.9   Space dependence of the amplitude and the phase of the fast and thermal removal adjoints for $x_0 = 0$ in a PWR SMR

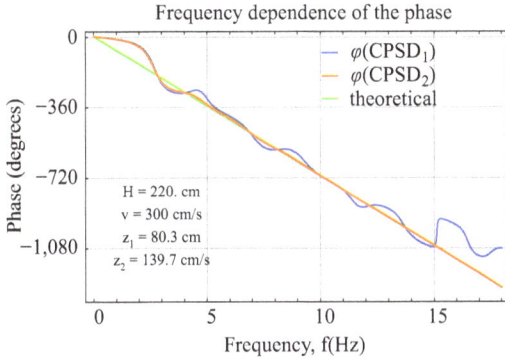

Frequency dependence of the phase

H = 220. cm
v = 300 cm/s
z_1 = 80.3 cm
z_2 = 139.7 cm/s

*Figure 4.10   Dependence of the phase of the cross-correlation on frequency
between two axially displaced detectors for propagating
perturbations in a PWR SMR*

However, subcooled boiling in the upper part of a PWR is an operational concern, whose detection is sought after by PWR utilities. One possibility to discover
subcooled boiling is noise analysis, developed in connection with the interpretation of
BWR in-core noise [47]. As it is known, the streaming of the two-phase flow, through
the presence of the local component, has the effect on correlations and cross-spectra
between axially displaced detectors that the propagating noise leads to a linearly, or
quasi-linearly decreasing phase as a function of the frequency. Pure water does not
exert sufficient perturbation that it can be detected, but in the presence of subcooled
boiling, the appearance of linear or quasi-linear phase could indicate the onset of subcooled boiling. This, on the other hand, assumes the existence of a local component
with a sufficiently high amplitude.

The dependence of the phase of the CPSD between two axially placed detectors,
both in the fast and the thermal groups, is shown in Figure 4.10.

As can be expected from a relatively small reactor, for small frequencies, the
phase deviates noticeably from linear, and stays close to zero, because the point
kinetic term dominates. With the increase of the frequency, the phase of the thermal noise soon relaxes to the theoretical linear phase. The phase of the CPSD of the
fast detectors shows a more irregular behaviour, but basically it also follows the linear
dependence on frequency.

4.3   Advanced light water SMR

The advanced light water technology SMR (ALWR SMR) is represented here with
a small modular pressurised reactor concept. The reactor core resembles a typical
western PWR with UO_2 fuel, but the size is somewhat reduced with a height of 280.0
cm and radius of 140 cm. This model was inspired by the Rolls-Royce SMR, based
on information available in the open literature.

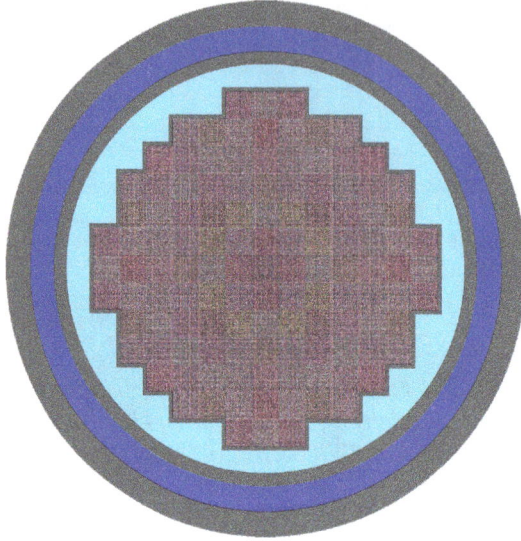

Figure 4.11   The geometry of the LWR SMR core

Table 4.2   Group constants and kinetic parameters for the LWR SMR core

Group	D	$\nu\Sigma_f$	Σ_a	Σ_R	χ_p	χ_d
1	1.52824	0.009093	0.010674	0.016067	1	1
2	0.38562	0.193929	0.107096	0	0	0.

β	λ	v_1 [cm/s]	v_2 [cm/s]	R [cm]	H [cm]
0.0051757	0.454141	$1.79667 \cdot 10^7$	$4.3316 \cdot 10^5$	140	280

The layout of the core is shown in Figure 4.11. Since the core diameter is equal with the height of the core, this means that both in the horizontal cross section of the core for the treatment of the variable strength absorber and vibrating fuel pin, as well as in the axial direction for the propagating perturbations, a slab with thickness $2\,a = H = 280$ cm will be used.

The group constants and the kinetic parameters used in the calculation are shown in Table 4.2.

The static fluxes in Groups 1 and 2 are shown in Figure 4.12. Like with other thermal systems, the flux in Group 1 is higher than in Group 2.

4.3.1   The Green's functions

The space and frequency dependence of the amplitude and the phase of the LWR SMR core along a horizontal cross section of the reactor are shown in Figure 4.13.

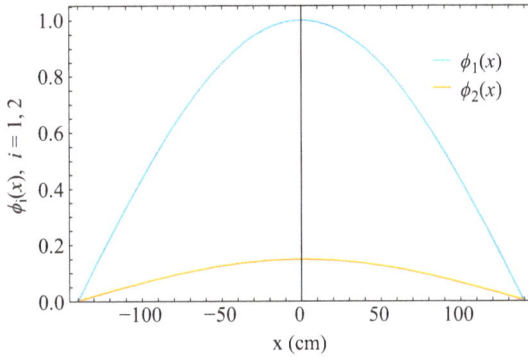

Figure 4.12 The static fluxes in the LWR SMR core

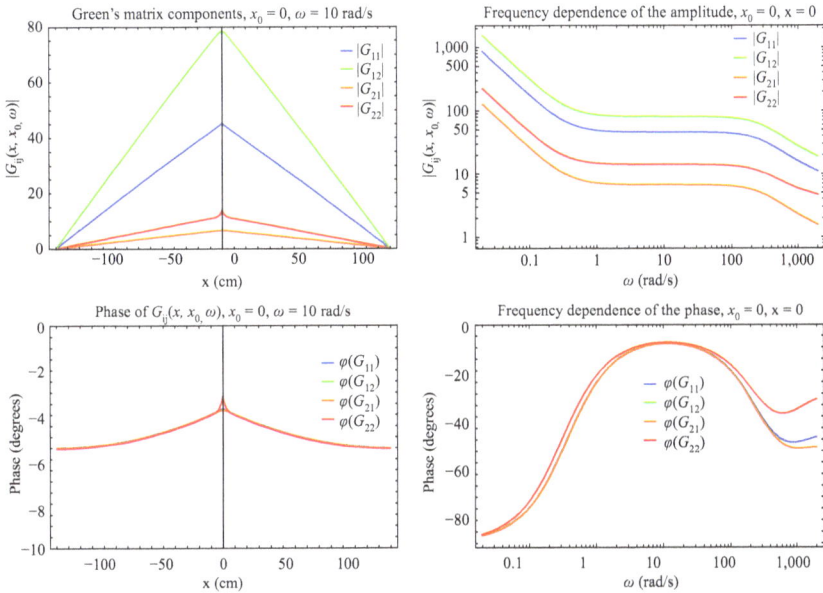

Figure 4.13 Space dependence of the amplitude and the phase of the components of the Green's function for $\omega = 10$ rad/s (left figures), and frequency dependence of the same for $x = x_0 = 0$ (right figures) in the LWR SMR core

Due to the relatively large size of the core, the Green's functions show a significant space-dependence, with a visible peak corresponding to the local component in G_{22}. The character of the space dependence of the amplitude shows a substantial resemblance to that of the large commercial BWR, discussed in Chapter 2, in that the space dependence is nearly linear. This represents a significant deviations from the point kinetic behaviour, which is promising from the point of view of localisation and diagnosing various perturbations.

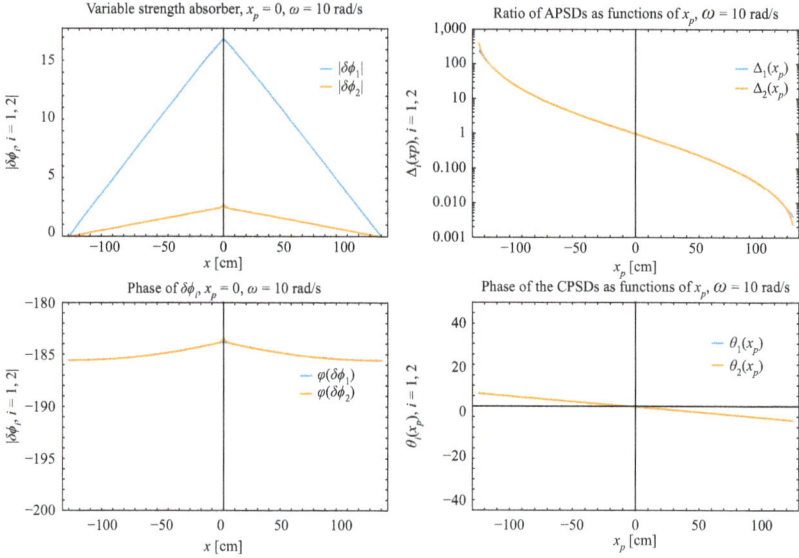

Figure 4.14   *Space dependence of the amplitude and the phase of the noise in the two energy groups, induced by an absorber of variable strength for $x_p = 0$ (left column) and the amplitude and phase localisation curves for the absorber of variable strength (right column) for the LWR SMR core*

4.3.2   Absorber of variable strength

As usual, the space dependence of the amplitude and the phase is very similar to those of the Green's function components, the thermal noise showing the local peak. As expected, the ratio of the peripherally placed two neutron detectors varies very significantly, making the localisation of the absorber of variable strength possible. The variations of the phase are not helpful even in this case.

4.3.3   Vibrating fuel pin

Based on the good localisation possibilities of the absorber of variable strength, it is expected that these will be just as good for the localisation of the vibrating fuel pin. Figure 4.15 shows the amplitude and phase of the neutron noise by two fuel pin positions: a central one and one 40 centimetres off the core centre. It demonstrates that the onset of the contribution of the point kinetic term starts at a much larger distance from the origin than e.g. for the Allegro core.

The localisation curves shown in Figure 4.16 confirm that the position of the vibrating fuel element can be achieved from the ratio of the APSDs of two peripherally placed detectors, and from the phase of their CPSD. Although the inverse of the amplitude localisation function is many-valued, together with the information in the phase, a unique position can be found from the information in the amplitude and in the phase.

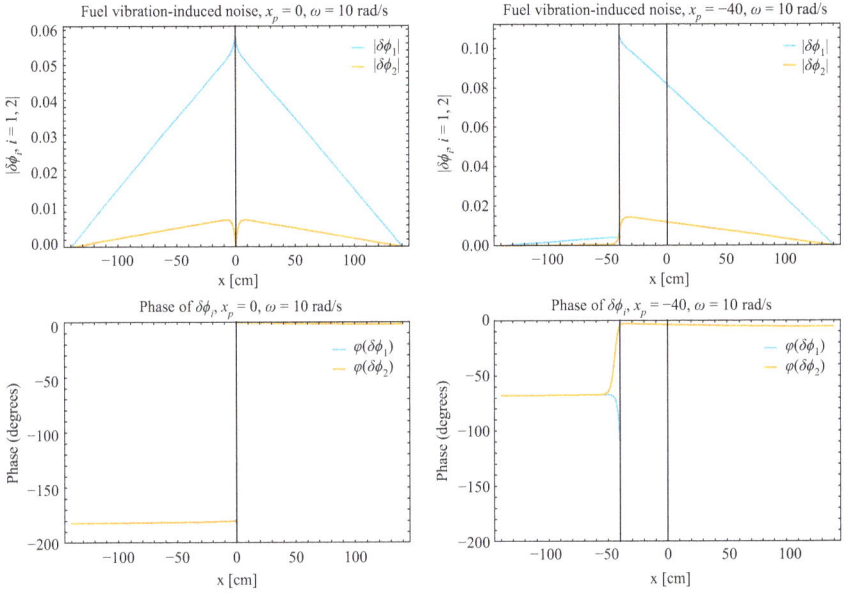

Figure 4.15 *Space dependence of the amplitude and the phase of the noise in the two energy groups, induced by a vibrating fuel rod, for $x_p = 0$ (left column) and $x_p = -40$ cm (right column), respectively, in the LWR SMR core*

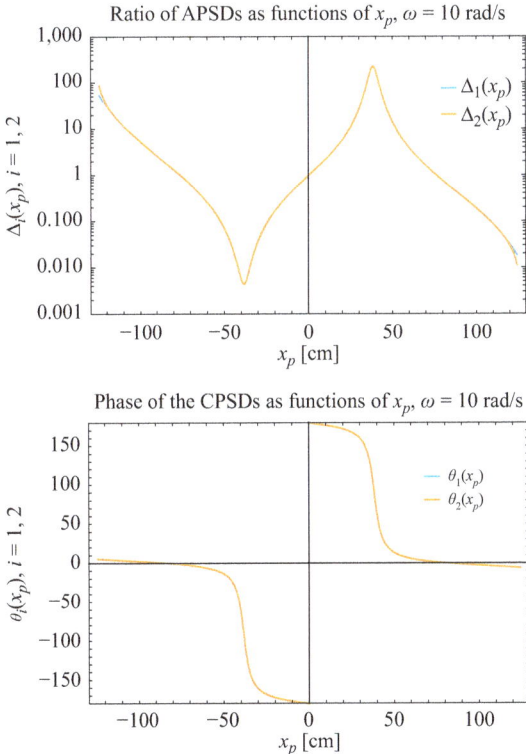

Figure 4.16 *Dependence of the amplitude and phase localisation functions on the position of the vibrating fuel rod in the LWR SMR core*

4.3.4   Propagating perturbations

Since the diameter of the LWR SMR is the same as its height, no recalculation of the leakage compensation is necessary. The neutronic transfer properties of the core will be the same in the z direction as they were in the x direction, and the properties of the removal adjoints can be expected to be similar to that of the Green's matrix components, but with an amplified relative weight of the local component in the thermal noise.

The spatial dependence of the amplitude and the phase of the removal adjoints are shown in Figure 4.17. Again, these look similar to the removal adjoints of the large commercial BWR, discussed in Chapter 2, although the magnitude of the peak of the local component is somewhat smaller here.

The investigations of the possibilities of two-phase flow diagnostics are motivated by the same arguments as in the small PWR SMR treated earlier in the previous Section. Namely, in-core neutron noise measurements may be useful for the detection

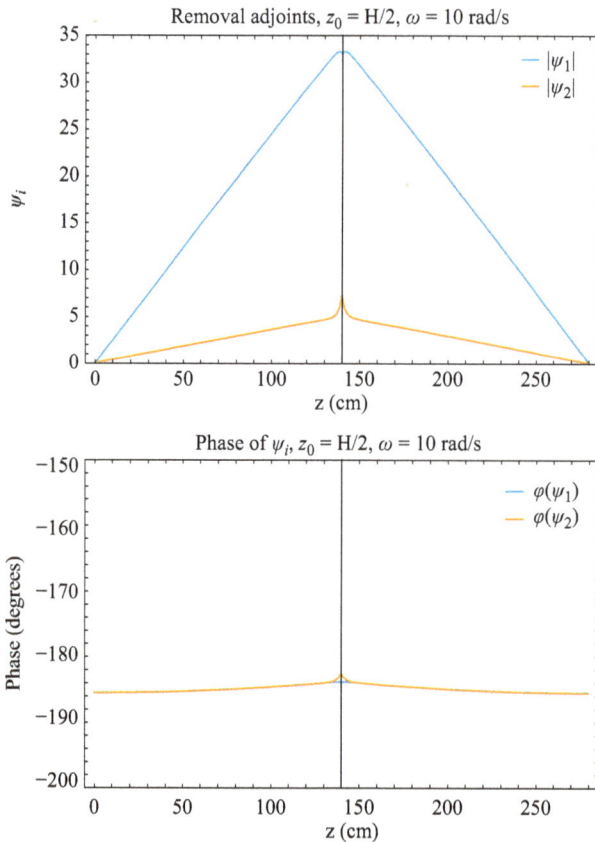

Figure 4.17   Space dependence of the amplitude and the phase of the fast and thermal removal adjoints for $x_0 = 0$ in the LWR SMR

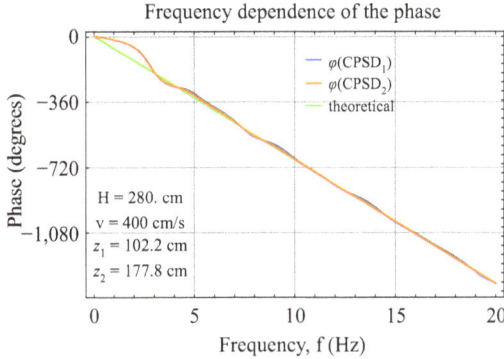

Figure 4.18   Dependence of the phase of the cross-correlation on frequency between two axially displaced detectors for propagating perturbations in the LWR SMR

of subcooled boiling in the upper part of the core. For this reason it is worth checking the suitability of the in-core detector signals to reproduce a linear phase for perturbations of the removal cross section.

The dependence of the phase between two axially placed detectors, both in the fast and the thermal group, is shown in Figure 4.18.

The figure shows that the phase has a clear linear dependence on the frequency with the proper slope, for the correlations between both the fast and the thermal neutron detector signals. The initial oscillations around the linear phase last longer than for the large BWR, up to approximately 5 Hz. Hence the determination of the transit time has to be based on the values of the phase above 5 Hz. On the other hand, in a real measurement, the signals become more "noisy" above 5 Hz, hence the determination of the transit time in practice will be less straightforward than in the case of the large BWR.

In summary, among the planned SMR types, the LWR SMR, despite its lower power and more compact size, in-core neuron noise analysis appears to be applicable essentially the same way as in large BWRs or PWRs.

4.4   Molten salt fast reactor

The molten salt fast reactor (MSFR) is a homogeneous molten salt reactor (MSR) concept. In terms of neutronics, it represents perhaps the most conceptually straightforward design imaginable. The core is a single compact cylinder, where the fluoride salt (LiF-ThF_4-UF_4) acts both as coolant and fuel. The core is surrounded by a breeding blanket containing thorium. The MSFR concept was originally developed by the SAMOFAR project [97]. However, here we used an implementation of the model made openly available by the Advanced Reactors and Fuel Cycles group of the University of Illinois at Urbana-Champaign. The active core has a height of 188 cm and a radius of 112.75 cm. The layout of the system is shown in Figure 4.19.

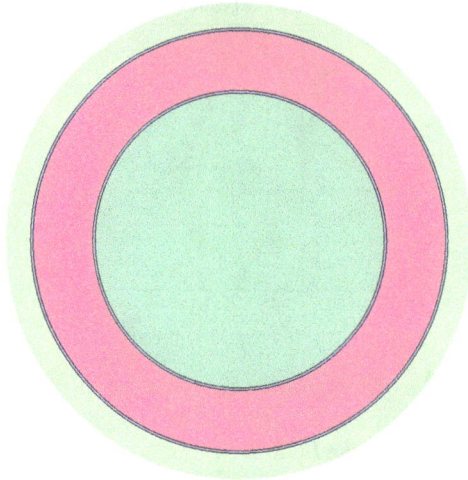

Figure 4.19   The geometry of the MSFR core

Table 4.3   Group constants and kinetic parameters for the MSFR core

Group	D	$\nu\Sigma_f$	Σ_a	Σ_R	χ_p	χ_d
1	2.167	0.005352	0.003248	0.05226	0.59606	0.03133
2	1.18506	0.007447	0.006924	0	0.40394	0.96866

β	λ	v_1 [cm/s]	v_2 [cm/s]	R [cm]	H [cm]
0.003093	0.322958	$2.09586 \cdot 10^9$	$1.3511 \cdot 10^8$	112.75	188

The group constants and the kinetic parameters used in the calculation are shown in Table 4.3.

The static fluxes in Groups 1 and 2 are shown in Figure 4.20. Similarly to other fast systems, the flux in Group 2 is higher than in Group 1.

4.4.1   The Green's functions

Because the MSFR is a molten salt system, in the dynamic calculations, as discussed in the foregoing, the calculations will be made by replacing β with the reduced effective delayed neutron fraction $\beta_{eff} = 0.4\,\beta$. The space and frequency dependence of the amplitude and the phase of the Green's functions for the MSFR core, calculated with this value of β_{eff}, along a horizontal cross section of the reactor are shown in Figure 4.21. Since this is a small, strongly coupled core, the Green's functions show a strong point reactor behaviour, with a practically invisible local component, which should be seen in G_{11}. As it is the case with the static flux, the Group 1 components G_{11} and G_{12} are much larger than the Group 2 ones, G_{21} and G_{22}. The plateau region

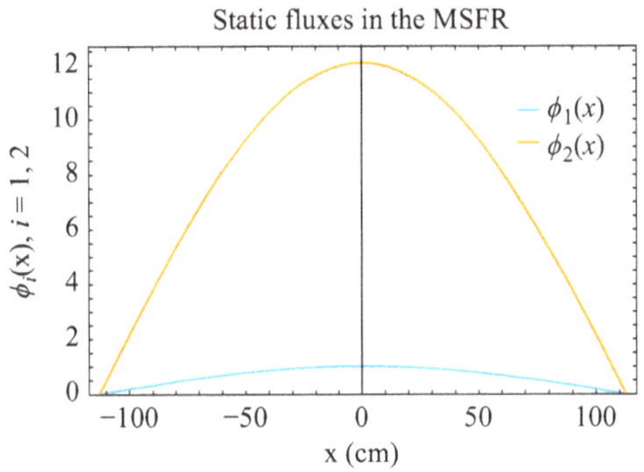

Figure 4.20   The static fluxes in the MSFR core

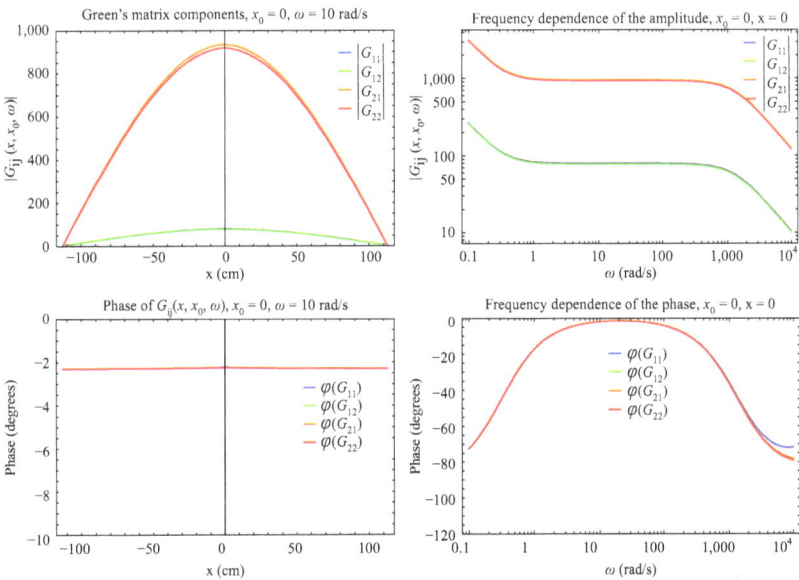

Figure 4.21   Space dependence of the amplitude and the phase of the components
of the Green's function for $\omega = 10$ rad/s (left figures), and frequency
dependence of the same for $x = x_0 = 0$ (right figures) in the MSFR
core

is rather wide, lies between about 0.5 to 2000 rad/s. In all aspects the behaviour is
very similar to that of the Allegro core. More similarities will also be seen in the
continuation.

4.4.2   Absorber of variable strength

As usual, the space dependence of the amplitude and the phase is very similar to
those of the Green's function components, the noise not showing the peak of the local

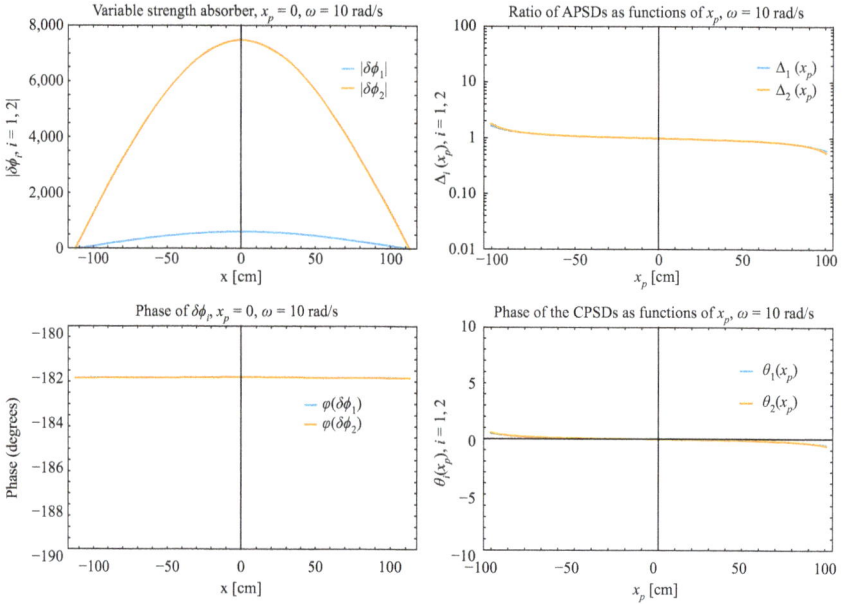

Figure 4.22   Space dependence of the amplitude and the phase of the noise in the two energy groups, induced by an absorber of variable strength for $x_p = 0$ (left column) and the amplitude and phase localisation curves for the absorber of variable strength (right column) for the MSFR core

component. They show a strong point kinetic character, indicating the difficulties with locating the position of the absorber rod.

This expectation is confirmed on the r.h.s. plots of Figure 4.22. As expected, the ratio of the APSDs and the phase of the CPSDs of two peripherally placed neutron detectors varies very little, making the localisation of the absorber of variable strength in principle not possible. One can though note that the variations in the ratios of the APSDs are somewhat larger than in the case of Allegro.

4.4.3   Vibrating absorber rod

Similarly to the case of the MSDR, in an MSR, considering the noise induced by a vibrating fuel pin is not actual. The possible effects of any perturbations related to fuel fluctuations are addressed while considering the propagating perturbations. Hence, again in similarity with the case of the MSDR, we consider instead the noise induced by a vibrating absorber.

The amplitude and the phase of the noise in the two energy groups for a vibrating absorber rod are shown in Figure 4.23 for a central (left figures) and a slightly off-central fuel rod (right figures). In contrast to the case of the absorber of variable strength, for the central absorber, the effect of the local component is now visible.

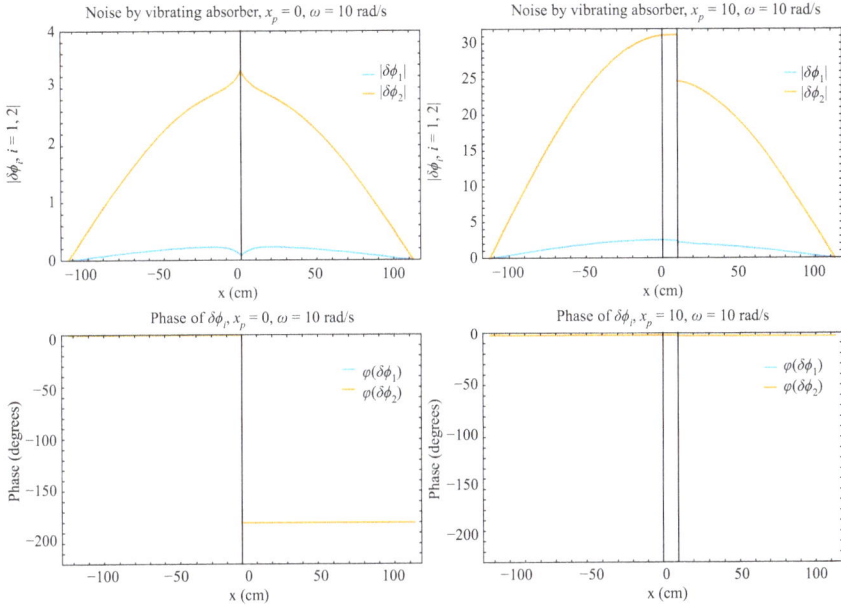

Figure 4.23   Space dependence of the amplitude and the phase of the noise in the two energy groups, induced by a vibrating absorber rod, for $x_p = 0$ (left column) and $x_p = -10$ cm (right column), respectively, in the MSFR SMR core

The spatial dependence is quite similar to that of the MSDR case, despite the spectral differences between the two cores.

For the off-central case, at the moderate distance of 10 cm from the core centre, the point kinetic component is already totally dominating, which is seen on both the small gap in the amplitude at the equilibrium rod positions, and even more on the phase, which is very close to zero at both sides. This is due to both the relatively small size of the system, the strong neutronic coupling in fast systems, and the reduced effective delayed neutron fraction of the MSR.

The localisation curves, shown in Figure 4.24 confirm that the position of the vibrating absorber rod cannot be determined from the ratio of the APSDs of two peripherally placed detectors, and from the phase of their CPSD, except for the special case of a central rod. The conclusion is that a vibrating component cannot be localised with neutron noise measurement in the MSFR SMR.

4.4.4   Propagating perturbations

For the study of the effect of propagating perturbations, the group constants were modified such that the size of the 1D model matches the core height, 188 cm.

For the calculation of the propagation Green's functions and the phase of the CPSD between two axially placed detectors, we need to again choose how to model density fluctuations of the fluoride salt. This will be similar to the case of the MSDR,

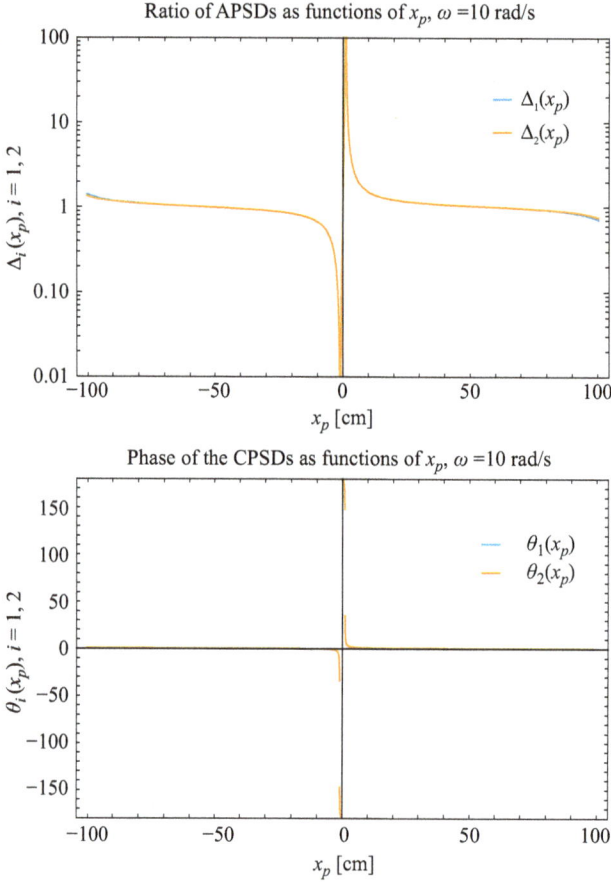

Figure 4.24   Dependence of the amplitude and phase localisation functions on the position of the vibrating absorber rod in the MSFR core

in that the density fluctuations will affect the fission cross sections. However, there will also be two differences. One is that, in contrast to the MSDR, the MSFR is a fast spectrum core, hence the perturbation of the fuel will appear in both groups (because $\chi_2 \neq 0$). As it will be seen soon, this will have significant consequences. The other difference is that although there is no thermalisation in this core, the scattering from Group 1 to Group 2 takes place in the fluoride salt, which means that the density fluctuations will affect also the removal cross section.

Assuming again that the effect of the perturbation will induce cross-section fluctuations proportional to their static values, one can define the (arbitrarily normalised) coefficients α_1 and α_2 as follows:

$$\Sigma_{tot} = \Sigma_{a1} + \Sigma_{a2} + \nu\Sigma_{f1} + \nu\Sigma_{f2} + \Sigma_R \tag{4.1}$$

and we define

$$\alpha_1 = \frac{\Sigma_{a1} + \Sigma_R - \chi_1 \left(\nu \Sigma_{f1} + c_\mu \, \nu \Sigma_{f2} \right)}{\Sigma_{tot}} \qquad (4.2)$$

and

$$\alpha_2 = \frac{-\Sigma_R - \chi_2 \left(\nu \Sigma_{f1} + c_\mu \, \nu \Sigma_{f2} \right) + c_\mu \, \Sigma_{a2}}{\Sigma_{tot}} \qquad (4.3)$$

In the aforementioned formulae, the very weak frequency dependence of the spectral parameters $\chi_1(\omega)$ and $\chi_2(\omega)$ was neglected, based on the same arguments as in the case of the MSDR. The numerical values are

$$\chi_1 = (1 - \beta) \chi_{p1} + \beta \chi_{d1} = 0.595 \qquad (4.4)$$

$$\chi_2 = (1 - \beta) \chi_{p2} + \beta \chi_{d2} = 0.405 \qquad (4.5)$$

With these definitions, and with the numerical values of the cross sections and the spectral parameters, one obtains

$$\alpha_1 = -0.00796 \qquad (4.6)$$

$$\alpha_2 = -0.05247 \qquad (4.7)$$

These values show that a similar fortuitous combination of the quantitative values of the cross section as in the case of the MSDR, which led to the situation of having $\alpha_1 \approx -\alpha_2$, seen in (3.5) and (3.6), is not valid here, since α_1 and α_2 have very different absolute values, and also have the same sign. The explanation can be seen in a comparison between (4.6) and (4.7) with (3.2) and (3.3). Since the MSDR is a thermal system, $\chi_2 = 0$, and α_2 is positive, whereas $\chi_1 = 1$, which leads to a negative α_1 with approximately the same absolute value. Since the MSFR is a fast system, both χ_1 and χ_2 are about 0.5, which decreases the absolute value of α_1, and turns α_2 into a negative value as well.

The result of this fact, combined with the already very minor relative weight of the local component, is seen in the propagation adjoints, shown in Figure 4.25. No trace of the local component is seen in these plots, and the phase is almost perfectly constant over the whole cross section of the core. This amounts to say that what regards the propagation adjoints, the MSFR resembles much more to the SFR than to the MSDR.

The dependence of the phase of the CPSD between two axially placed detectors, shown in Figure 4.26, only amplifies this statement. Like it was the case with the SFR, it is actually surprising that although the phase curve shows large oscillations, globally it still follows the straight line of the theoretical phase curve which would be seen with a large and sharp local component. Actually, the character of the curve also resembles to that of the SFR. Nevertheless, the same statement is valid also here, namely that although the phase globally follows the correct linear dependence of the theoretical value, it is doubtful whether in a real measurement with such oscillations could be used to determine the transit time of the flow from curve fitting, when the effects of other noise sources, background noise, etc., are also included in the measured signal.

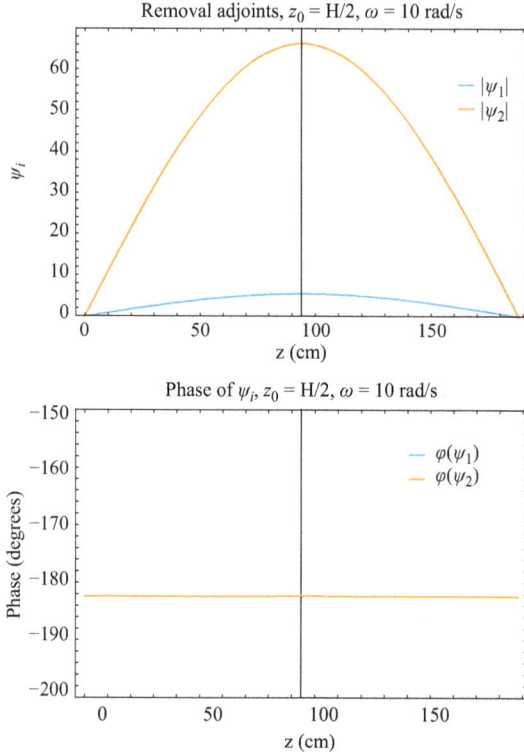

Figure 4.25 *Space dependence of the amplitude and the phase of the fast and thermal propagation adjoints for $x_0 = 0$ in the MSFR*

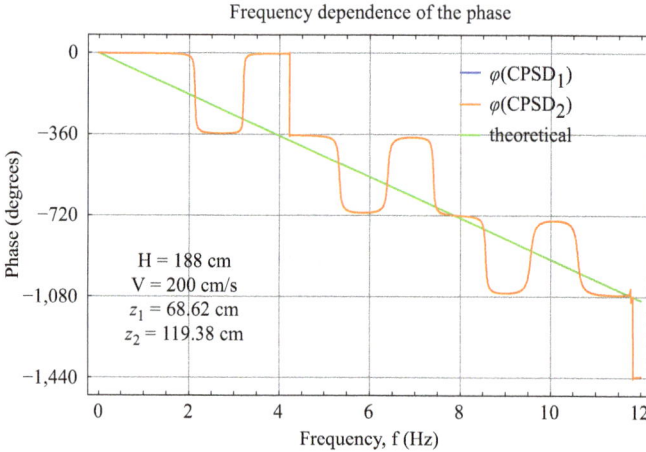

Figure 4.26 *Dependence of the phase of the cross-correlation on frequency between two axially displaced detectors for propagating perturbations in the MSFR*

4.5 Lead-cooled fast reactor SMR (LFR SMR)

A small, 200 MWth lead-cooled thorium-fuelled fast reactor (LFR SMR) was modelled based on [98]. The 33 fuel assembly and 4 control assembly positions are surrounded by an MgO reflector. The fuel consists of 15 w% enriched UN. The active core height of the reactor is 120 cm and the radius is 73.74 cm. The layout of the core is shown in Figure 4.27.

The group constants and the kinetic parameters used in the calculation are shown in Table 4.4.

The static fluxes in Groups 1 and 2 are shown in Figure 4.28. Similarly to the other fast spectrum cores, the flux in Group 2 is higher than in Group 1.

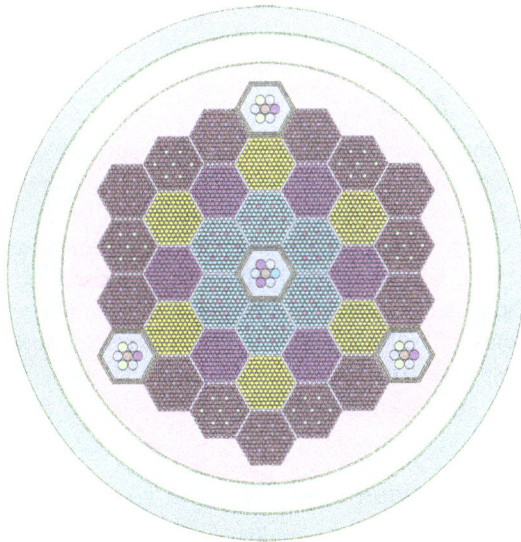

Figure 4.27   The geometry of the LFR SMR core

Table 4.4   Group constants and kinetic parameters for the LFR SMR core

Group	D	$\nu\Sigma_f$	Σ_a	Σ_R	χ_p	χ_d
1	2.08596	0.024141	0.009643	0.051161	0.58901	0.03953
2	1.19035	0.008247	0.060805	0	0.41099	0.96047

β	λ	v_1 [cm/s]	v_2 [cm/s]	R [cm]	H [cm]
0.008748	0.53006	$2.07941 \cdot 10^9$	$4.05158 \cdot 10^8$	73.74	120

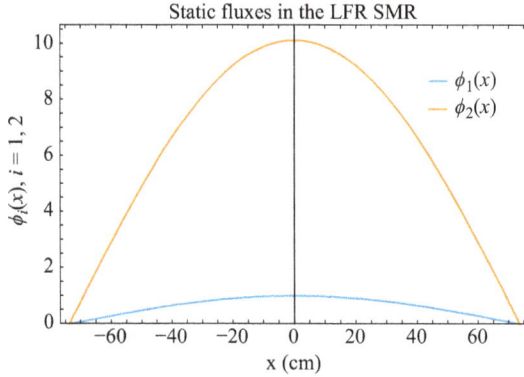

Figure 4.28   The static fluxes in the LFR SMR core

4.5.1   The Green's functions

The space and frequency dependence of the amplitude and the phase of the LFR SMR core along a horizontal cross section of the reactor are shown in Figure 4.29.

Due to the small size and tight neutronic coupling in the fast core, the Green's functions show a strong point kinetic character, with the amplitudes following the shape of the static flux, and having a nearly constant phase. The local component is not visible in the plots. Corresponding to the short prompt neutron lifetime, the frequency dependence of both the amplitude and the phase shows a wide plateau, the phase being close to zero over most of the plateau. The upper break frequency is particularly high, the highest among all the cores considered.

4.5.2   Absorber of variable strength

As usual, the space dependence of the amplitude and the phase is very similar to those of the Green's function components, and there is no local peak in the noise. As could be expected, the ratio of the APSDs of the peripherally placed two neutron detectors varies very mildly as a function of the position of the absorber, and the phase of the CPSDs is nearly constant. This confirms that localisation of this type of noise source is not practical in the LFR SMR.

4.5.3   Vibrating fuel pin

The chances of localisation may be better for the vibrating components. Figure 4.31 shows the amplitude and phase of the neutron noise by two fuel pin positions: a central one and one at 10 centimetres off the core centre. Corresponding to the situation in similar small cores, a small distance away from the centre the reactivity term already dominates in the induced noise. The gap in the amplitude at the position of the fuel rod is quite small, and the phase is very close to zero at both sides of the fuel pin, and does not change much in the entire core, except very close to the fuel pin.

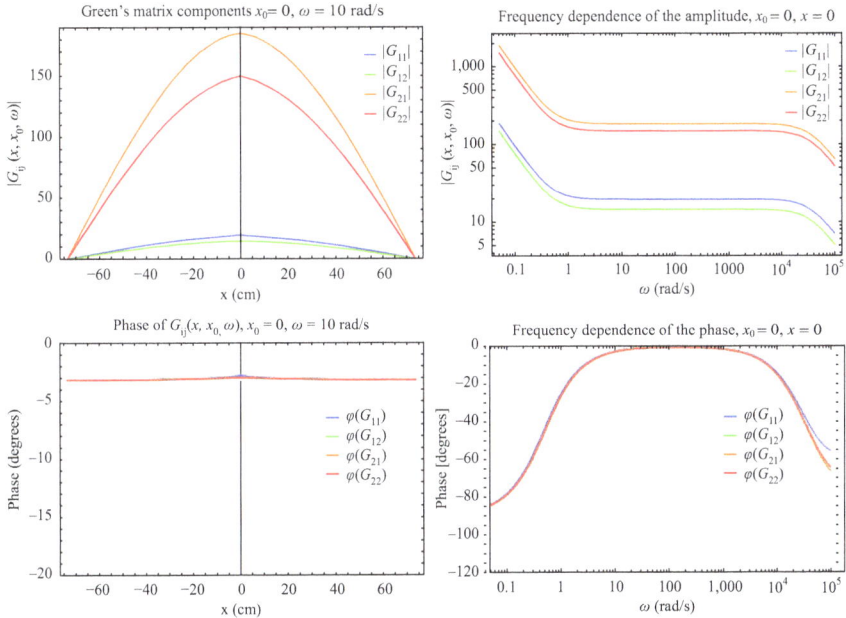

Figure 4.29 Space dependence of the amplitude and the phase of the components of the Green's function for $\omega = 10$ rad/s (left figures), and frequency dependence of the same for $x = x_0 = 0$ (right figures) in the LFR SMR core

On the other hand, as the top left in Figure 4.31 shows, for the central absorber, i.e. when the reactivity term is missing, the effect of the local component becomes visible. This is because the absence of the point kinetic component increases the relative weight of the local component in the total noise (which in this case consists of only the space-dependent and the local components), and the type of the perturbation amplifies this relative weight further, since the spatial derivative of the Green's function appears in the formulae.

Because of this, the noise induced by a vibrating absorber or fuel rod is suitable to show the characteristics of the local component. From the top left plot of Figure 4.31, it is seen that the local component in an LFR has a relatively large range or "field of view", about 20 cm. However, the weight of this component in the total noise is rather small; it is only visible when the reactivity term of the noise is missing, and this happens only in the special locations where the derivative of the static flux is zero. As the top right plot of Figure 4.31 shows, in practically all other cases the contribution from the local component is negligible even for the vibrating fuel rod.

From the strongly point kinetic behaviour of the induced noise, which is valid for all fuel pin positions except the central one, it is not surprising that the localisation of a vibrating fuel pin (or vibrating absorber) is not feasible with neutron noise methods in

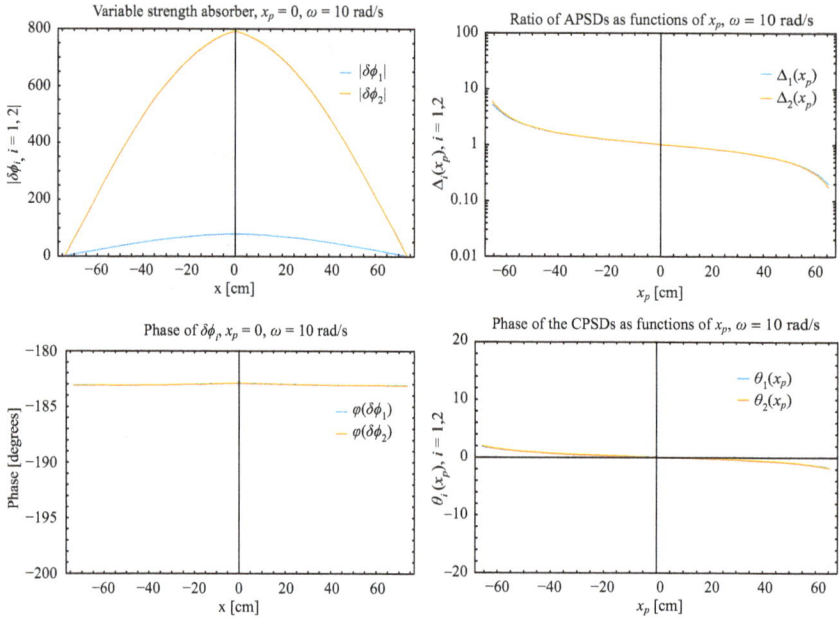

Figure 4.30   Space dependence of the amplitude and the phase of the noise in the two energy groups, induced by an absorber of variable strength for $x_p = 0$ (left column) and the amplitude and phase localisation curves for the absorber of variable strength (right column) for the LFR SMR core

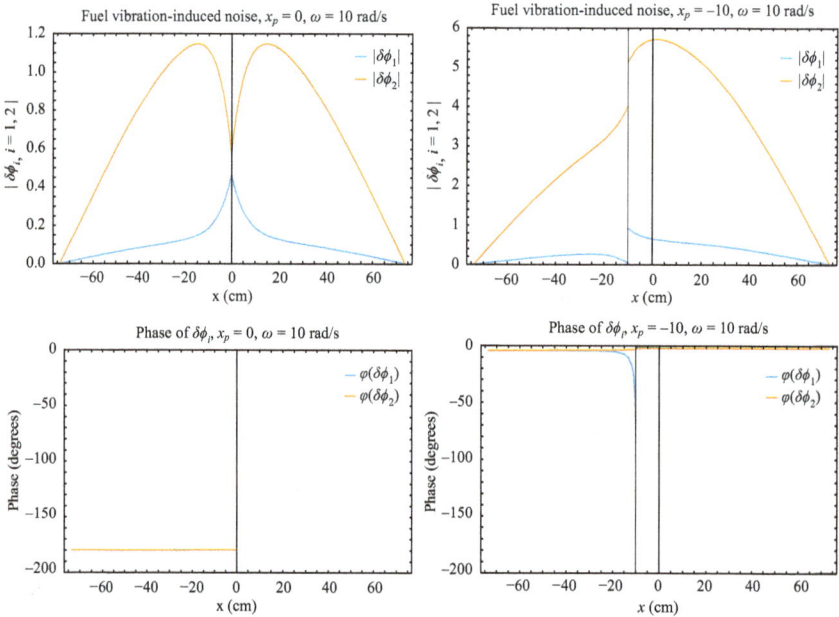

Figure 4.31   Space dependence of the amplitude and the phase of the noise, induced by a vibrating fuel rod, for $x_p = 0$ (left column) and $x_p = -10$ cm (right column), respectively, in the LFR SMR core

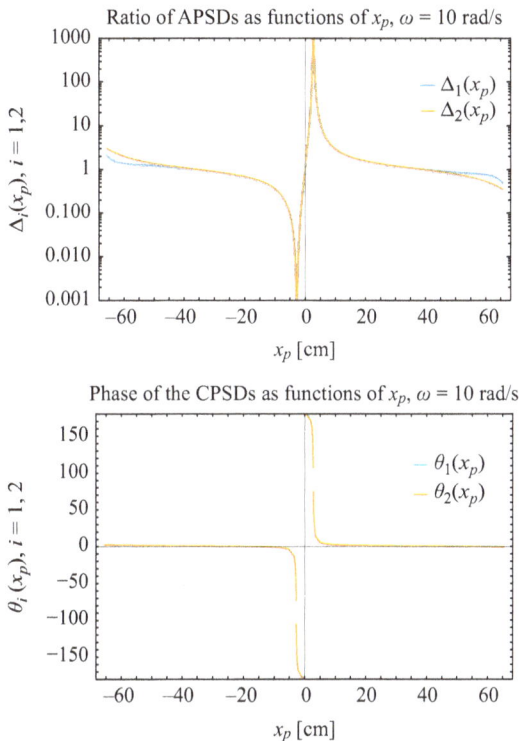

Figure 4.32   Dependence of the amplitude and phase localisation functions on the position of the vibrating fuel rod in the LFR SMR core

the LFR SMR. This is seen on the localisation curves, shown in Figure 4.32. It is only in the neighbourhood of the core centre, at small distances, where the interference of the space-dependent and point kinetic component leads to a huge discontinuity in the amplitude such that the noise becomes very small on one side, that the location of the vibrating component can be guessed. Outside this small area, the amplitude and phase localisation curves are practically insensitive to the position of the vibrating component.

4.5.4   Propagating perturbations

As it was mentioned earlier, neutron noise induced by propagating perturbations is interesting in liquid metal-cooled reactors, both for the noise induced by small temperature/density variations, and also to detect and quantify the occurrence and propagation of bubbles.

For the investigation of the neutron noise induced by propagating perturbation, the core will be modelled in its axial direction. Since the height of the LFR SMR is somewhat smaller than its diameter, a slightly less space-dependent behaviour can be expected. As usual, the absorption cross sections were adjusted to the core height of 120 cm.

With the data of the core, and assuming that only the absorption and removal cross sections are perturbed, with the same relative weight as the static cross sections (see (2.159)), one obtains

$$\alpha_1 = 0.876449$$

and

$$\alpha_2 = 0.520515$$

As with the SFR, the removal cross section does not dominate, thus the two coefficients have the same sign, and will not amplify the local component in the propagation adjoints. This is confirmed by Figure 4.33, which shows the spatial dependence of the amplitude and the phase of the propagation adjoints. They look similar to the individual components of the Green's function, and no local component is visible in any of the two adjoints. Although the space dependence of the propagation adjoints is not very pronounced, there is a chance that they show sufficient "peaking" such that they can be used for determining the transit time between axially displaced detectors.

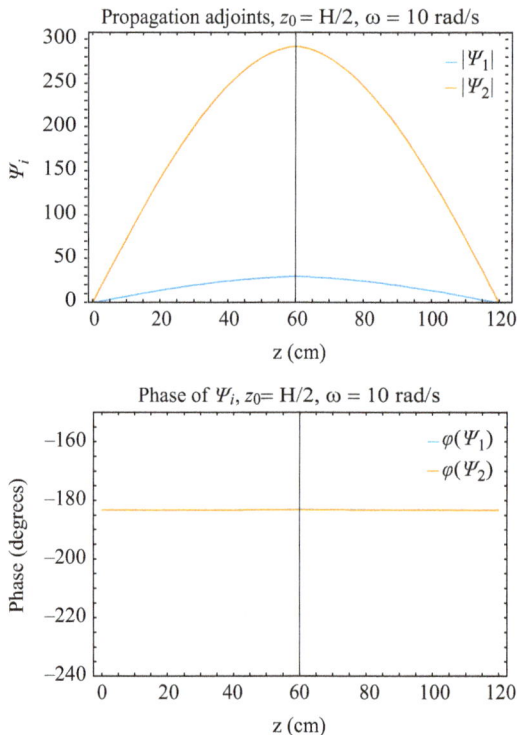

Figure 4.33   Space dependence of the amplitude and the phase of the fast and thermal propagation adjoints for $x_0 = 0$ in the LFR SMR

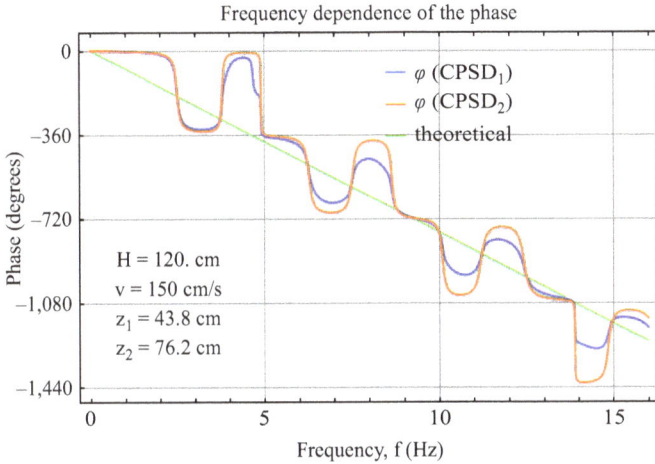

Figure 4.34   Dependence of the phase of the cross-correlation on frequency between two axially displaced detectors for propagating perturbations in the LFR SMR

The dependence of the phase between two axially placed detectors is shown in Figure 4.34. The figure shows that the phase oscillates strongly around the theoretical linear dependence. Actually, it shows considerable resemblance to the phase oscillations seen with the SFR and the MSFR cores. The judgment is also the same, namely that although in this theoretical study it seems as if globally, the phase follows the slope of the linear dependence of the theoretical value, in a real measurement it will not be possible to extract the transit time of the coolant between two in-core neutron detectors. On the other hand, in-core neutron detectors may be useful in detecting and quantifying voids. As mentioned in Chapter 2, Section 2.3.8, the spectral properties of the NRMS of the fast and epithermal detectors might be used to quantify the void content.

It is thus seen that in the LFR SMR, most of the diagnostic tasks performed in current reactors, localising a perturbation or measuring transit time of the flow, cannot be performed with in-core neutron noise measurements. On the other hand, there are several methods suitable for measuring flow rate of liquid metal coolant, which will be mentioned in Chapters 7 and 8.

As it was also mentioned in Chapter 2, Section 2.3.8, stability of lead-cooled reactors may be a concern with reduced flow rate in natural circulation. This in itself is a strong argument that despite the difficulties of diagnosing perturbations, there is an incentive to implement in-core neutron detectors for surveillance of the stability properties of the core.

4.6   Generic fluoride salt-cooled high-temperature reactor

The generic fluoride-salt-cooled high-temperature reactor core (gFHR) model functions as a benchmark for core designers, methods developers, safety analysts, and researchers to better understand the physics behaviour of Kairos Power's FHR core.

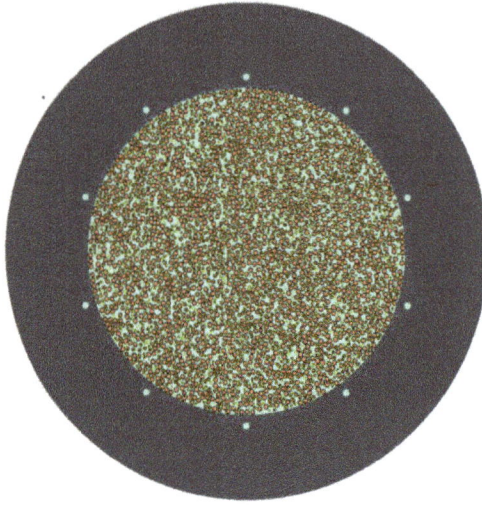

Figure 4.35   The geometry of the FHR core

The gFHR (in the continuation referred to as FHR) represents a full-scale FHR, inspired by Kairos Power's global design but incorporates several simplifications and is made in Serpent 2. The model is made with the intention that the methodology is easily adaptable to other, more specific FHR designs (i.e. Hermes low power demonstration reactor, KP-FHR) [99–101].

The concept consists of a pebble bed core with TRISO particles embedded into the pebbles, cooled with molten FLiBe salt. The use of pebbles allows for continuous refuelling of the core. Furthermore, the continuous cycling of the pebbles results in a steady-state isotopic distribution, i.e. an equilibrium core condition. The active core height of the reactor is 310 cm and the radius is 120 cm. A layout of the core is shown in Figure 4.35.

Such a reactor is designed currently by Kairos Power [102], and a prototype is planned to be built at Oak Ridge National Laboratory (ORNL). Despite of its thermal spectrum and that it is not designed or suitable for breeding or transmutation of waste, it counts as a Generation-IV (Gen-IV)-type SMR.

The group constants and the kinetic parameters used in the calculation are shown in Table 4.5.

The static fluxes in Groups 1 and 2 are shown in Figure 4.36. Like with other thermal systems, the flux in Group 1 is higher than in Group 2.

4.6.1   The Green's functions

The space and frequency dependence of the amplitude and the phase of the FHR core along a horizontal cross section of the reactor are shown in Figure 4.37. Although the diameter of the core is medium large, the Green's function components exhibit a distinctly point kinetic behaviour. A very small amplitude local component can be

Table 4.5 Group constants and kinetic parameters for the FHR core

Group	D	$\nu\Sigma_f$	Σ_a	Σ_R	χ_p	χ_d
1	1.15102	0.000671	0.001625	0.002805	1	1
2	0.97649	0.013140	0.008269	0	0	0.

β	λ	v_1 [cm/s]	v_2 [cm/s]	R [cm]	H [cm]
0.005438	0.412054	$1.02468 \cdot 10^7$	$4.98095 \cdot 10^5$	120	310

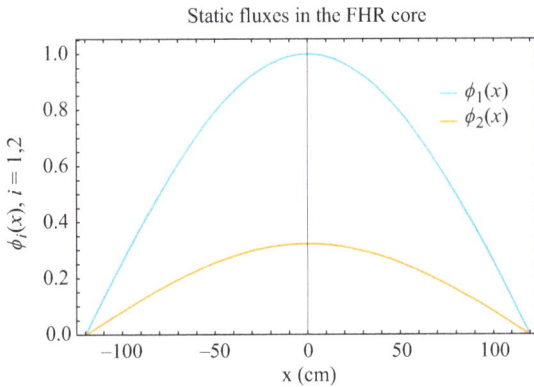

Figure 4.36 The static fluxes in the FHR core

observed in G_{22}, which is best visible in the phase. Apart from the moderate size of the core, a contributing factor to this behaviour is the relative smallness of the delayed neutron fraction $\beta = 0.00544$, which is smaller than that of LWRs.

The phase change of the Green's functions across the core is also very small, in agreement with the point kinetic behaviour of the amplitude. An interesting feature of the phase is that at the plateau frequency of 10 rad/s, it lies significantly more below zero, over $-20°$, which is much more than most of the other cores, especially the fast spectrum cores. It is only the MSDR, which is also graphite moderated, but with a much more massive amount of graphite, in which the phase delay is even larger, and in which the phase is also noticeably space-dependent. Like in the MSDR, the large negative phase is due to the sluggishness of the graphite moderated system. It also means that at plateau frequencies the Green's function have a non-negligible imaginary part.

The frequency dependence of the amplitudes shows that due to the thermal spectrum and low neutron velocities, the upper break frequency of the amplitude is relatively low, around 20–30 rad/s, and the amplitude is not constant in the plateau region. The FHR resembles to the MSDR even in this respect.

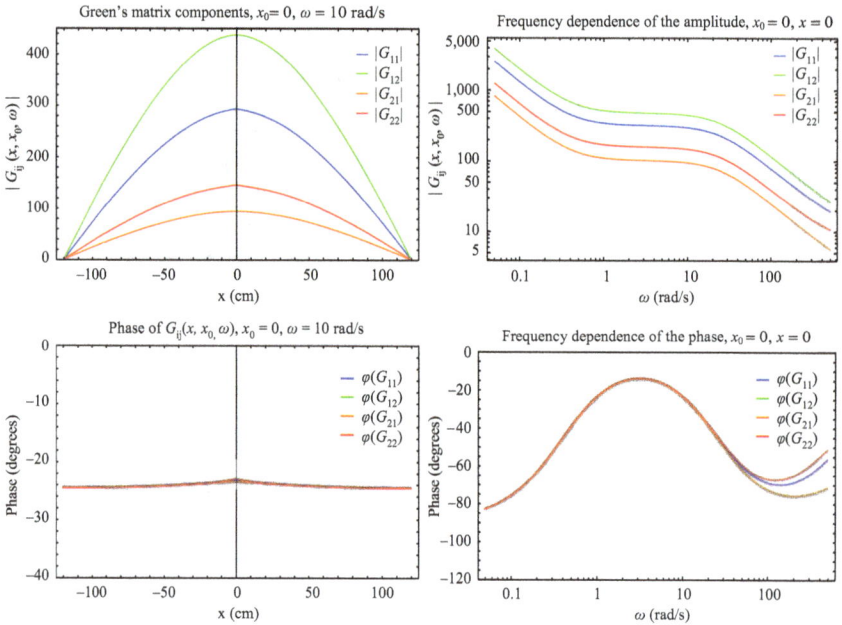

Figure 4.37   Space dependence of the amplitude and the phase of the components of the Green's function for $\omega = 10$ rad/s (left figures), and frequency dependence of the same for $x = x_0 = 0$ (right figures) in the FHR core

4.6.2   Absorber of variable strength

As usual, the space dependence of the amplitude and the phase is very similar to those of the Green's function components, which in this case means point kinetic response, as seen in Figure 4.38. Accordingly, the variation of the localisation curves is rather small, which indicates that only detectors in the vicinity of such a perturbation can be used to locate it. On the other hand it is interesting that the dependence of the phase of the CPSD of two peripherally positioned neutron detectors shows a stronger dependence on the position of the absorber than in most of the other cores. This indicates that the information in the phase in the thermal FHR is more helpful than in the other cores, again except in the MSDR.

4.6.3   Vibrating fuel pin

Based on the poor localisation possibilities of the absorber of variable strength and the strong point kinetic character of the Green's function, the prospects of localising a vibrating fuel rod are not promising. This is also a somewhat conceptual type of problem, since the FHR does not have fuel pins, and not even control rods in the present design. However, for the study of the properties of the system, it is still interesting to investigate this problem.

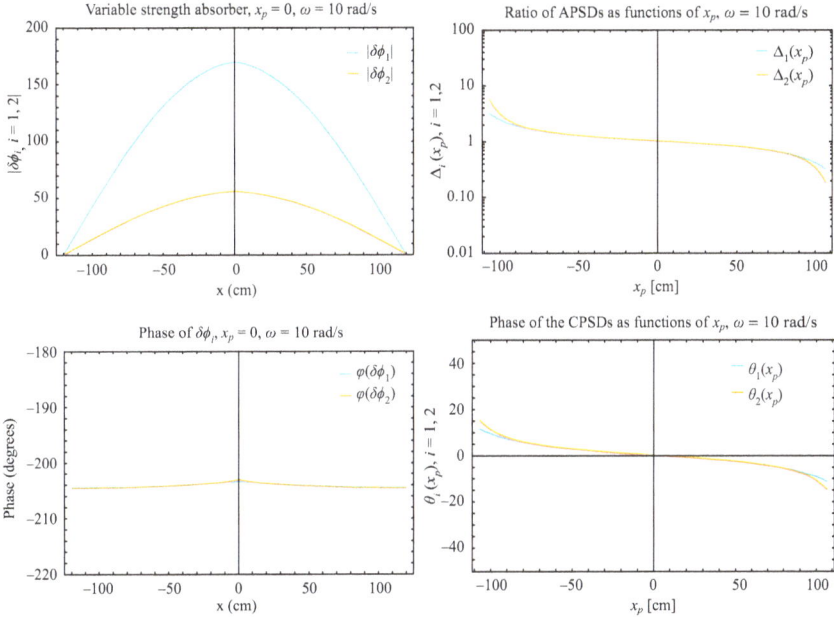

Figure 4.38   Space dependence of the amplitude and the phase of the noise in the two energy groups, induced by an absorber of variable strength for $x_p = 0$ (left column) and the amplitude and phase localisation curves for the absorber of variable strength (right column) for the FHR core

Figure 4.39 shows the amplitude and phase of the neutron noise by two fuel pin position: a central one and one 10 centimetres off the core centre. The amplitude of the noise shows the characteristic features of thermal systems with a strong local peak in the fast noise, and a large dip in the thermal noise. As usual, the phase is opposite at the opposite sides of the vibrating element.

Because of the large contribution of the reactivity component in the Green's functions, as soon as the equilibrium position of the vibrating pin is moved off the core centre where it does not have a reactivity effect, the point kinetic component appears with a relatively large weight. Moving the equilibrium position off with 10 cm from the centre, the point kinetic component already dominates over the space-dependent one, which is seen on the phase which is nearly the same at the two sides of the vibrating component.

In agreement with these characteristics, the localisation curves, shown in Figure 4.40 confirm that it is difficult, if not impossible, to recover the position of the vibrating fuel element from the ratio of the APSDs of two peripherally placed detectors, and from the phase of their CPSD. Apart from the central equilibrium position of the vibrating fuel pin, the dependence of the ratio of the APSDs and the phase of the CPSD on the position of the vibrating pin is weak. In practice, this is probably not a huge disadvantage, since there are no axially extended fuel pins of absorber rods in the core that could vibrate.

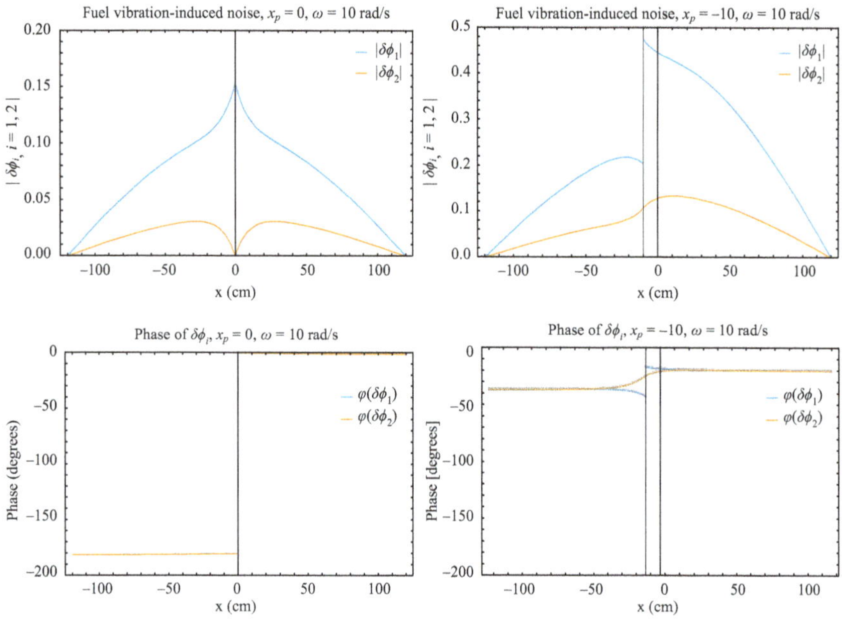

Figure 4.39   Space dependence of the amplitude and the phase of the noise in the two energy groups, induced by a vibrating fuel rod, for $x_p = 0$ (left column) and $x_p = -40$ cm (right column), respectively, in the FHR core

4.6.4   Propagating perturbations

For the propagating perturbations, the cross sections were readjusted to yield the critical width of the 1D model to be equal to the height of the core, 310 cm. A somewhat more pronounced space dependence can be expected in the axial direction.

As with the other reactor types, one question is the representation of the transfer function combination for the given the noise source, i.e. the weight factors α_1 and α_2, which compose the so-called fast and thermal propagation adjoints, see (2.162) and (2.163). For the FHR, it appears reasonable that only the fluctuations of the absorption cross sections are taken into account. This is because both the fission and the moderation of the neutrons in this reactor takes place in the graphite of the pebbles; the fluoride salt has a much less moderation capacity, and it does not contain fissile material. A change in the density of the coolant will thus primarily affect the absorption cross sections in the two groups.

The coefficients α_1 and α_2 used for the FHR will therefore be calculated according to the formulae next, with the numerical values for this system also shown:

$$\alpha_1 = \frac{\Sigma_{a1}}{\Sigma_{a1} + \Sigma_{a2}} = 0.18846 \tag{4.8}$$

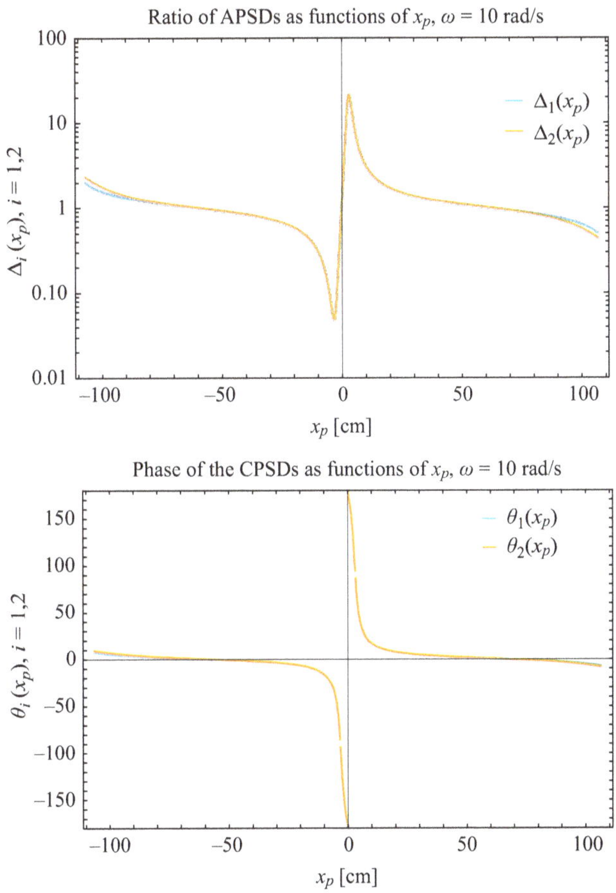

Ratio of APSDs as functions of x_p, $\omega = 10$ rad/s

Phase of the CPSDs as functions of x_p, $\omega = 10$ rad/s

Figure 4.40 *Dependence of the amplitude and phase localisation functions on the position of the vibrating fuel rod in the FHR core*

and

$$\alpha_2 = \frac{c\mu \, \Sigma_{a2}}{\Sigma_{a1} + \Sigma_{a2}} = 0.2622 \tag{4.9}$$

According to the definition of the coefficients, they need to have the same sign, and thus in the FHR no amplification of the local component similar to that in the LWRs or the MSDR can be expected.

The spatial dependence of the amplitude and the phase of the propagation adjoints, shown in Figure 4.41, reflect this fact. These look similar to the other small SMRs, whether thermal or fast. A very small local component can be noticed in the thermal transfer function, both in the amplitude and the phase. However, this local component appears with a smaller weight than what can be observed in G_{22}, due to the fact that the propagation adjoints are the weighted sum of two Green's functions elements.

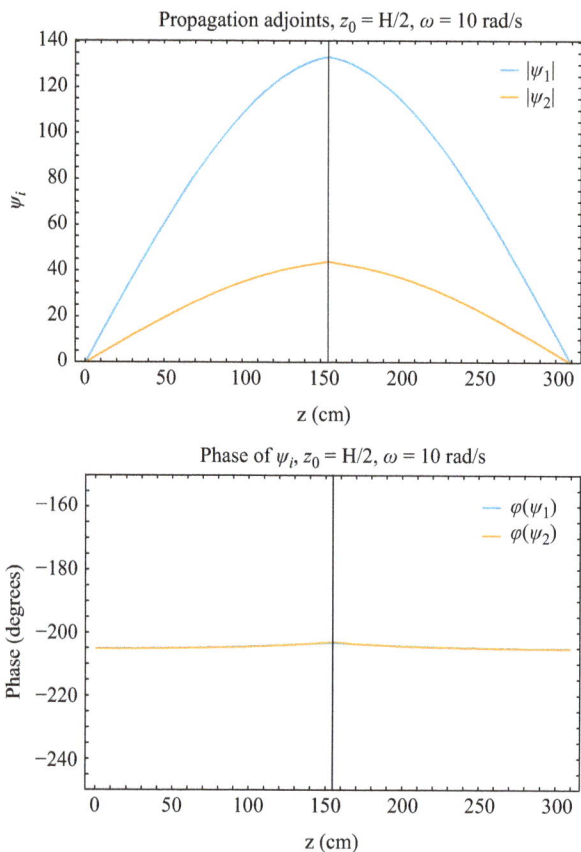

Figure 4.41   *Space dependence of the amplitude and the phase of the fast and*
thermal removal adjoints for $x_0 = 0$ in the FHR

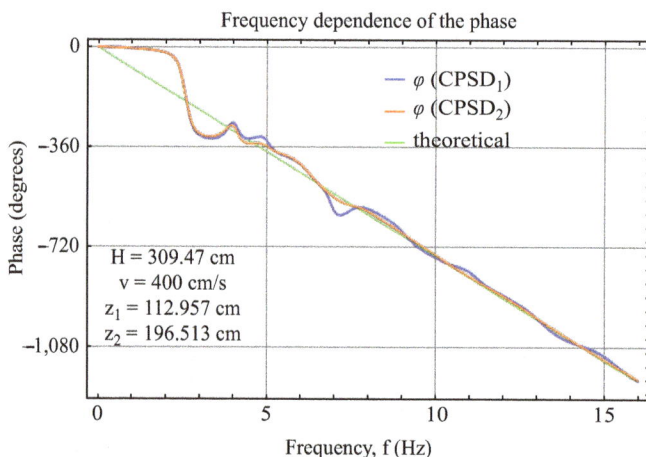

Figure 4.42   *Dependence of the phase of the cross-correlation on frequency*
between two axially displaced detectors for propagating
perturbations in the FHR

The dependence of the phase between two axially placed detectors, both in the fast and the thermal group, is shown in Figure 4.42. The figure shows that, somewhat surprisingly, the phase has a clear linear dependence on the frequency with the proper slope, for both the fast and the thermal neutron detector signals. The initial oscillations around the linear phase last up to approximately 4 Hz. Hence the determination of the transit time has to be based on the values of the phase above 4 Hz. On the other hand, in a real measurement, the signals become more "noisy" above 5 Hz, hence the determination of the transit time in practice will be less straightforward than the figure would suggest.

In summary, localisation of perturbations in the core from in-core neutron noise does not seem feasible in the FHR, but one should be able to extract information on the flow of the coolant with in-core neutron detectors.

Chapter 5

A survey of the dynamic core transfer calculational codes by the various groups worldwide, their principles, potentials for applications in Gen-IV systems and SMRs

Hoai-Nam Tran[1]

5.1 Introduction

Neutron noise, i.e. the fluctuation of neutron fluxes, is induced by the local/global fluctuations of neutronic and/or thermal–hydraulic (TH) characteristics, or by the mechanical vibrations of reactor components [7]. The neutron noise can be detected by neutron detectors located at in-core and/or ex-core positions, which could provide useful information for core monitoring, diagnostics and surveillance. The neutron noise technique is considerably advantageous and is being applied for online monitoring of the operation of most of the nuclear reactors, since it can be conducted without the disruption of reactor operation [7].

Numerical methods for neutron noise analysis with respect to the fluctuations of core properties by solving the noise equations are being continuously developed to provide the knowledge of neutron noise behaviour. In general, numerical analysis can provide a reactor transfer function between the noise sources and the induced noise, where the noise sources are defined via the fluctuations in macroscopic cross sections. The reactor transfer function is powerful for noise calculations, since it needs to be solved only once, and the space- and frequency-dependent noise induced by any source type can be calculated. This technique has been well applied to simplified reactor core models or in analytical calculations. For a large heterogeneous system with a multi-energy group model, solving for the reactor transfer function is a challenge, since it is time consuming and usually requires a large memory. In such a case, numerical methods tend to solve for the noise equations with a predefined noise source, i.e. a fixed-source problem. Though many research groups worldwide have recently devoted themselves to the development of numerical tools for neutron noise calculations, numerical simulation still remains a challenge for interpreting detector signals and improving core surveillance.

[1] Phenikaa Institute for Advanced Study, Phenikaa University, Hanoi, Vietnam

Two approaches are applied to analyse the neutron noise based on the simulations in the time-domain [103–107] and in the frequency-domain [7,46,108–112]. In the case of periodic perturbations, i.e. when the frequency of the perturbations is known, the space-dependent noise can be solved in the frequency domain using dedicated tools. The frequency-domain simulations are advantageous by simplifying the description of noise sources and avoiding the complexity of time discretisation. Thus, the stability of the solutions can be guaranteed. The noise simulations can be performed at different frequencies to obtain the space-dependent frequency spectrum of the noise. From this, the space and time-dependent noise can be derived via inverse Fourier transform [108]. However, a limitation of frequency-domain simulations is their inability to define random and/or non-periodic fluctuations. Moreover, coupling frequency-domain simulators with existing core physics codes in the time-domain presents challenges, as it requires an interface for transferring data between the two domains. In the time-domain, simulations can be conducted using existing neutron kinetic codes, where the static and fluctuating components are not separated. However, most of these codes were not originally developed for simulating neutron noise. Recently, several attempts to address this issue have been reported in the literature [104,107,113–115].

The noise equations in the frequency domain are derived from the time-dependent neutron balance equations by assuming small fluctuations around the static values of all quantities, e.g. the cross sections and the neutron flux, and by taking the Fourier transform of time-dependent terms. The outcome of the Fourier transform, after subtracting all static terms and neglecting higher-order terms, is the neutron noise equations with a fixed-source and complex quantities. Several attempts have been conducted for the development of numerical tools by simulating the space- and frequency-dependent noise. The neutron noise equations are solved in the frequency domain with the noise source defined via the fluctuations of macroscopic cross sections. Through the definition of the noise sources, the neutron noise induced by various phenomena can be simulated, e.g. perturbations of cross sections, vibrations of control rod or fuel assemblies. The development of a numerical tool was initiated at Chalmers University of Technology for the noise calculations of light water reactors (LWRs) based on two-group diffusion theory and a finite difference method (FDM) [108,110]. Other attempts have been conducted to develop the noise simulators using an analytical nodal method (ANM) for LWRs [116], and hexagonal geometrical systems such as VVERs [109,117], or using finite element method [117,118].

The diffusion theory was successfully deployed to simulate the neutron noise in a number of numerical tools. However, in some cases, localised perturbations in heterogeneous systems may lead to anisotropic sources in both static and noise equations. Thus, the accurate simulations of both static flux and noise may need to be done using the transport approximation. Although the transport approximation requires higher computational time, it provides a good option to verify the transfer function/neutron noise calculated based on the diffusion theory. Several developments were conducted for solving the neutron noise in heterogeneous systems applying advanced approximations, such as P1 approximation [119], SN and SP3 theories [120,121], discrete ordinates method [120,122], method of characteristics (MOC) [106] and Monte Carlo methods [111,112,123,124].

This chapter aims at surveying the worldwide attempts in the development of computational codes for calculating the dynamic core transfer/neutron noise in various reactor systems, as well as the potential application in the next generation reactors, including Gen-IV reactors and small modular reactors (SMRs).

5.2   A survey of the core transfer calculation codes

5.2.1   *Time-domain neutron noise simulators*

Most of the attempts dedicated to the development of neutron simulations in the time domain are aimed at extending the ability of existing reactor dynamic codes to calculate the neutron noise induced by in-core fluctuations [104,107,113–115,125]. For example, SIMULATE-3K (S3K), a commercial reactor dynamic code in the time domain not originally developed for noise calculations, has been extended to calculate the neutron noise induced by fuel vibrations and TH fluctuations. The CMSYS platform, developed at Paul Scherrer Institute, comprises several codes such as CASMO-5, SIMULATE-3 and SIMULATE-3K [113,115], which can simulate the neutron noise induced by fuel assembly vibrations based on the delta gap model of CASMO-5 and the fuel vibration model of SIMULATE-3K [113]. Other commercial codes in the time domain, such as DYN3D and PARCS, have been modified to simulate neutron noise induced by perturbations in absorbers of variable strength and by a travelling (propagating) perturbation [105,125]. These codes model perturbations as either the fluctuations of macroscopic cross sections or as time-dependent vibrations of fuel assemblies.

The time-domain neutron noise equations are obtained from the space-time-dependent neutron diffusion and time-dependent precursor equations. The time-dependent diffusion equations in two-energy groups and six-groups of delayed neutron precursors used in the PARCS code are written as [125]:

$$\frac{1}{v_1}\frac{\partial \phi_1(\boldsymbol{r},t)}{\partial t} = -\nabla\left[-D_1(\boldsymbol{r},t)\nabla\phi_1(\boldsymbol{r},t)\right] - \Sigma_{a,1}(\boldsymbol{r},t) + \Sigma_{12}(\boldsymbol{r},t)\phi_1(\boldsymbol{r},t)$$
$$+(1-\beta)\nu\Sigma_{f,1}(\boldsymbol{r},t)\phi_1(\boldsymbol{r},t) + (1-\beta)\nu\Sigma_{f,2}(\boldsymbol{r},t)\phi_2(\boldsymbol{r},t)$$
$$+\sum_{k=1}^{6}\lambda_k C_k(\boldsymbol{r},t) \tag{5.1}$$

$$\frac{1}{v_2}\frac{\partial \phi_2(\boldsymbol{r},t)}{\partial t} = -\nabla\left[-D_2(\boldsymbol{r},t)\nabla\phi_2(\boldsymbol{r},t)\right] - \Sigma_{a,2}(\boldsymbol{r},t)\phi_2(\boldsymbol{r},t)$$
$$+\Sigma_{12}(\boldsymbol{r},t)\phi_2(\boldsymbol{r},t), \tag{5.2}$$

and

$$\frac{\partial C_k(\boldsymbol{r},t)}{\partial t} = \beta_k\nu\Sigma_{f1}\phi_1(\boldsymbol{r},t) + \beta_k\nu\Sigma_{f1}\phi_1(\boldsymbol{r},t) - \lambda_k C_k(\boldsymbol{r},t);$$
$$k = 1,\ldots,6 \tag{5.3}$$

In the aforementioned equations, all variables have their usual meaning. v, λ, β and C are neutron velocity, time constant, effective delayed neutron fraction, and density of the delayed neutron precursors, respectively. The diffusion coefficients, macroscopic cross sections, neutron fluxes and the density of the six groups of neutron precursors are space- and time-dependent. The bold letter r refers to as a position vector. To derive the time-dependent noise equations, a perturbation is assumed as $X(t) = X_0 + \delta X(t)$ to all time-dependent quantities, and then the equations without the perturbations are subtracted. Here X_0 and $\delta X(t)$ are the static part and the time-dependent part (perturbation) of the variable, respectively. The PARCSv3.2 code was developed to solve the time-dependent neutron diffusion equations in 3D models. The code can handle both Cartesian and/or hexagonal geometries of fuel assemblies. Several methods for spatial discretisation were implemented in PARCSv3.2, such as FDM, a hybrid method combining ANM and nodal expansion method (NEM). A coarse mesh finite difference (CMFD) technique was also implemented to accelerate the convergence speed. To simulate the neutron noise corresponding to the fluctuations in time-domain, several developments were made in the PARCSv3.2 source code: introduction of perturbations, editing the corresponding cross sections and the induced noise.

The verification of neutron noise calculations using PARCSv3.2 in the time domain, compared to CORE SIM calculations in the frequency domain, was performed and reported [125]. CORE SIM was specifically developed for neutron noise simulations in LWRs. The comparison was conducted based on two types of noise sources: a point-like source corresponding to the fluctuation of absorption cross section and a perturbation in the coolant flow travelling along a fuel channel. The results demonstrated that PARCSv3.2 is able to simulate the neutron noise in some conditions, particularly a point-like source at the frequencies of 0.1–10 Hz. In the case of a travelling perturbation, the results are reliable at a frequency lower than 1 Hz and with the perturbation close to the core centre. The results also suggested that the method used for noise calculations in the PARCSv3.2 code needs further improvement and verification [125].

A neutron noise simulator in the time-domain was developed using the NEM method for spatial discretisation and the implicit-difference method for time discretisation [107]. The simulator solves the neutron noise equations in two energy groups and in 3D rectangular geometries. The static solution was verified based on the KWU and SMART integrated pressurised water reactor (iPWR) benchmark problems in comparison to the PARCS code. Then, the noise calculations were performed for the two cores and compared to the CORE SIM and SD-HACNEM codes in the frequency-domain. The perturbations were assumed as a random vibration of a control rod and a vibration with combining frequencies. The approach proved the advantage in simulating random perturbations. Consequently, the tool is considered suitable for dealing with random perturbations/vibrations [107]. Several other codes for neutron noise simulation in the time-domain can be found in [104,114,115].

5.2.2   CORE SIM

The CORE SIM code has been developed continuously at Chalmers University of Technology for simulating the space- and frequency-dependent neutron noise induced by spatially distributed or localised sources in the frequency domain. The neutron noise equations in the two-group diffusion theory are derived from the time-dependent diffusion equation. By neglecting the higher order of fluctuations, the first-order neutron noise equation is written as [108,110]:

$$\left[\nabla \cdot \overline{\overline{D}}(r) \nabla + \overline{\overline{\Sigma}}_{dyn}(r, \omega) \right] \times \left[\begin{array}{c} \delta\phi_1(r, \omega) \\ \delta\phi_2(r, \omega) \end{array} \right] = \left[\begin{array}{c} S_1(r, \omega) \\ S_2(r, \omega) \end{array} \right] \tag{5.4}$$

where

$$\overline{\overline{D}}(r) = \left[\begin{array}{cc} D_1(r) & 0 \\ 0 & D_2(r) \end{array} \right] \tag{5.5}$$

$$\overline{\overline{\Sigma}}_{dyn}(r, \omega) = \left[\begin{array}{cc} -\Sigma_1(r, \omega) & \nu\Sigma_{f2}(r, \omega) \\ \Sigma_{rem}(r) & -\Sigma_{a2}(r, \omega) \end{array} \right] \tag{5.6}$$

$$\Sigma_1(r, \omega) = \Sigma_{a1}(r, \omega) + \Sigma_{rem}(r) - \nu\Sigma_{f1}(r, \omega) \tag{5.7}$$

$$\nu\Sigma_{f1,2}(r, \omega) = \frac{\nu\Sigma_{f1,2}(r)}{k_{eff}} \left(1 - \frac{i\omega\beta_{eff}}{i\omega + \lambda} \right) \tag{5.8}$$

$$\Sigma_{a1,2}(r, \omega) = \Sigma_{a1,2}(r, \omega) + \frac{i\omega}{\upsilon_{1,2}} \tag{5.9}$$

The r.h.s. vector in (5.4) represents the noise source in the fast and thermal groups, which can be modelled through the fluctuations of macroscopic cross sections. The fluctuations can be a result of mechanical or thermal processes, such as absorption perturbation, core barrel vibrations, fuel assembly vibrations, etc. The source term is written as

$$\begin{aligned} \left[\begin{array}{c} S_1(r, \omega) \\ S_2(r, \omega) \end{array} \right] &= \overline{\phi}_{rem}(r)\delta\Sigma_{rem}(r) + \overline{\overline{\phi}}_a(r) \left[\begin{array}{c} \delta\Sigma_{a1}(r, \omega) \\ \delta\Sigma_{a2}(r, \omega) \end{array} \right] \\ &+ \overline{\overline{\phi}}_f(r) \left[\begin{array}{c} \delta\nu\Sigma_{f1}(r, \omega) \\ \delta\nu\Sigma_{f2}(r, \omega) \end{array} \right] \end{aligned} \tag{5.10}$$

where the δX_i, $i = \{rem, a1, a2\}$, etc., stand for the fluctuations of the macroscopic cross sections, corresponding to the actual perturbation, and further,

$$\overline{\phi}_{rem}(r) = \left[\begin{array}{c} \phi_1(r) \\ -\phi_1(r) \end{array} \right] \tag{5.11}$$

$$\overline{\overline{\phi}}_a(r) = \left[\begin{array}{cc} \phi_1(r) & 0 \\ 0 & \phi_2(r) \end{array} \right] \tag{5.12}$$

and

$$\overline{\overline{\phi}}_f(r) = \left[\begin{array}{cc} -\phi_1(r) & -\phi_2(r) \\ 0 & 0 \end{array} \right] \tag{5.13}$$

All symbols in the aforementioned equations have their usual meaning.

Continuous development is ongoing, including the CORE SIM+ version, which has the capability to handle finer meshes [126]. CORE SIM+ features enhancements to optimise both steps of neutron noise simulation, enabling the modelling of a broad range of neutron noise sources (Figure 5.1). Key features of CORE SIM+ include

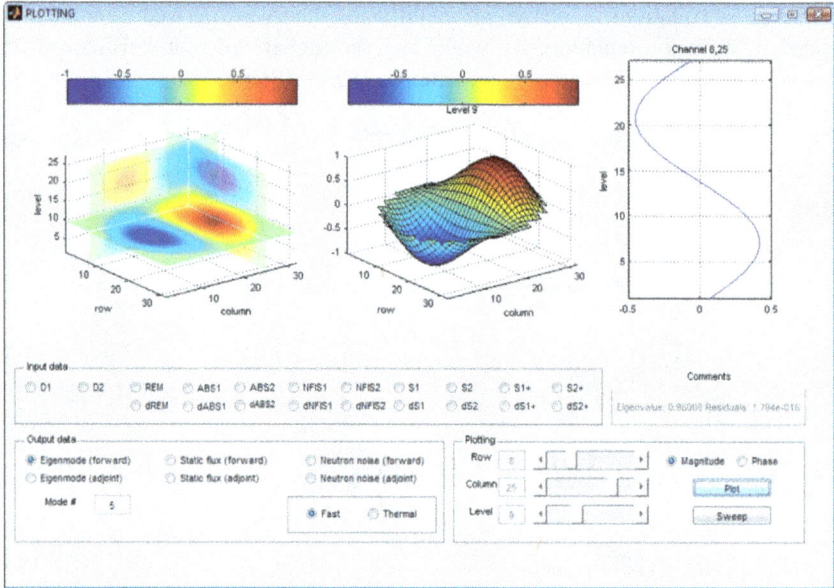

Figure 5.1   The graphical user interface (GUI) for the visualisation of input and output data in CORE SIM [110]

a non-uniform spatial mesh, efficient matrix/vector construction, effective numerical solvers combined with acceleration methods, and automated generation of the reactor transfer function [110,126]. Three options are available in CORE SIM+ for solving the static forward and adjoint problems: the standard power method (PM), the PM accelerated with Chebyshev polynomials, and PM combined with a non-linear acceleration based on the Jacobian Free Newton Krylov algorithm [126]. Extensive verification and validation efforts have been undertaken for this tool. The capability to estimate the point kinetic component of a perturbed system was verified. The Green's function generator and the direct solver were confirmed to produce very similar results when simulating fuel cell vibration in a simple system. CORE SIM+ was utilised to simulate an experiment involving a vibrating cluster of fuel rods in a research reactor, and the results showed reasonable agreement with measurements. Additional verification exercises were successfully conducted and reported in [126].

The CORE SIM versions have proven to be flexible and efficient in generating neutron noise databases for nuclear power reactors. The core transfer function of a reactor system can be determined using CORE SIM+. These databases are valuable for studying the reactor response to various perturbations and for training machine-learning algorithms aimed at core monitoring and diagnostics. An illustrative example is provided with a representative database generated for a pressurised water reactor (PWR). This database encompasses scenarios induced by absorbers of varying strength, perturbations travelling with the coolant flow, and vibrations of the core barrel and fuel assemblies, across different frequencies and locations.

5.2.3   DYN3D

DYN3D is a reactor dynamic code, which was continuously developed by the HZDR (Helmholtz-Zentrum Dresden-Rossendorf), Germany for more than 20 years [127,128]. The code is considered as a 3D best-estimate tool for static and transient analysis of LWRs. DYN3D was originally developed to analyse accidents initiated by reactivity in VVERs. The improvement of physical models and numerical methods in the DYN3D code is continued in coupling TH with fuel performance codes. DYN3D has become an advanced numerical tool for transient analysis of LWRs and innovative designs. DYN3D solves the static and time-dependent multi-group neutron diffusion equations or simplified transport equations by deploying the NEM method. Cartesian, hexagonal or trigonal geometries can be handed by DYN3D. The axial height of the reactor core is divided into a number of layers, so that a 3D node represents a prism determined by fuel assembly geometries. In transient calculations, the time derivative terms are approximated by applying implicit first-order Euler method combined with the frequency transformation.

A one-dimensional (1D) TH solver and fuel rod behaviour module were also implemented in the DYN3D code. The TH model solves the balance equations of mass, momentum and energy of one or two-phase flow, heat transfer regime mapping and a thermo-mechanical fuel rod model. Recent development of DYN3D extended its ability to simulate the steady-state and transients of SFRs based on the multi-group cross-section set generated by using the Monte Carlo code Serpent [127].

DYN3D solves the two-group diffusion equations in the time-domain using the NEM method with transverse integration for the spatial dependence. Six groups of delayed neutron precursors were taken into account. DYN3D is compared with different approaches for the treatment of the time and space dependency. The analytical solutions of the 1D problems are evaluated to compare with DYN3D. Comparison of DYN3D and a linearised method is only possible if the magnitudes of the perturbations are sufficiently small, so that the higher-order terms can be neglected.

Many European institutions have recently attempted to investigate the magnitudes and patterns of flux fluctuations primarily observed in the KWU PWRs. Among the numerical tools used to study neutron flux fluctuations is the time-domain reactor dynamics code. Since DYN3D and other similar codes were not specifically developed to simulate low-amplitude neutron flux fluctuations, their applicability in this area needs verification. To aid in the verification of DYN3D for simulating neutron flux fluctuations, two cases of perturbations are considered: a localised absorber with variable/oscillatory strength and a travelling oscillatory perturbation. In this analysis, both DYN3D and CORE SIM, along with analytical approaches in the frequency domain, are used. The results obtained are compared in terms of the distributions of amplitude and phase of the induced neutron flux fluctuations. These comparisons are repeated with varying amplitudes and frequencies of the perturbations. The results demonstrate a good agreement both qualitatively and quantitatively. The deviations observed between the DYN3D results and the reference results show a dependency on the perturbation magnitude, which is attributed to the omission of higher-order terms in the calculation of the reference solutions.

5.2.4   SD-HACNEM

Numerous efforts have been devoted to developing computational tools for neutron noise simulations in both rectangular and hexagonal reactor systems at Sharif University of Technology, initiated by N. Vosoughi. A noise simulator was created for the VVER with hexagonal-structured geometry, using the FDM for the spatial discretisation of hexagonal meshes [109]. The Galerkin Finite Element Method (GFEM) was employed with unstructured triangular elements to solve the two-group neutron noise equations for both rectangular and hexagonal reactor cores [117,118]. A noise simulator was developed using the average current nodal expansion method (ACNEM) in hexagonal geometry [129]. This tool has the potential to diagnose the displacement of two fuel assemblies, where the noise source is scaled according to assembly sizes. The zeroth-order NEM with a higher-order flux expansion was implemented to increase the accuracy. This method incorporates average partial currents on surfaces, providing advantages over the FDM and FEM methods with a higher accuracy and reduced computational times. Additionally, the NEM method demonstrates superior convergence.

The SD-HACNEM code was developed for neutron noise simulations in rectangular geometries by solving the two-group diffusion noise equations using the high-order NEM [130]. In particular, the static calculations were developed using the ACNEM and the high-order ACNEM (HACNEM), utilising a nodal mesh equal to assembly size. The static results were compared with that from the BIBLIS-2D benchmark problem. Following this, the neutron noise calculations were developed in the frequency domain using the HACNEM method. The results were corroborated through zero-frequency simulations and adjoint calculations. The numerical findings revealed that the HACNEM method exhibited more accuracy in neutron noise simulation without reducing the mesh size compared to the ACNEM method.

5.2.5   Noise calculations using transport theories

Some extensions were developed for calculating the neutron noise using the transport theory. The derivation of the neutron noise transport equation in the frequency-domain can be started from the time-dependent neutron transport equation by applying a linear approximation and Fourier transformation of all time-dependent fluctuations. The final form of the neutron noise transport equation with one delayed neutron precursor is written as follows [131]:

$$\Omega \cdot \nabla \delta\phi\left(r, \Omega, E, \omega\right) + \Sigma_{t0}\delta\phi\left(r, \Omega, E, \omega\right)$$

$$= \int_{4\pi} d\Omega' \int dE' \Sigma_{s0}\left(r, \Omega' \to \Omega, E' \to E, \omega\right) \delta\phi\left(r, \Omega', E', \omega\right)$$

$$+ \frac{\chi(E)}{4\pi k_{eff}} \left(1 - \frac{i\omega\beta}{i\omega + \lambda}\right) \int_{4\pi} d\Omega' \int dE' \nu\Sigma_{f0}\left(r, E'\right) \delta\phi\left(r, \Omega', E', \omega\right)$$

$$- \frac{i\omega}{v(E)} \delta\phi\left(r, \Omega, E, \omega\right) + S\left(r, \Omega, E, \omega\right) \tag{5.14}$$

where

    Σ_t is the macroscopic total cross section;

    Σ_s is the macroscopic scattering cross section;

    Σ_f is the macroscopic fission cross section;

    $\chi(E)$ the fission neutron spectrum;

    $\nu(E)$ the number of neutrons per fission;

    $v(E)$ is the neutron speed;

    β is the fraction of the delayed neutrons;

    λ is the time decay constant of the delayed neutron precursors;

    $S(r, \Omega, E, \omega)$ is the neutron noise source.

The noise source $S(r, \Omega, E, \omega)$ calculated from the fluctuations of the macroscopic cross sections in the transport theory is written as

$$S(r, \Omega, E, \omega) = -\delta\Sigma_t(r, E, \omega)\,\phi_0(r, \Omega, E)$$

$$+ \int_{4\pi} d\Omega' \int dE'\, \delta\Sigma_s(r, \Omega' \to \Omega, E' \to E, \omega)\,\phi_0(r, \Omega', E') \qquad (5.15)$$

$$+ \frac{\chi(E)}{4\pi k_{eff}}\left(1 - \frac{i\omega\beta}{i\omega + \lambda}\right) \int_{4\pi} d\Omega' \int dE'\, \nu\delta\Sigma_f(r, E', \omega)\,\phi(r, \Omega', E')$$

The form of the neutron noise transport equation in the frequency-domain is somewhat similar to a conventional fixed source transport equation. The main differences include the complex-valued variables of the noise $\delta\phi(r, \Omega, E, \omega)$ and fluctuations of cross sections $\delta\Sigma(r, E, \omega)$, the complex source term $S(r, \Omega, E, \omega)$ with a negative sign, the production term multiplied by $1 - i\omega\beta/(i\omega + \lambda)$, and the additional term of $-i\omega\,\delta\phi(r, \Omega, E, \omega)/v$. While the conventional Monte Carlo methods handle only real and positive particle weights.

Neutron noise calculations based on the P1 theory were developed and applied to a 1D two-region model, comparing the results with diffusion theory [119]. The neutron noise was calculated using the Green's function technique, and the computed Green's functions were successfully benchmarked against semi-analytical calculations. Due to the limited number of detector positions available for comparing detector readings to numerical estimates of neutron noise in measurements, the shape of the numerical solution becomes crucial. The results presented for an LWR and a heavy water reactor (HWR) indicate that both diffusion and P1 theories produce nearly identical outcomes, with minor differences mainly observed in the local component and the reflector region. It is important to note that the static flux used is cosine-shaped due to the homogeneous core composition; in real heterogeneous commercial reactors, the static flux is flatter in the middle, resulting in even smaller differences. Conversely, the static flux displays a sharper gradient near the reflector. Since the discrepancies between P1 and diffusion theories relate to terms proportional to the gradient of the static flux, larger differences may be expected close to the reflector.

The study also demonstrated that neglecting the effects of fluctuations in cross sections on transport cross sections yields results nearly identical to those obtained by accounting for these effects. This implies that calculations can be performed more quickly and with lower memory requirements, as calculating the Green's function involves less complex solution procedures. For HWRs, the induced neutron noise significantly differs from that of LWRs, but the same conclusions were applied, indicating no practical advantage in using a higher-order theory. Transitioning from 1D to 3D calculations, relevant to power reactor problems, leads to a steeper/narrower local component of neutron noise. This increased steepness could result in a larger difference between P1 and diffusion theory in the vicinity of the local component. Nevertheless, the same conclusions remain valid, as the global component of neutron noise is the most critical for practical applications.

The simplified PN method has garnered significant attention in reactor physics, as it allows for partial capture of neutron transport effects without greatly increasing computational time and code complexity. Gong *et al.* developed a neutron noise simulator based on two-group transport SP3 theory and investigated the differences between the noise derived from SP3 and diffusion theory [121]. They derived the theory for first-order neutron noise within the multi-group SP3 approximation. The analysis considers the symmetric form of the SP3 equation for neutron flux. Neutron dynamics are examined in a limited 2D or 3D domain with boundary conditions. The process begins by expressing the time- and space-dependent flux of a critical system within the multi-group SP3 equations. By subtracting the static equations from the time-dependent neutron diffusion equations and eliminating the precursor density using a temporal Fourier transform, the first-order neutron noise equation in SP3 theory is formulated as [121]:

$$
\begin{aligned}
&\left(\frac{i\omega}{v_g} + \Sigma_{r,g}\right) \delta\phi_{0,g}(\mathbf{r},\omega) - \nabla D_{0,g} \nabla \delta\phi_{0,g}(\mathbf{r},\omega) \\
&\quad -2\nabla D_{0,g} \nabla \delta\phi_{2,g}(\mathbf{r},\omega) - \sum \Sigma_{s,g'\to g} \delta\phi_{0,g'}(\mathbf{r},\omega) \\
&\quad -\frac{1}{k_{eff,0}} \left(\chi_{p,g}(1-\beta) + \chi_{d,g} \sum_j \frac{\lambda_j \beta_j}{\lambda_j + i\omega} \right) \sum_{g'=1}^{G} \nu\Sigma_{f,g'} \delta\phi_{0,g'}(\mathbf{r},\omega) \\
&= S_{0,g}(\mathbf{r}),
\end{aligned}
\tag{5.16}
$$

$$
\begin{aligned}
&\left(\frac{i\omega}{v_g} + \Sigma_{tr,g}\right) \delta\phi_{2,g}(\mathbf{r},\omega) - \frac{2}{5}\nabla D_{0,g} \nabla \delta\phi_{0,g}(\mathbf{r},\omega) \\
&\quad -\frac{4}{5}\nabla D_{0,g} \nabla \delta\phi_{0,g}(\mathbf{r},\omega) - \frac{3}{5}\nabla D_{2,g} \nabla \delta\phi_{2,g}(\mathbf{r},\omega) \\
&= S_{2,g}(\mathbf{r})
\end{aligned}
\tag{5.17}
$$

and

$$
\begin{aligned}
S_{0,g}(\mathbf{r},\omega) &= \delta\Sigma_{r,g}\phi_{0,g}(\mathbf{r},\omega) + \sum \delta\Sigma_{s,g'\to g}\phi_{0,g'}(\mathbf{r},\omega) \\
&\quad +\frac{1}{k_{eff,0}} \left(\chi_{p,g}(1-\beta) + \chi_{d,g} \sum_j \frac{\lambda_j \beta_j}{\lambda_j + i\omega} \right) \sum_{g'=1}^{G} \delta\nu\Sigma_{f,g'}\phi_{0,g'}(\mathbf{r},\omega)
\end{aligned}
\tag{5.18}
$$

$$
S_{2,g}(\mathbf{r},\omega) = -\delta\Sigma_{tr,g}\phi_{2,g}(\mathbf{r}),
\tag{5.19}
$$

where

   g is the energy group;
   $\phi_{0,g}$ is the scalar flux;

$\phi_{2,g}$ is the second moment of angular flux;

$\Sigma_{tr,g}$ is the transport cross section;

$\Sigma_{r,g}$ is the removal cross section;

$\Sigma_{s,g'\to g}\phi_{0,g'}$ is the scattering cross section from energy group g' to group g;

χ_p is the prompt neutron spectrum;

χ_d is the delayed neutron spectrum;

$\nu\Sigma_{f,g}$ is the fission production cross section;

$D_{0,g} = \dfrac{1}{3\Sigma_{tr,g}}$ is the scalar diffusion constant;

$D_{2,g} = \dfrac{3}{7\Sigma_{tr,g}}$ is the second moment of the diffusion constant;

$k_{eff,0}$ is the neutron multiplication factor;

ω is the frequency of the perturbation.

The perturbation may be the fluctuations of neutronic and TH parameters, and the mechanical vibrations of structural components of the reactor.

A noise simulation code, CORCA-NOISE, has been developed and integrated into the SP3 solver CORCA-PIN. Noise calculations using CORCA-NOISE were performed to demonstrate the validity of CORCA-NOISE, to investigate a shell-mode core-barrel vibration in comparison with the calculations using the diffusion theory, and to investigate the transfer function of the 2D mixed oxide (MOX) benchmark core. A shell-mode core-barrel vibration benchmark and a 2D two-group MOX benchmark were applied to demonstrate the validity of the numerical solution code. The effect of cross sections and frequency on the induced noise were investigated, the amplitude and phase of the noise were compared between the SP3 and diffusion theories. The results exhibited considerable difference between the two theories in power reactor noise at plateau frequencies, particularly in heterogeneous cores. Thus, further investigations are suggested to answer if it is worth to introduce the SP3 approximation in the noise simulation.

NOISE-SN was developed for noise simulation of LWRs by solving the transport equation in the frequency domain. Spatial discretisation was implemented based on a finite difference, discrete ordinates, and multi-energy method. In the NOISE-SN code, the solutions of both the static transport and neutron noise equations were performed by an inner-outer iterative procedure. In the inner iteration, a transport sweep with the discrete angular directions was conducted to evaluate the neutron fluxes in a group and construct the self-scattering term. After the inner iterations are completed, the down-scattering term is evaluated based on the neutron fluxes. The fission source is updated in the outer iteration and the inner iteration is repeated again. In the outer iteration of the static calculation, the effective multiplication factor is also updated. The iterative procedure is repeated until obtaining the convergence. The CMFD method was implemented in the NOISE-SN code to accelerate the static fluxes and noise calculations in the frequency-domain. The criticality calculations in the static module of NOISE-SN rely on a standard CMFD method. A similar algorithm is followed for the CMFD acceleration of the noise calculations. After one inner-outer iteration, the estimated neutron noise is used to construct a low-order diffusion-like equation.

In order to verify the code, it was used for calculating the neutron noise based on two 2D benchmark problems: the C4V problem and the C5G7 problem. In the C4V problem, NOISE-SN was compared against CORE SIM+ to investigate the possible differences between the two codes [132]. In addition, the effect of the order of discrete ordinates is analysed and the algorithm based on the fictitious source method is tested. In the C5G7 problem, the impact of the ray effect on NOISE-SN is shown and the two mitigation strategies (i.e. increasing the order of discrete ordinates and introducing a fictitious source in the discrete ordinates equation) are evaluated.

5.2.6   Monte Carlo methods for noise calculations

A Monte Carlo method was developed for solving the transport equation of neutron noise in the frequency-domain [111,123,124]. Yamamoto developed a strategy for solving the neutron noise transport equation based on the Monte Carlo method by introducing (1) complex-valued weights to handle the complex quantities, (2) weight cancellation of positive and negative weights, and (3) an additional term of $i\omega\delta\phi(\mathbf{r}, \Omega, E, \omega)/v(E)$ during the random walk processes [111]. The method was applied for the noise calculations in an infinite homogeneous model and compared to analytical diffusion solutions. A good agreement between the Monte Carlo method and the analytical solution was reported. However, significant differences were found near the noise source at a high frequency.

A new Monte Carlo method was proposed for neutron noise calculation without using a weighting technique [112]. The general neutron noise theory and the new Monte Carlo method were compared against that proposed by Yamamoto [111] and to deterministic methods for isotropic and anisotropic noise sources. The comparison was conducted with the perturbations in a wide range of frequencies [0.01–100 Hz], where the conventional algorithm for fixed-source problems can be used for both Monte Carlo methods.

A challenge of the Monte Carlo methods is to control the growing number of particles generated due to low- or high-frequency noise. The aforementioned Monte Carlo techniques could potentially address this issue [111,112]. However, the weight cancellation method is complicated to implement in a production-level Monte Carlo code. In contrast, the method proposed by Rouchon *et al.* [112] is expected to be viable for production-level codes, although the calculation efficiency deteriorates outside the plateau frequency range. Yamamoto proposed a method for suppressing the divergence [131]; however, this divergence does not occur within the plateau frequency range. Implementing an algorithm to address the divergence issue into a production-level Monte Carlo code may require extensive modifications. A specific algorithm for solving the neutron noise transport equation for the plateau region has been successfully integrated into the MCNP4C code.

The modified MCNP code can be utilised across a frequency range from 0.02 Hz to several tens of Hz. It was employed for neutron noise calculations in a 1D homogeneous fuel solution and the results were compared with those obtained using a two-energy group in-house research code. The favourable agreement between the two codes supports the validity of the modified MCNP code. This code was then applied to neutron noise calculations for a benchmark boiling water reactor (BWR) model. Close to the neutron noise source, the frequency dependence of amplitude and phase shift significantly differs from that predicted by point kinetics. However, as the detector moves further from the noise source, the frequency dependence begins to resemble that of point kinetics. This spatial dependence of neutron noise is believed to arise from higher mode effects. Neutron noise detected near the source is particularly sensitive to the neutron source itself, which can cause deviations from expected point kinetics behaviour. Therefore, it is essential to consider higher-order mode effects and the algorithm for controlling the number of particles in the calculations.

5.2.7   Noise simulator for fast reactors

A previous analytic investigation of neutron noise induced by fuel vibration indicated that the neutron noise in a fast group could provide useful information for identifying the vibrations of fuel rods in LWRs. The neutron noise in a fast reactor could be useful for assessing the core dynamic condition. Measurements of neutron noise in fast reactors and a test facility were conducted early. Thus, similarly to the analytical two-group extensions described in Chapter 2, it is highly motivated to extend the noise methods to fast reactors, one of the Gen-IV systems.

For this purpose, a neutron noise simulator for fast reactor with hexagonal geometries was developed based on the multi-groups diffusion theory [133]. The tool solved both the static and noise diffusion equations, where the neutron noise equation was solved in the frequency-domain with several groups of delayed neutron precursors. Since the multi-group neutron noise equations in diffusion theory have not been presented else-where in this book, they are summarised here for completeness. To construct the noise equations, it is necessary to define a noise source via the static flux and the fluctuations of cross sections. This means that the solution of the k_{eff} and the static fluxes was also required and implemented in this tool.

The multi-group diffusion equation for the static state is written as

$$-\nabla \cdot [D_g \nabla \phi_g (r)] + \Sigma_{t,g} \phi_g (r) = \frac{1}{k_{eff}} \chi_g \sum_{g'} \nu \Sigma_{f,g'} \phi_{g'} (r) + \sum_{g' \neq g} \Sigma_{s,g' \to g} \phi_{g'} (r)$$

(5.20)

where
   $g = 1, 2, \ldots, G$ denotes the energy group,
   ϕ_g is the neutron flux in group g,
   D_g is the diffusion coefficient in group g,
   $\nu \Sigma_{f,g}$ is the production cross section in group g,
   $\Sigma_{s,g' \to g}$ is the scattering cross section from group g' to group g,
   $\Sigma_{t,g}$ is the total cross section, which is defined as

$$\Sigma_{t,g} = \Sigma_{a,g} + \sum_{g' \neq g} \Sigma_{s,g \to g'},$$

(5.21)

   $\Sigma_{a,g}$ is the absorption cross section in group g,

χ_g is the fission energy spectrum, which is expressed via the prompt, χ_g^p and delayed, $\chi_{g,j}^d$ spectra as follows (see (2.106)):

$$\chi_g = (1 - \beta) \chi_g^p + \sum_j \beta_j \chi_{g,j}^d \tag{5.22}$$

Further, $j = 1, 2, \ldots, J$ denotes the group of delayed neutron precursors, and β is the total fraction of delayed neutrons

$$\beta = \sum_j \beta_j \tag{5.23}$$

It is noticed that all cross sections in (5.20) and (5.21) are space-dependent.

The multi-group noise equation are obtained from the space- and time-dependent diffusion equations by assuming that all time-dependent terms, $X(\boldsymbol{r}, t)$, can be split into a stationary component, $X_0(\boldsymbol{r})$, which corresponds to the value at the steady state, plus a small fluctuation, $\delta X(\boldsymbol{r}, t)$ as

$$X(\boldsymbol{r}, t) = X_0(\boldsymbol{r}) + \delta X(\boldsymbol{r}, t) \tag{5.24}$$

By assuming that the fluctuations are small so that only the first-order noise needs to be taken into account, products of fluctuating terms can be neglected and the result is a linear equation for the fluctuation of the flux. Subtracting the static equation and after performing a Fourier transform of all time-dependent terms, $\delta X(\boldsymbol{r}, t)$, as

$$\delta X(\boldsymbol{r}, \omega) = \int_{-\infty}^{\infty} \delta X(\boldsymbol{r}, t) e^{-i\omega t} dt \tag{5.25}$$

with the assumption that the system was in the unperturbed (critical) state at $t = -\infty$, the first-order space- and frequency-dependent neutron noise in multi-group diffusion theory is written as follows:

$$-\nabla \cdot [D_g \nabla \delta\phi_g(\boldsymbol{r}, \omega)] + \Sigma_{t,g}(\omega) \delta\phi_g(\boldsymbol{r}, \omega) = \frac{1}{k_{eff}} \chi_g(\omega) \sum_{g'} \nu\Sigma_{f,g'} \delta\phi_{g'}(\boldsymbol{r}, \omega)$$

$$+ \sum_{g' \neq g} \Sigma_{s,g' \to g} \delta\phi_{g'}(\boldsymbol{r}, \omega) + S_g(\boldsymbol{r}, \omega) \tag{5.26}$$

where $\delta\phi_g(\boldsymbol{r}, \omega)$ denotes the space- and frequency-dependent noise in group g. The frequency-dependent total cross section in (5.26) is written as

$$\Sigma_{t,g}(\omega) = \Sigma_{a,g}(\omega) + \sum_{g' \neq g} \Sigma_{s,g \to g'} \tag{5.27}$$

with

$$\Sigma_{a,g}(\omega) = \Sigma_{a,g} + \frac{i\omega}{v_g} \tag{5.28}$$

The k_{eff} in the noise equation is the eigenvalue obtained from the static calculation. $\chi_g(\omega)$ denotes the frequency-dependent fission energy spectrum, which is obtained from the equation of delayed neutron as

$$\chi_g(\omega) = \chi_g - \sum_j \chi_{g,j}^d \frac{i\omega\beta_j}{\lambda_j + i\omega} \qquad (5.29)$$

which is the generalisation of the two-group version with one average delayed neutron group, (2.115), to several energy groups and delayed neutron precursor groups. The last term of (5.26) is the noise source, which is calculated from the fluctuations of macroscopic cross sections as

$$S_g(\boldsymbol{r},\omega) = -\delta\Sigma_{a,g}(\omega)\,\phi_g(\boldsymbol{r}) - \sum_{g'\neq g}\delta\Sigma_{s,g\to g'}(\omega)\phi_g(\boldsymbol{r}) + \sum_{g'\neq g}\delta\Sigma_{s,g'\to g}(\omega)\phi_{g'}(\boldsymbol{r})$$

$$+\frac{1}{k_{eff}}\chi_g(\omega)\sum_{g'}\delta\left[\nu\Sigma_{f,g'}(\omega)\right]\phi_{g'}(\boldsymbol{r}) \qquad (5.30)$$

In this equation, the fluctuation of the diffusion coefficient in the source term is neglected. The static equation (5.20) is an eigenvalue problem, where the k_{eff} and the static fluxes correspond to the fundamental mode, while the balance equation for the neutron noise, as given by (5.26), is an inhomogeneous equation with an external source. Another important difference is that all quantities in (5.26) are frequency-dependent, i.e. complex quantities.

The numerical tool was developed for fast reactors with hexagonal geometries by solving both the static and noise equations based on the multi-groups diffusion theory (Figure 5.2). The noise equation was solved in the frequency-domain with several groups of delayed neutron precursors. Spatial discretisation was implemented based on the FDM, in which the hexagonal geometry of the fuel assembly was radially divided into six equilateral triangles, and if needed for more accuracy, each triangle can be further divided into four subtriangles. By this spatial discretisation each fundamental node in a 3D core model consists of two equilateral triangular bases and three rectangular sides. The finite difference approximation was then applied to calculate the leakage term in the static and noise equations for each node. A power iterative solution process was implemented that consisted of an outer and an inner iteration loop to solve the static and noise equations. The noise solution module of this tool only solves for a fix-source problem, since the problem size is too big to solve for the core transfer function. This means that the noise source is required to be prior defined for solving the induced noise. To accelerate the convergence of both the static and noise calculations, a CMFD acceleration technique was implemented in this tool, where a coarse mesh is defined radially as a hexagonal assembly.

Verification calculations for this tool were conducted based on two benchmark problems: 1) a 2D thermal reactor core model, and 2) a 3D SFR. In the first problem, static and noise calculations were performed on the basis of a 2D two-group homogeneous core model in comparison with analytical solutions. Point-like noise sources were assumed to appear at the central position or non-central position of the core. Good agreement was found between the numerical and analytical results in both forward and adjoint problems. In the second benchmark problem, the tool was

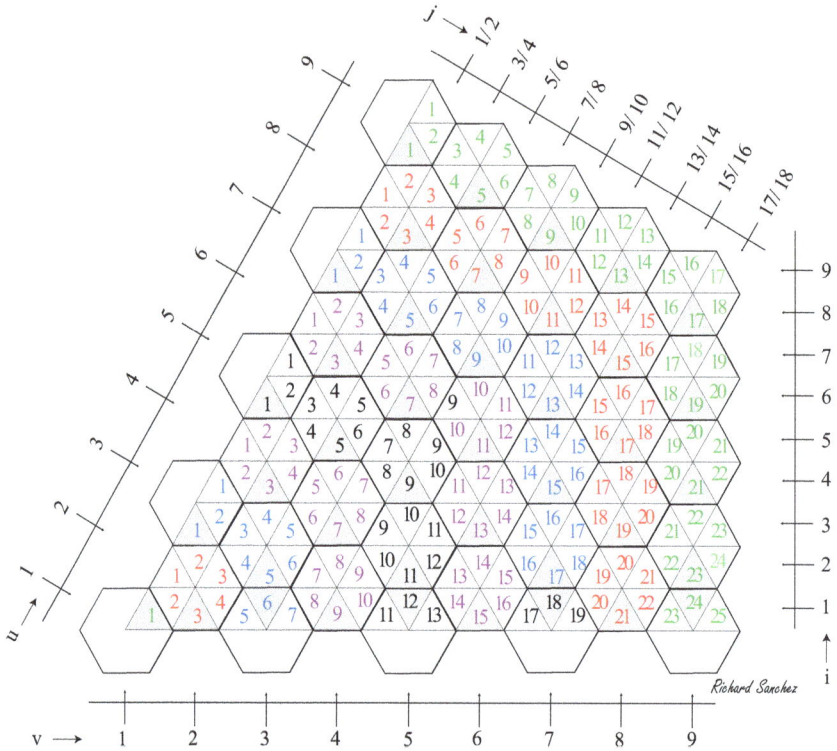

Figure 5.2   *Triangular discretisation in the 1/6th model of a hexagonal system used in the noise simulator. (i/j) coordinates are used to handle the triangular meshes, while (u/v) coordinates are used to handle the hexagonal meshes [133]*

used to perform static calculations for a large SFR core and benchmarked against the ERANOS code [134]. A cross sections in 33 energy groups and eight groups of delayed neutron precursors for the SFR core were generated using the ERANOS code and processed for use in this noise simulator. There is no available reference for benchmarking the frequency-dependent noise in a multi-group model; nevertheless, the noise calculations could still be considered as reliable if the static calculations give adequate accuracy. Calculations of the space- and frequency-dependent noise induced by the perturbation of the absorption cross section near the core centre have been performed.

5.3   Potentials for applications in Gen-IV systems and SMRs

In 2001, the framework for international collaboration on Gen-IV nuclear technologies was established, with the aim of achieving sustainability, economic efficiency, safety and reliability, resistance to proliferation, and physical protection [135,136].

Six reactor systems have been chosen as the Gen-IV technologies, which include the gas-cooled fast reactor (GFR), lead-cooled fast reactor (LFR), molten salt reactor (MSR), sodium-cooled fast reactor (SFR), supercritical water-cooled reactor (SCWR), and very high-temperature reactor (VHTR).

Recently, SMR concepts have gained increasing attention due to their advantages in terms of design, economic viability, construction timelines, and site costs. According to International Atomic Energy Agency (IAEA), SMRs are defined as nuclear reactors with a power output ranging from 10 to 300 MWe [137]. These reactor components can be designed using modular technology, allowing them to be manufactured in factories and then transported to the installation site. As a result, SMRs present a more cost-effective alternative to traditional large power reactors. They are particularly suitable for remote areas where energy demand is lower and located far from the national electricity grid.

SMRs are being developed for several key reactor types, including advanced LWRs, HWRs, and Gen-IV reactors. They are engineered with innovative fuel concepts, high discharge burnup, and long cycle lifetimes, which help reduce proliferation risks and radiation emissions. SMRs incorporate passive or inherent safety features; for example, the VHTR employs inherent safety design by utilising a low power density and high thermal conductivity of its core materials to ensure passive decay heat removal. This design allows the reactor to remain safe even during a loss of coolant accident. Another benefit of SMRs is that they do not require many of the safety provisions and engineered mechanisms that conventional reactors need. Their low capital costs and short construction timelines make financing less burdensome. Additionally, many SMRs feature proliferation resistance due to designs that eliminate on-site refuelling and enable long fuel cycles, being capable of using advanced fuel cycles that incorporate recycled materials.

More than 80 SMR designs are currently being developed or implemented, catering to various applications such as electricity generation, hybrid energy systems, heating, water desalination, and steam production. In 2020, the floating nuclear power plant (FNPP) known as Akademik Lomonosov, which consists of two KLT-40S modules, began commercial operations in Russia. In December 2021, China's HTR-PM reactor (250 MWt) was connected to the grid. An iPWR is one of the SMR designs to be deployed in the near future. A prototype iPWR design, CAREM-25 (27 MWe), is under construction in Argentina. Other iPWR designs include SMART (Korea), NuScale (USA), mPower (USA), and Westinghouse SMR (USA). The NuScale reactor received standard design approval in 2020, and other concepts are either in the design phase or seeking construction licenses. Consequently, the potential application of dynamic core transfer and neutron noise calculations in next-generation reactors has garnered significant interest.

To date, several attempts have been made to apply neutron noise simulators to analyse noise behaviours in Gen-IV reactors and SMRs. For example, noise analysis for the SMART iPWR design was conducted using a time-domain noise simulator [107]. The tool employs the NEM method for spatial discretisation, utilising two groups and 3D rectangular geometry, along with implicit-difference discretisation for time evolution. Noise analysis was conducted for the SMART core, which is the

System-integrated Modular Advanced PWR design with the power output of 107 MW developed by the Korea Atomic Energy Research Institute. The SMART core consists of 57 fuel assemblies of the 17 × 17 type, with an active fuel height of 2 m. Simulations were performed to analyse the neutron noise induced by control rod vibrations in various scenarios: single rod vibration, regulating bank vibration, and random vibration. The results demonstrated the advantages of the tool and confirmed its ability to simulate vibrating perturbations. Furthermore, the evaluations indicated the potential application of the tool for both large-scale reactors and SMRs.

A noise simulator for fast reactors was utilised to examine the neutron noise generated by periodic core deformation, known as core flowering, in a large SFR [84]. The SFR core configuration is illustrated in Figure 5.3 [84]. The analysis was carried out using a 3D model of the SFR, with a cross-section set created for the model that included 33 energy groups, generated by the ERANOS code [134]. In the SFR, fuel assemblies are arranged with spacer pads and secured by a lower grid base, which can lead to geometric core deformations, such as flowering and compaction, particularly in the upper section of the reactor. Consequently, core deformation was modelled to occur from the midplane to the top of the core (see Figure 5.4 [84]). Observations on vibration-induced neutron noise in SFRs indicated that mechanical vibration frequencies are likely in the range of 1 to 10 Hz. Therefore, the noise simulations were executed at a frequency of 5 Hz [84].

The results of the modelling of the SFR core deformations revealed that optimal detection positions for the phenomena using noise instruments would be located in the outer radial reflector, with several detectors positioned along the height [84]. The noise amplitude was found to increase linearly with the deformation amplitude,

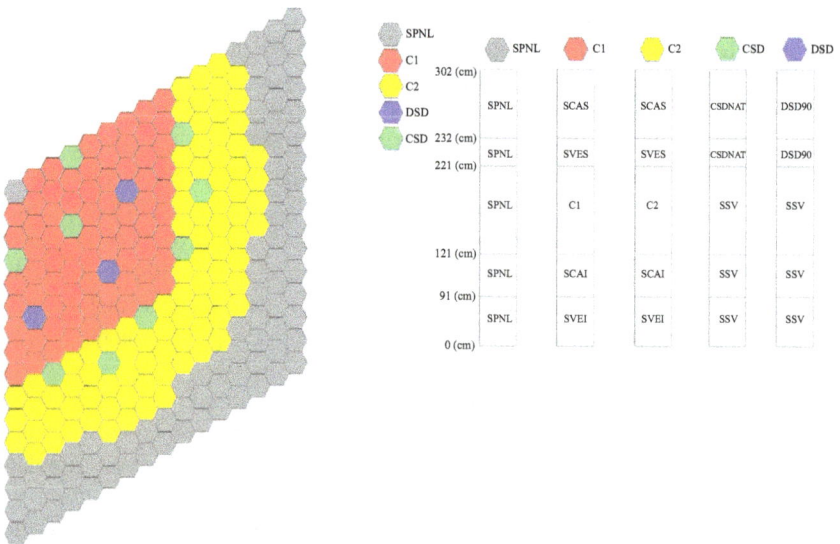

Figure 5.3   Configuration of a large SFR core [84]

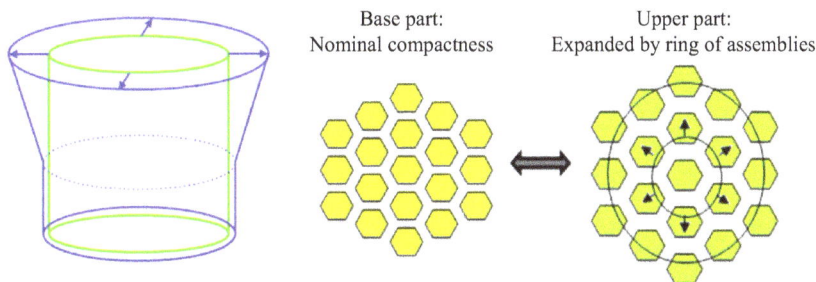

Figure 5.4   Model of core deformation on the upper half of an SFR core [84]

suggesting that detecting effects from small vibration amplitudes could be challenging. Furthermore, it was observed that in the SFR, a fission chamber containing fissile materials such as U-235 or Pu-239 coated on its inner wall has potential utility for in-core fast neutron detection. However, the energy-dependent cross section of the detector is complex, presenting challenges for numerical simulations aimed at predicting or interpreting the measurement phenomena.

The properties of neutron noise induced by localised perturbations were explored using a noise simulator based on a simplified 2D model of the SFR [138]. The study assumed three representations of noise sources related to localised perturbations in the absorption, fission, and scattering cross sections, each positioned at the core's centre. The investigation focused on space- and energy-dependent noise in a broad frequency range of 0.1 to 100 Hz.

The coordinate transformation method (CTM) was integrated into the DYN3D code to simulate small structural deformations in SFRs [139]. DYN3D previously utilised this method to model uniform radial core expansions. The CTM was extended to accommodate non-uniform radial deformations, such as core flowering. The method's viability was assessed through several steps. First, functionality was verified using the rectangular solver. Second, DYN3D's capability to model core expansions was tested on an SFR, employing hexagonal geometry for flowering scenarios with transverse isotropic flowering profiles. Additionally, the continuous bending of assemblies was effectively represented by vertical segments gradually displaced in the radial direction, resulting in a stair-stepped geometry. DYN3D was subsequently tested on the core flowering of the Phénix reactor, demonstrating its feasibility for simulating scenarios that extend beyond transverse isotropic shapes.

5.4   Concluding remarks

A survey has been conducted on the global development of computational codes designed to simulate dynamic core transfer function/neutron noise in various reactor systems. Numerous efforts have been dedicated to developing numerical tools for modelling neutron noise in LWRs. Two distinct approaches are employed in these tools: time-domain and frequency-domain methods. The most common and effective approach uses diffusion theory to solve space-dependent neutron noise in the frequency domain. Various noise sources can be simulated by defining fluctuations in macroscopic cross sections. Several extensions of these tools have been demonstrated for next-generation reactors, including Gen-IV reactors and SMRs, indicating strong interest and potential for diagnosing advanced reactor systems.

Chapter 6

Developments of zero-power noise methods for fast reactor systems

Imre Pázsit[1] and Yasunori Kitamura[2]

6.1 Introduction

This chapter deals with the extension of the pulse counting-based reactivity measurement methods, such as the Feynman- and Rossi-alpha methods for next generation systems. As mentioned in the previous chapters, one important feature of the Gen-IV reactors is that they will have a fast-neutron spectrum. The consequences of this fact and the need for extending the theoretical basis of the reactivity measurement methods for two-group theory will be discussed in Sections 6.2 and 6.3. Also, other aspects, such as availability of detectors and the need of including gamma photon counting into these methods, will also be discussed (Section 6.4). Section 6.5 summarises the principles of the two-detector Feynman-alpha measurement. Further, although they are not at the moment the main priority for next generation reactors, accelerator-driven systems (ADS) were a very promising candidate a couple of decades ago, and they are still potentially interesting. These systems will be driven by neutron sources whose statistical properties are very different from those of the traditional simple Poisson sources, hence the Feynman- and Rossi-alpha methods need to be modified for using them in such systems. This extension has already been made [6], but it will be briefly summarised in Section 6.6. Finally, one development, recently initiated for present reactors, as well as for safeguards problems, but of which the Gen-IV systems and small modular reactors (SMRs) would also largely benefit, is a new method to extract the same statistical information from the continuous signals of ionisation and fission chambers than from the pulse counting methods. Due to it being a new method with a limited number of publications, it will be described in Section 6.7 with some real examples of its use.

6.2 Traditional Feynman-alpha for fast reactors

Since several of the next generation systems will be fast ones, one challenge is whether the pulse counting-based reactivity determination methods, which were

[1]Division o Subatomic, High Energy and Plasma Physics, Department of Physics, Chalmers University of Technology, Sweden
[2]Division of Nuclear Engineering Science, Kyoto University Institute for Integrated Radiation and Nuclear Science, Japan

elaborated for thermal systems, remain applicable for such systems. This is not so much a question of instrumentation, since fast detectors and electronics are easily available for much shorter prompt neutron decay times (except the problems with dead time, see also Section 6.7). The more crucial question is whether the traditional pulse counting methods, elaborated in one-group theory, remain applicable in a fast spectrum.

This latter question was investigated in recent experimental work, and there appears to exist evidence that the traditional Feynman- and Rossi-alpha formulae can be used to extract the prompt neutron decay constant [140–143]. The focus in these works lie on the Rossi-alpha formula, and the results showed that the prompt neutron decay constant could be extracted for systems for systems as deeply subcritical as $k_{eff} = 0.42$ [142]. In order to extract the reactivity, the main challenge is to determine the neutron generation time Λ, which was achieved by Monte Carlo calculations.

From the aforementioned works, it appears that the traditional Rossi- and Feynman-alpha methods are applicable for fast systems with detection of fast neutrons with plastic scintillators, but the evaluation of the measurements requires a detailed numerical simulation of the measurement. This was conceptually simple for the measurements, which were performed on well-defined homogeneous systems with simple geometry. It remains to see the applicability of the method in real-life measurements for next generation systems with a fast spectrum.

6.3 Feynman-alpha for fast reactors: two-group theory

The aforementioned indications of the potential applicability of the traditional Feynman- and Rossi-alpha techniques in systems with a fast spectrum notwithstanding, it is interesting to investigate the alternative of the extension of these techniques to two-group theory. Some incentive for going into this direction came from the differential die-away analysis (DDAA) [144], and the differential die-away self-interrogation (DDSI) methods [145] of nuclear safeguards. In these measurements two separate exponential terms appear, which can be explained by the use of two-group theory. The DDSI method is, in principle, equivalent to a two-group version of the Rossi-alpha method, and the interest in the DDAA and DDSI methods spawned work trying to explore the feasibility of a two-group version of the Feynman- and Rossi-alpha methods for determining the subcritical reactivity in systems where a two-group methodology is necessary.

There exist several publications on the subject [146–149], containing a formal derivation of the two-group Feynman-alpha formulae. In [146], the two-group Feynman-alpha formula was derived by using the backward master equation approach, with the assumption of only prompt neutrons in the system, and using the counts of fast neutrons. In [147], the same model was used but solved by the forward master equation approach, leading to the same result. In [148], the backward approach was used to derive the two-group Feynman-alpha formula for fast neutrons, including also delayed neutrons. The most general treatment is given in [149], which uses the forward master equation approach to calculate the two-group Feynman-alpha formula with the inclusion of delayed neutrons, for both fast, thermal, and fast-thermal cross-Feynman-alpha formulae.

However, none of the aforementioned results have been tested on measurements yet, only were compared with simulations. The simulations merely proved that by using the proper parameters in the analytical formulae, taken from Monte Carlo simulations, one can obtain the same curve from the analytical model as from the Monte Carlo simulations. The problem is that this does not solve the diagnostic task; that is, it does not give any recipe on how to extract the subcritical reactivity from the measurement.

The reason is generic to a large class of problems. A simple model, like the traditional one-group treatment, uses approximations, but contains only a small number of parameters, and is explicitly dependent on the sought quantity (in our case the subcritical reactivity). A more advanced approach, such as two-group theory, may be a more realistic and better model of the process, but has the disadvantage that it depends on a larger number of parameters, and the quantity of interest, the subcritical reactivity, is hidden, or does not appear at all in a form which makes it possible to extract it from the measurements. For instance, in the two-group model with delayed neutrons, and accounting for detection of both fast and thermal neutrons, the roots of the characteristic equation, appearing as exponents in the time-dependence of the Feynman-alpha formula, are given as solutions of a fourth-order algebraic equation, which cannot even be given in an analytic form. Hence the extraction of the reactivity from the measurement is not possible by analytical methods, and one has to resort to non-parametric unfolding methods, such as machine learning (artificial neural networks).

To fill up the gap between the theory and its application, we thus investigate the possibility of extracting the reactivity from the simplest two-group model still with parametric methods. This is the model accounting for prompt neutrons only, and calculating the Feynman-alpha from the counts of fast neutrons. Following Reference [147], first we outline the derivation of the Feynman-alpha formula for fast neutrons. Then the possibility of extracting the subcritical reactivity from a measurement will be investigated.

6.3.1   Derivation of the two-group Feynman-alpha formula for fast neutrons

We start with deriving a forward master equation for the probability $P(N_1, N_2, Z_1, t)$ for having N_1 fast, N_2 thermal neutrons at time t in the system and having detected Z_1 fast neutrons in the interval $(0, t)$. Fast fission is neglected, and it is assumed that the extraneous source injects fast neutrons into the system with simple Poisson statistics with intensity S. The reaction intensities and the fission number distribution are denoted as follows:

λ_{c_1} intensity of capture of a fast neutron in the system
λ_d intensity of capture of a fast neutron in the detector
λ_R intensity of removal (downscattering) of a fast neutron
λ_{c_2} intensity of capture of a thermal neutron
λ_{f_2} intensity of thermal fission
$p_f(k)$ the number distribution of fission neutrons

The total reaction intensities in the fast and thermal groups, denoted as λ_1 and λ_2, respectively, are defined as

$$\lambda_1 = \lambda_{c_1} + \lambda_R + \lambda_d \tag{6.1}$$

and

$$\lambda_2 = \lambda_{c_2} + \lambda_{f_2} \tag{6.2}$$

With these preliminaries, the following forward-type master equation can be written down for $P(N_1, N_2, Z_1, t)$:

$$\begin{aligned}
\frac{dP(N_1, N_2, Z_1, t)}{dt} &= \lambda_{c_1}(N_1 + 1)P(N_1 + 1, N_2, Z_1, t) \\
&\quad + \lambda_R(N_1 + 1)P(N_1 + 1, N_2 - 1, Z_1, t) \\
&\quad + \lambda_d(N_1 + 1)P(N_1 + 1, N_2, Z_1 - 1, t) \\
&\quad + \lambda_{c_2}(N_2 + 1)P(N_1, N_2 + 1, Z_1, t) \\
&\quad + \lambda_{f_2}(N_2 + 1)\sum_{k}^{N_1} p_f(k)P(N_1 - k, N_2 + 1, Z_1, t) \\
&\quad + SP(N_1 - 1, N_2, Z_1, t) - (\lambda_1 N_1 + \lambda_2 N_2 + S)P(N_1, N_2, Z_1, t)
\end{aligned} \tag{6.3}$$

Defining the generating functions

$$G(x, y, z, t) = \sum_{N_1}\sum_{N_2}\sum_{Z_1} x^{N_1} y^{N_2} z^{Z_1} P(N_1, N_2, Z_1, t) \tag{6.4}$$

and

$$q_f(x) = \sum_{k} p_f(k)\, x^k \tag{6.5}$$

we obtain for the generating function the equation

$$\begin{aligned}
\frac{\partial G(x, y, z, t)}{\partial t} &= (\lambda_{c_1} + \lambda_R y + \lambda_d z - \lambda_1 x)\frac{\partial G(x, y, z, t)}{\partial x} \\
&\quad + (\lambda_{c_2} + \lambda_{f_2} q_f(x) - \lambda_2 y)\frac{\partial G(x, y, z, t)}{\partial y} + S(x - 1)G(x, y, z, t)
\end{aligned} \tag{6.6}$$

The procedure of taking the various moments, and solving for the expectation and the variance of the number of fast neutron detections as a function of the detection time goes along the general routines, and it will not be detailed here. The final result can be written in a condensed form as

$$\frac{\sigma_Z^2(T)}{\langle Z_1 \rangle} = 1 + Y_1\left(1 - \frac{1 - e^{-\omega_1 T}}{\omega_1 T}\right) + Y_2\left(1 - \frac{1 - e^{-\omega_2 T}}{\omega_2 T}\right) \tag{6.7}$$

Here,

$$-\omega_1 = -\frac{1}{2}(\lambda_1 + \lambda_2) + \frac{1}{2}\sqrt{(\lambda_1 - \lambda_2)^2 + 4\lambda_1\lambda_2\nu_{\text{eff}}} \tag{6.8}$$

$$-\omega_2 = -\frac{1}{2}(\lambda_1 + \lambda_2) - \frac{1}{2}\sqrt{(\lambda_1 - \lambda_2)^2 + 4\lambda_1\lambda_2\nu_{eff}} \tag{6.9}$$

where the notation

$$\nu_{eff} = \frac{\nu\lambda_{2f}\lambda_R}{\lambda_1\lambda_2} \tag{6.10}$$

was introduced, with ν being the first moment of $p_f(k)$, i.e. the mean number of neutrons per fission. The coefficients Y_1 are Y_2 are quite lengthy and complicated functions of the intensities, the exponents ω_1 and ω_2, and the second factorial moment of the number of neutrons per fission, $\langle \nu (\nu - 1) \rangle$. The full expressions for Y_1 are Y_2 are found in [146] and [147], and will not be given here. Like in the traditional Feynman-alpha methods, the exponents ω_1 and ω_2 can be extracted from a measurement without the knowledge of the expressions for Y_1 are Y_2.

Equation (6.7) looks formally similar to the traditional Feynman-alpha expression. But whereas in the traditional method the first exponent, the prompt neutron decay constant α is a simple explicit function of the reactivity, expressions (6.8) and (6.9) for the exponents ω_1 and ω_2 are a much more implicit functions of the reactivity. The possibility of subtracting the reactivity from the two-group Feynman-alpha expression will be investigated in the next subsection.

6.3.2 Extracting the reactivity from a two-group Feynman-alpha measurement

We need to express the reactivity in terms of the parameters of the system, similarly to the one-group definition (2.22). In an infinite homogeneous system, in the absence of fast fission, the multiplication constant $k \equiv k_\infty$ is given as

$$k = \frac{\nu\lambda_{f_2}\lambda_R}{\lambda_1\lambda_2} \tag{6.11}$$

which is the same as the ν_{eff} of (6.10).

An inspection of (6.8) and (6.9) shows that

$$\omega_2 > 0 \quad \text{and} \quad \omega_1 < \omega_2 \tag{6.12}$$

It is also seen that for $k = 1$, one has $\omega_1 = 0$, as expected.

One can rewrite (6.8) for ω_1 as

$$-\omega_1 = -\frac{1}{2}(\lambda_1 + \lambda_2) + \frac{1}{2}\sqrt{(\lambda_1 + \lambda_2)^2 + 4\lambda_1\lambda_2 (k - 1)} \tag{6.13}$$

and by using (6.11) this can further be rewritten as

$$\omega_1 = \frac{1}{2}(\lambda_1 + \lambda_2) - \frac{1}{2}\sqrt{(\lambda_1 + \lambda_2)^2 + 4\nu\lambda_{f_2}\lambda_R \rho} \tag{6.14}$$

From (6.14) the reactivity can be expressed as

$$\rho = \frac{\omega_1^2 - \omega_1(\lambda_1 + \lambda_2)}{\nu\lambda_{f_2}\lambda_R} \tag{6.15}$$

Equation (6.15) shows that unlike in the one-group case, the reactivity is not a linear function of the decay constant ω_1. Extracting the reactivity from the measured ω_1 is still possible in principle, but one has to know a substantially larger set of parameters to extract ρ than the simple parameter prompt neutron generation time Λ. In addition, from this form it is not obvious that extrapolating the Feynman- and Rossi alpha formulae from the form obtained by accounting for prompt neutrons only to the case when delayed neutrons are also taken into account, can be obtained such that one simply replaces ρ with $\rho - \beta$. This latter question can be investigated from the results of [148]. Hence the feasibility of the two-group Feynman-alpha method in practical applications is yet to be investigated.

Finally, we note that for systems close to critical, ω_1^2 can be neglected in (6.15). This will lead to

$$\omega_1 \approx -\rho \frac{\nu \lambda_{f_2}}{\lambda_1 + \lambda_2} \frac{\lambda_R}{\lambda_1 + \lambda_2} = \frac{\rho}{\Lambda_T} \frac{\lambda_R}{\lambda_1 + \lambda_2} \qquad (6.16)$$

where

$$\Lambda_T = \frac{1}{\nu \lambda_{f_2}} \qquad (6.17)$$

is the prompt neutron reproduction time in a thermal system. It is thus seen that ω_2 can be written as

$$\omega_1 \approx -\frac{\rho}{\Lambda^*} \qquad (6.18)$$

if Λ^* is defined as

$$\Lambda^* = \Lambda_T \frac{\lambda_1 + \lambda_2}{\Lambda_R} \qquad (6.19)$$

That is, for systems close to critical, there will be a linear relationship between the reactivity and ω_1. Nevertheless, to extract the reactivity, one still needs to know the same amount of system parameters (reaction intensities) as in the general case.

6.4   Feynman-alpha with gamma counting

As mentioned earlier, the possibility of using gamma photons, in addition to neutrons, both in pulse counting for zero-power noise [73] as well as in current mode for power reactor noise [77], was suggested as early as in the late 1960s. On the power reactor noise side, as described earlier, some singular pilot experiments were made in the past, but the technique has not yet been tested at operating power reactors. On the zero-power diagnostic side, the feasibility and potential advantages of monitoring subcritical reactivity with gamma pulse counting has been noted relatively recently.

First, partly by inspiration from multiplicity counting in nuclear safeguards where the use of gamma detection got already started, some extension of the original works on gamma count statistics were made. The extensions concerned accounting both for neutron and gamma total detection without discrimination, as well as for including gamma reactions other than prompt gammas from fission, such as photoneutrons, whose existence establishes a two-way coupling between the evolution of the neutron and photon chains [150,151]. Approximately at the same time, the feasibility of using gamma counting for the Feynman-alpha method was investigated in several experiments [152–154]. In these experiments it was verified that one can extract the prompt neutron decay constant α by fitting the measured gamma data to the time-dependent part of the traditional Feynman-alpha expression for neutrons.

In these measurements also some advantages of the gamma-count-based method were emphasised; namely, that it can be used in ex-core measurements farther away from the core than with a neutron detector, and partly that in some cases it gave more accurate estimation of the prompt neutron α than the one obtained from the neutron counting. These circumstances make the gamma-count-based reactivity measurements a promising candidate for several of the next generation reactors, due to the fact that in several designs, there will be no room for in-core detectors, hence one will be restricted to ex-core measurements, possibly not even close to the core. This fact motivates an account of the method and the underlying theory.

The number of papers on the theory of gamma-count-based reactivity measurements is very small. The aforementioned experimental works all refer to the original publication of Gelinas and Osborn [73]. This work, similarly to the later publications [150,151] is based on a one-neutron energy-group point model for prompt neutrons only, i.e. delayed neutrons are neglected. Both the calculations, as well as the results are rather involved, using notations in a very formal style, making it difficult to relate them to simple physical parameters, such as the reactivity. The reason for this is that the master equations used are written in a style similar to the case of the Feynman-alpha expression with delayed neutrons, namely that they follow up the number of all species (neutrons and delayed neutron precursors for the neutron case, and neutrons and photons for the gamma-counting case).

The description can be substantially simplified by recognising that due to the speed of the photons, the intensity of their reaction is so high, that their lifetime can be neglected. This makes it possible to drop the number of the photons in the system, and only keep the number of photon detections as a random variable. In this model it is assumed that each photon born in a fission process will be detected instantly with a probability (detector efficiency) ε_η, or not detected with a probability $(1 - \varepsilon_\eta)$. The fate of the non-detected photons is not followed up, since they will not be detected later. This assumption is similar to the concept of "superfission", introduced in nuclear safeguards, which means that the lifetime of a (deeply) subcritical chain in a fissile item is so short, that the internal multiplication of the neutrons before leaving the item can be considered as instantaneous. This assumption simplifies the treatment and improves its transparency substantially, and will lead to the same result as the more complicated theories while giving more insight. Since this is a new concept, we describe its main aspects in the following, in the same framework as the description of the traditional method in Section 2.2.2.

6.4.1   General principles

We will seek the probability distribution $P(N, Z, W, t)$ that at time t, there are N neutrons in the subcritical system driven by a stationary Poisson source of intensity S, and in the time interval $[0, t)$, Z neutron and W gamma counts were registered. As before, neutrons can be captured in the system, captured in the detector, or cause fission with intensities λ_c, λ_d and λ_f, respectively. In a fission event, prompt neutrons and gammas are generated with number distributions $p_f(n)$ and f_k, respectively, and in the moment of the fission, each gamma can be detected with a probability ε_η. In general, certain quantities, such as expectations and detector efficiency will be denoted by the letter ν for neutrons and η for gamma photons (the latter in order to conform with the notations of [73]).

With these preliminaries, the following master equation can be derived for $P(N, Z, W, t)$, considering the possible mutually exclusive events between t and $t + dt$:

$$
\begin{aligned}
P(N, Z, W, t + dt) &= \lambda_c \, dt \, P(N + 1, Z, W, t)(N + 1) \\
&+ \lambda_d \, dt \, P(N + 1, Z - 1, W, t)(N + 1) \\
&+ \lambda_f \, dt \sum_{n=0}^{\infty} p_f(n) \sum_{k=0}^{\infty} f_k \sum_{\ell} \binom{k}{\ell} \varepsilon_\eta^\ell (1 - \varepsilon_\eta)^{k-\ell} \\
&\times P(N + 1 - n, Z, W - \ell, t)(N + 1 - n) \\
&+ S \, dt \, P(N - 1, Z, W, t) \\
&+ P(N, Z, W, t) \left[1 - \{ (\lambda_f + \lambda_c + \lambda_d) N + S \} \, dt \right] \quad (6.20)
\end{aligned}
$$

Rearranging yields

$$
\begin{aligned}
\frac{d P(N, Z, W, t)}{dT} &= \lambda_c P(N + 1, Z, W, t)(N + 1) \\
&+ \lambda_d P(N + 1, Z - 1, W, t)(N + 1) \\
&+ \lambda_f \sum_{n=0}^{\infty} p_f(n) \sum_{k=0}^{\infty} f_k \sum_{\ell} \binom{k}{\ell} \varepsilon_\eta^\ell (1 - \varepsilon_\eta)^{k-\ell} \\
&\times P(N + 1 - n, Z, W - \ell, t)(N + 1 - n) \\
&+ P(N, Z, W, t) \left[N(\lambda_f + \lambda_c + \lambda_d) + S \right] \quad (6.21)
\end{aligned}
$$

Introducing the generating functions of $P(N, Z, W, t)$, $p_f(n)$ and f_k the usual way, i.e.

$$
G(x, z, w, t) = \sum_N \sum_Z \sum_W x^N z^Z w^W P(N, Z, W, t) \quad (6.22)
$$

as well as

$$
g(x) = \sum_{n=0}^{\infty} p_f(n) x^n \quad \text{and} \quad h(w) = \sum_{k=0}^{\infty} f_k w^k \quad (6.23)
$$

we obtain from (6.21) the following master equation for $G(x, z, w, t)$:

$$
\begin{aligned}
\frac{\partial G(x, z, w, t)}{\partial t} &= \{ \lambda_f [g(x) \, h(c(w)) - x] - \lambda_c (x - 1) - \lambda_d (x - z) \} \\
&\times \frac{\partial G(x, z, w, t)}{\partial x} + (x - 1) S \, G(x, z, w, t) \quad (6.24)
\end{aligned}
$$

where

$$
c(w) = \varepsilon w + 1 - w \quad (6.25)
$$

is the generating function of the detection process [6].

Since (6.24) does not contain derivatives w.r.t. to w, with $w = 1$ it reverts to the equation for the neutron number and neutron counts only. Hence the moments of the latter, such as its variance and mean, can be determined independently of the moments of the gamma counts. The converse is also true, in that since (6.24) does not contain derivatives w.r.t. to z either, the moments of the gamma counts can also be derived, basically with the same amount of effort, as those of the neutron counts. Finally, it is also possible to derive the joint moments of the neutron and gamma counts, and thereby derive a "two-detector" Feynman-alpha, which can be measured with the simultaneous use of a neutron and a gamma detector, respectively.

6.4.2 First moments

Since the moments of the neutron counts have been derived a long time ago, it is in principle redundant to derive them again. However, in order to get better insight into the fact that the moments of the neutron and photon counts have the same functional dependence, including the Feynman-alpha formula, the main steps of the derivation of the prompt neutron Feynman-alpha will be also given here.

The derivation of the moments of the counts cannot be made without the determination of the expectation of the neutron number and the joint expectation of the neutron number and neutron counts. The expectation $\langle N \rangle \equiv N$ is obtained by differentiating (6.24) w.r.t. x and substituting $x = z = w = 1$, yielding

$$\frac{dN}{dt} = \frac{\rho}{\Lambda} N(t) + S \tag{6.26}$$

where the reactivity ρ and the prompt neutron generation time Λ were already defined in (2.22) and (2.23). The stationary solution of (6.26) is

$$N = \frac{\Lambda S}{-\rho} = \frac{N}{\alpha} \tag{6.27}$$

where the prompt neutron decay constant α without delayed neutrons

$$\alpha = -\frac{\rho}{\Lambda} > 0 \tag{6.28}$$

was also introduced.

The expectations $\langle Z \rangle \equiv Z$ and $\langle W \rangle \equiv W$ of the neutron and gamma counts, respectively, can be easily obtained in a similar way, resulting in the well-known result for the neutron counts as

$$Z(t) = \lambda_d N t = \varepsilon_\nu \lambda_f N t \tag{6.29}$$

where the neutron detection efficiency ε_ν as introduced in the standard way as

$$\varepsilon_\nu = \frac{\lambda_d}{\lambda_f} \tag{6.30}$$

and similarly for the gamma counts as

$$W(t) = \lambda_f \varepsilon_\eta \, \eta \, N t \tag{6.31}$$

where $\eta \equiv \langle \eta \rangle$ is the mean number of gamma photons per fission.

6.4.3   Variance to mean and covariance to mean

For the second-order moments, which are needed in order to calculate the variance, it is practical to introduce the modified variances μ_{ab} [4,6] as

$$\mu_{aa} = \langle a(a-1) \rangle - \langle a \rangle^2 = \sigma_a^2 - \langle a \rangle \tag{6.32}$$

and

$$\mu_{ab} = \langle a\,b \rangle - \langle a \rangle \langle b \rangle = \text{Cov}\{a\,b\} \tag{6.33}$$

From (6.24), the following equation system is obtained for the modified moments:

$$\frac{d\,\mu_{NN}(t)}{dt} = \lambda_f \langle \nu(\nu-1) \rangle N - 2\,\alpha\,\mu_{NN} \tag{6.34}$$

$$\frac{d\,\mu_{NZ}(t)}{dt} = \varepsilon_\nu \lambda_f \mu_{NN} - \alpha\,\mu_{NZ} \tag{6.35}$$

$$\frac{d\,\mu_{ZZ}(t)}{dt} = 2\,\varepsilon_\nu\,\lambda_f \mu_{NZ} \tag{6.36}$$

$$\frac{d\,\mu_{NW}(t)}{dt} = \lambda_f \eta\,\varepsilon_\eta \left[\nu N + \mu_{NN} \right] - \alpha\,\mu_{NW} \tag{6.37}$$

$$\frac{d\,\mu_{WW}(t)}{dt} = \lambda_f \langle \eta\,(\eta-1) \rangle\,\varepsilon_\eta^2\,N + 2\,\lambda_f \eta\,\varepsilon_\eta\,\mu_{NW} \tag{6.38}$$

$$\frac{d\,\mu_{ZW}(t)}{dt} = \lambda_f \eta\,\varepsilon_\eta\,\mu_{NZ} + \varepsilon_\nu\,\lambda_f \mu_{NW} \tag{6.39}$$

In the stationary case, μ_{NN} is constant, hence (6.34) reduces to a simple algebraic equation with the known solution. Equations (6.35) and (6.36) can be solved sequentially independently from (6.37) and (6.38), i.e. independently of the moments including gammas, and likewise (6.37) and (6.38) can be solved independently from the neutron counts. The last equation for the covariance μ_{ZW} of neutron and gamma counts can be solved in possession of μ_{NZ} and μ_{NW}.

The solutions are quite straightforward, from which the gamma variance to mean and the neutron-gamma covariance to mean are of interest. However, for the sake of comparison, we give here the well-known traditional Feynman-alpha formula for prompt neutrons, which reads as

$$\frac{\sigma_Z^2(t)}{Z(t)} = 1 + \varepsilon_\nu \frac{D_\nu}{\rho^2} \left(1 + \frac{1 - e^{-\alpha t}}{\alpha t} \right) \tag{6.40}$$

where D_ν is the Diven factor of prompt fission neutrons,

$$D_\nu = \frac{\langle \nu(\nu-1) \rangle}{\nu^2}. \tag{6.41}$$

The gamma variance to mean is obtained as follows:

$$\frac{\sigma_W^2(t)}{W(t)} = 1 + \frac{\mu_{WW}}{W} = 1 + \varepsilon_\eta\,\eta \left[D_\eta + \left(\frac{2}{|\rho|} + \frac{D_\nu}{\rho^2} \right) \left(1 + \frac{1 - e^{-\alpha t}}{\alpha t} \right) \right] \tag{6.42}$$

where similarly to (6.41), D_η is the gamma Diven factor

$$D_\eta = \frac{\langle \eta(\eta - 1) \rangle}{\eta^2} \tag{6.43}$$

The time-dependent part of (6.42) is identical with that of the traditional neutron-based Feynman-alpha formula, and agrees with that derived by Gelinas and Osborn (Equation (38a) in [73]) and the corresponding formulae in [150]. The term

$$\varepsilon_\eta \, \eta \, D_\eta \tag{6.44}$$

corresponds to the asymptotic (saturation) value of first term on the r.h.s. of Equation (38a) in [73] for $t \to \infty$ (see also [152]), which is consistent with the fact that our assumption of instantaneous detection of the photons at their birth shortcuts the time evolution between the birth and the detection of gamma photons. This means that the variance to mean does not start from unity at $t = 0$ as it does in the traditional case (or in other words, the $Y_\eta(t)$ function, represented by the last term of (6.42), is not equal to zero at $t = 0$, rather it is equal to $\varepsilon_\eta \, \eta \, D_\eta$). If one takes into account the non-zero lifetime of the photons before detection, then $Y_\eta(t)$ still starts from zero, but it jumps to its asymptotic value $\varepsilon_\eta \, \eta \, D_\eta$ during a very short time, which cannot be resolved in the experiment.

The term D_ν / ρ^2 in the factor multiplying the time-dependent function is also the same as that in ([73]), but we also have an additional term $2/|\rho|$. It is not clear whether the deviation between our formulae and those in [73] are due to the assumption of the instantaneous detection in our model, or to something else. Close to criticality, this extra term can be neglected besides the term proportional to ρ^{-2}. Besides, its presence or absence does not affect the recovering the parameter α from measurements by the curve fitting procedure. On the other hand, in deeply subcritical systems, the presence of this term is advantageous, because it extends the applicability of the method to deeper subcriticalities.

An alternative method to the gamma variance to mean is the neutron-gamma covariance to mean method. Since the covariance concerns two different signals, one has the choice of choosing either the neutron or the gamma counts for the mean. The difference is only in some multiplication factors, and since using the expectation of the gamma counts leads to a simpler formula, we will display the neutron-gamma covariance to the gamma mean. This is obtained from the aforementioned as

$$\frac{\mu_{ZW}(t)}{W(t)} = \varepsilon_\nu \left[\frac{1}{|\rho|} + \frac{D_\nu}{\rho^2} \right] \left(1 + \frac{1 - e^{-\alpha t}}{\alpha t} \right) \tag{6.45}$$

This expression shows a striking similarity to the variance-to mean of the neutron counts; not only the time dependence, but all factors except the extra $1/|\rho|$ term are the same. In addition, similarly to the two-detector neutron-neutron covariance to mean formula (see the next subsection), the factor unity is missing, and the covariance to mean contains only the part in excess to the Poisson value of unity, which might be advantageous in measurements. The extra term $\varepsilon_\eta \, \eta \, D_\eta$ is also missing, hence the neutron - gamma covariance to mean starts from zero at $t = 0$. Therefore, when both neutron and gamma detection is available, the covariance to mean may have advantages over both the pure neutron or pure gamma variance to mean formula. In cases where the measurement cannot be made close to the core, and hence gamma detection is more effective than neutron detection, which is expected to be the case in several Gen-IV and SMR systems, the use of the gamma variance to mean method might prove to be substantially more advantageous.

The simplified model introduced here was able to reconstruct the same results as the more complicated models in which the number of gamma photons is kept as a separate random variable. Because of its simplicity and transparency, this model is suitable to be extended to include also delayed neutrons, or two neutron energy groups. Due to the promising features of the gamma variance to mean or the neutron-gamma covariance to mean method, this is definitely worth following up in future work, as also suggested by Darby *et al.* [153].

6.5   Two-detector Feynman-alpha method

An interesting variant of the Feynman-alpha method is to use two detectors, and instead of calculating the variance to mean, using the covariance to mean* for extracting the prompt neutron decay constant α. That is, instead of the variance of the single detector counts $\sigma_Z^2(t)$, to use

$$\langle Z_1 Z_2 \rangle - \langle Z_1 \rangle \langle Z_2 \rangle \equiv \mu_{Z_1,Z_2}(t) \tag{6.46}$$

Early experience showed that the use of the two-detector method had advantages over the traditional variance to mean method, in that it gave a better estimate of the searched parameter, because the estimation was less sensitive to uncorrelated events. This observation and its experimental verification was reported already at one of the very first Florida conferences in 1966 [155,156]. Reference [155] states that "Detector efficiency is mainly responsible for the ratio of correlated to uncorrelated reactor noise. The advantage of the cross-correlation method is particularly important if only low detector efficiency is available, as is the case, for instance, in fast reactors." It is therefore clear that this method will gain increased interest for next generation nuclear systems.

The advantages of the two-detector cross-correlation method can be slightly amplified with the fact that it yields the Feynman Y function directly, instead of using a small deviation from the unity of the Poisson variance to mean. Another advantage of the method could be that having two detectors in two suitable chosen different spatial positions could in principle compensate for the deviation of point kinetics (often expressed as the presence of higher modes).

The theoretical considerations behind the two-detector cross-correlation method in [155] were based on phenomenological arguments, similarly to the first versions

---

*The word "covariance" is used in several different meanings in this book, and care needs to be exercised to use the correct interpretation. The prefix "co-" in the present use of "covariance to mean" refers to using the joint statistics of the detector counts during the same measurement time, but with two *different detectors*. The Rossi-alpha method is also often called "covariance to mean", but it refers to the joint statistics of detections with the same detector, but in *two different infinitesimal time intervals*, for which the correct terminology is auto-covariance to mean, as it is also called in [6] and in most parts of this book. Hopefully the terminology will not lead to confusion and it will be clear from the context what the word "covariance" refers to.

of the Feynman- and Rossi-alpha methods. It is desirable to provide a derivation from first principles, based on the master equation method, as it was the case with the Feynman- and Rossi-alpha methods [157]. Although the two-detector method is widely used in practice, strangely enough no such derivation of the method was found in the literature. Therefore, despite its simplicity, we give an account of its derivation here.

The derivation will follow closely the one given in Section 6.4 for the Feynman-alpha formula for gamma detections, but only for the neutronic part. The difference is that instead of one detector, we consider two detectors. For generality, we assume that the two detectors have different detection intensities λ_{d_1} and λ_{d_2}, respectively. This is not so much to account for the difference in the detector properties, rather to the detection efficiencies which enter the formula, and which in reality also account for the different detection sensitivities at different core positions, which is the case even if completely identical detectors are used.

Thus, we will be looking for an equation for the probability

$$P(N, Z_1, Z_2, t) \tag{6.47}$$

that at time t, there are N neutrons in the subcritical system driven by a stationary Poisson source of intensity S, and that there were Z_1 and Z_2 counts registered between $[0, t)$ in detector 1 and 2, respectively. As usual, the master equation for $P(N, Z_1, Z_2, t)$ will be converted to an equation for its generation function, defined as

$$G(x, z_1, z_2, t) = \sum_N \sum_{Z_1} \sum_{Z_2} x^N z_1^{Z_1} z_2^{Z_2} P(N, Z_1, Z_2, t) \tag{6.48}$$

Using the same notations as in Section 6.4 (except for the detection intensities), one can immediately write down the master equation for the generating function (6.48) as

$$\begin{aligned}
\frac{\partial G(x, z_1, z_2, t)}{\partial t} = {} & \{\lambda_f[g(x) - x] - \lambda_c(x - 1) - \lambda_{d_1}(x - z_1) \\
& - \lambda_{d_2}(x - z_2)\} \frac{\partial G(x, z_1, z_2, t)}{\partial x} + (x - 1)S\,G(x, z_1, z_2, t)
\end{aligned} \tag{6.49}$$

The derivation of $\mu_{Z_1, Z_2}(t)$ goes along the same lines as with the gamma-based Feynman-alpha method. We use the same notations as in Section 6.4, except for defining the reactivity and the detector efficiencies as

$$\rho = \frac{\langle \nu \rangle \lambda_f - (\lambda_f + \lambda_c + \lambda_{d_1} + \lambda_{d_2})}{\langle \nu \rangle \lambda_f} \tag{6.50}$$

and

$$\varepsilon_1 = \frac{\lambda_{d_1}}{\lambda_f}; \qquad \varepsilon_2 = \frac{\lambda_{d_2}}{\lambda_f} \tag{6.51}$$

With these notations, the equation for $N \equiv \langle N \rangle$ will be the same as before,

$$\frac{dN}{dt} = \frac{\rho}{\Lambda} N(t) + S \tag{6.52}$$

whereas the equations for the expectations of the detector counts $\langle Z_1 \rangle \equiv Z_1$ and $\langle Z_2 \rangle \equiv Z_2$ will read, respectively, as

$$\frac{dZ_1}{dt} = \varepsilon_1 \lambda_f N(t) \tag{6.53}$$

$$\frac{dZ_2}{dt} = \varepsilon_2 \lambda_f N(t) \tag{6.54}$$

The stationary solutions will also be the same. Regarding the second moments, the equation for the modified second moment μ_{NN} and its solution will also be the same, which will be written out here as

$$\mu_{NN} = \frac{\lambda_f \langle \nu (\nu - 1) \rangle N}{-2 \nu \rho} \tag{6.55}$$

The equations and their solutions for the joint modified moments $\mu_{N,Z_i}, i = 1, 2$ will also be formally the same as in the case of one detector, except for the indexing:

$$\mu_{N,Z_i}(t) = \frac{\varepsilon_i \lambda_f^2 \langle \nu (\nu - 1) \rangle N (1 - e^{-\alpha t})}{2 \alpha^2}; \qquad i = 1, 2 \tag{6.56}$$

For the determination of the covariance $\mu_{Z_1,Z_2}(t)$ first we differentiate (6.49) w.r.t. Z_1 and Z_2 to obtain

$$\frac{d \langle Z_1 Z_2 \rangle}{dt} = \varepsilon_1 \lambda_f \langle N Z_2 \rangle + \varepsilon_2 \lambda_f \langle N Z_1 \rangle \tag{6.57}$$

The key point here, which will ensure the symmetry of the final result, is that the joint expectations of the counts and the neutron number are multiplied with the efficiency of the other detector and vice versa. Making use of (6.53) and (6.54) to express the time derivative of the product $\langle Z_1 \rangle \langle Z_2 \rangle$, one arrives at

$$\frac{d \mu_{Z_1,Z_2}(t)}{dt} = \lambda_f \left(\varepsilon_1 \mu_{N,Z_2}(t) + \varepsilon_2 \mu_{N,Z_1}(t) \right) \tag{6.58}$$

from where it follows that

$$\mu_{Z_1,Z_2}(t) = \lambda_f \int_0^\infty \left(\varepsilon_1 \mu_{N,Z_2}(t') + \varepsilon_2 \mu_{N,Z_1}(t') \right) dt' \tag{6.59}$$

Substituting (6.56) into the aforementioned yields the result

$$\mu_{Z_1,Z_2}(t) = \varepsilon_1 \varepsilon_2 \lambda_f^3 \langle \nu (\nu - 1) \rangle N t \left(1 - \frac{1 - e^{-\alpha t}}{\alpha t} \right) \tag{6.60}$$

Since now two different detectors with corresponding mean counts are available, from (6.60) there are different possibilities to derive a covariance to mean. Namely, the covariance can be divided with either of the counts, or with the geometrical mean. All those possibilities affect only the combination of the detector efficiencies multiplying the expression. Choosing the geometrical mean in the formula will lead to the covariance to mean written in standard Feynman notations as

$$\frac{\mu_{Z_1,Z_2}(t)}{\sqrt{Z_1 Z_2}} = \sqrt{\varepsilon_1 \varepsilon_2} \frac{D_\nu}{\rho^2} \left(1 + \frac{1 - e^{-\alpha t}}{\alpha t} \right) \tag{6.61}$$

This formula shows the heuristically predictable result that the neutron-neutron covariance to mean has the same dependence on the measurement time length as the traditional method, but also that it does not start from unity at time $t = 0$, rather from zero, i.e. yielding the Feynman $Y(t)$ function (cf. (2.2)). This is also a generic property of the covariance, such as in the neutron-gamma covariance function (6.45).

From the derivation it is also clear by extrapolation, that when including also delayed neutrons, the two-detector covariance to mean formula can be obtained from the variance to mean of the traditional formula with the same slight modifications as without delayed neutrons, namely that the covariance to mean is equal to the $Y(t)$ function of the traditional formula, but the multiplying factor concerning the detector efficiency depending on which quantity is used for the mean (either of the two detectors' count rate, or their geometrical mean).

6.6   Reactivity measurements in accelerator driven systems

At present, accelerator-driven subcritical systems (ADS) do not constitute a prime candidate for next generation systems, at least not in the near future. However, the idea is kept alive, and even if no large-scale experiments or commercial deployment of ADS is planned, there are on-going research projects which will supply useful scientific information and technical know-how, should a large-scale deployment of the concept become actual [158]. Therefore, it is worth reminding that when the ADS concept was intensively investigated about two decades ago, the extension of the pulse counting-based reactivity measurement methods was explored at depth. The concepts and the methods elaborated were summarised in [6], where also the methodology and the derivations are described in great detail. For the sake of completeness, the main findings and the formulae applicable for the different type of measurements will be briefly summarised here.

The difference between the reactivity measurement in traditional systems (either Gen-II, Gen-III, Gen-IV or SMR) and in an ADS lies primarily in the difference of the extraneous neutron sources used. In the traditional systems, the extraneous neutron source has a constant intensity and simple Poisson statistics, such as Am–Be, Sb–Be or Pu–Be source. For an ADS, which should not operate close to criticality, in order to have a high enough power, a much stronger extraneous source is needed. Such a source is an accelerator-based spallation source. Such a source deviates from the traditional sources in the following ways:

- Irrespective of whether the accelerator is run in the continuous or pulsed mode, its statistics will deviate from the simple Poisson one. Since a large random number of neutrons are generated by each proton impinging on the target, if the accelerator is run in a stationary mode, i.e. with a constant current, the source statistics will be compound Poisson.
- In most of the current designs, high-yield accelerator-based sources will be run in pulsed mode. This means that the source is not stationary in time, which is another deviation from the traditional sources. Further, one has qualitatively

different behaviour of the variance to mean for wide pulses, or very short ones, which can be approximated by temporal Dirac delta functions.
*   Although the pulsing of the system will be deterministic, one has the possibility of treating the pulses either as deterministic, starting the measurement always at a fixed time (such as the beginning of a pulse) or random, by starting the measurement at a random time.

The theory of all the aforementioned cases has been elaborated by several authors [159–161], and a list of the methods and the corresponding expressions for the Feynman- and Rossi-alpha expressions is given, together with the derivations, in [6]. Here we only give a list of the final results for the Feynman-alpha formulae for the various cases. As it was described in Section 2.2.1, the Rossi-alpha formulae can be derived from the Feynman-alpha by a twofold differentiation, hence those will not be displayed.

6.6.1   Variance to mean in a stationary ADS with spallation source

A spallation-based stationary neutron source (similarly to a spontaneous fission source, such as ^{252}Cf), has a so-called compound Poisson statistics. Similarly to the simple Poisson process, the source emission time intervals have an exponential distribution, but at each event a random number of source particles are injected into the system. The source neutrons have a number distribution $p_q(n)$, and the first two factorial moments of the number of emitted particles will be denoted as $q \equiv \langle q \rangle$ and $\langle q(q-1) \rangle$. For such a source, the variance to mean of the neutron counts has been calculated with both one and six delayed neutron groups [6]. Hence, for the derivation and the full formulae, we refer to that publication. The variance to mean for the steady spallation source then reads as

$$\frac{\sigma_Z^2(t)}{Z(t)} = 1 + \frac{\varepsilon D_{\nu_p}}{(\rho - \beta)^2} (1 + \delta) \left(1 - \frac{1 - e^{-\alpha t}}{\alpha t}\right) + Y_2 \left(1 - \frac{1 - e^{-\alpha_d t}}{\alpha_d t}\right) \quad (6.62)$$

Here the parameter δ is defined as

$$\delta = \frac{q D_q}{\nu D_{\nu_p}} (-\rho) \qquad (6.63)$$

as it was introduced in [162]. The parameter α_d is the decay constant related to the delayed neutrons. The values of Y_2 and α_d are given e.g. in [6].

A comparison with the traditional formula (2.2) shows that the Y_1 factor of the prompt term is enhanced by an additional term δ of (6.63). This term is negligible close to critical, because then the prompt chains are long, and the probability of detecting two neutrons from the same chain, started by a single source neutron, is high. On the other hand, in deep subcritical systems the chains started by a single neutron are short, and the probability of detecting two neutrons from the same chain is low. However, since several source neutrons are born simultaneously, detecting two neutrons from the same *source event* remains still reasonably high if the source multiplicity is high, and this is expressed by (6.62).

6.6.2 Pulsed sources

As it was mentioned in the introduction to this subsection, the accelerator driven spallation sources will be run in pulsed model. Regarding the pulsing, there are essentially
four subclasses possible, due to the fact that there are two possibilities for the pulse
width, wide and narrow, and the measurement can be evaluated either deterministically, i.e. starting the measurements always at the same point of the pulse train, such
as its start, or, randomly. Although the pulsing itself is deterministic, it is the way
of evaluation the measurement which differs, for simplicity these cases are called
"deterministic pulsing" and "random pulsing", respectively.

The significance of the pulse width is that if it is sufficiently wide, the individual
source events (the arrival of the individual protons on the target) can be regarded as
independent.[†] If the pulse is very short, then all neutrons in the pulse can be regarded
as emitted and injected into the core simultaneously, i.e. they are not independent.
Since both wide and narrow pulses can be both deterministically or randomly pulsed,
there are four main cases which will be very briefly summarised here. For wide pulses,
there is a further freedom of choosing a pulse shape, but the pulse shape can be handled in a simple manner, and we do not consider subclasses for the various pulse
shapes. All formulae refer to the case of prompt neutrons only; there are results available with the inclusion of delayed neutrons, but they are too involved to be included
in this book.

With the details of the calculations and the resulting formulae, we refer to the
original publications [6,159,160]. To give a flavour of the type of formulae, here we
only quote the results for finite width pulses; the case of narrow pulses (instantaneous
injection) are too lengthy to cite here.

Assuming a pulse repetition time T_0, a pulse shape $f(t)$ which is non-zero only
between $[0, T_0]$, we have the following results for the Feynman $Y(t) = \mu_{ZZ(t)}/Z(t)$
function, given with its two factors.

For the case of deterministic (synchronised) pulsing, one obtains

$$Z(t) = S_0 \lambda_d \left[a_0 t + \sum_{n=1}^{\infty} \frac{a_n \{1 - \cos(\omega_n t)\} + b_n \sin(\omega_n t)}{\omega_n} \right] \tag{6.64}$$

and

$$\mu_{ZZ}(t) = \frac{S_0 \lambda_d^2 \lambda_f \langle \nu(\nu - 1) \rangle}{\alpha^2} \left\{ a_0 t \left(1 - \frac{1 - e^{-\alpha t}}{\alpha t} \right) \right.$$
$$\left. + \sum_{n=1}^{\infty} \left[a_n \mathscr{A}_n(t) + b_n \mathscr{B}_n(t) + (1 - e^{-\alpha t})^2 \sum_{n=1}^{\infty} \frac{-\omega_n a_n + 2\alpha b_n}{\omega_n^2 + (2\alpha)^2} \right] \right\} \tag{6.65}$$

Here, α is the prompt neutron decay constant $\alpha = -\rho/\Lambda$ as before,

$$\omega_n = \frac{2n\pi}{T_0}; \quad n = 1, 2, \ldots \tag{6.66}$$

[†] As Degweker and Rana pointed out [161], this is not always the case.

$\mathscr{A}_n(t)$ and $\mathscr{B}_n(t)$ are combinations of simple trigonometric functions, depending on ω_n and α, and finally a_n and b_n are the coefficients of the Fourier series expansion of the periodically repeated pulse shape [6].

For the stochastic (non-synchronised) pulsing, the result is somewhat simpler:

$$Z(t) = S_0 \, \lambda_d a_0 T \tag{6.67}$$

and

$$Y(T) = \frac{\lambda_d \lambda_f \langle \nu(\nu - 1) \rangle}{\alpha^2} \left(1 - \frac{1 - e^{-\alpha T}}{\alpha T} \right)$$
$$+ \frac{2 S_0 \lambda_d}{a_0 T} \sum_{n=1}^{\infty} \frac{a_n^2 + b_n^2}{\omega_n^2} \sin^2 \left(\frac{\omega_n}{2} T \right) \tag{6.68}$$

Examples of measured and fitted results of both deterministic and stochastic pulsing are found in [6] and the references therein.

6.7   Zero-power noise analysis from continuous detector signals

There has been an interesting development relatively recently in the area of pulse counting-based stochastic reactivity measurement methods, the Feynman- and Rossi-alpha methods. The essence of this development is to shift the detection methodology from pulse counting to the analysis of time-resolved continuous detector signals, primarily that of fission chambers. Instead of evaluating the statistics of the discrete pulses, one calculates the (auto- or cross) covariance of the continuous detector signal. In order to distinguish the method based on the analysis of continuous signals from the pulse counting method with a compact terminology which is easy to refer to, it will be called the "time-resolved method".

The incentive came from the so-called Campbelling techniques, which are applied to the moments of the continuous detector signals, The essence is to estimate the expectation of the stationary signal from its higher-order moments, primarily from its variance, in order to suppress the minority component coming from gamma detection. The Campbelling techniques were originally derived with the assumption of independent detection events, which have Poisson statistics. Hence the original Campbelling technique cannot be used to determine temporal correlations between the detections arising from the fission chains, which lead to the deviation from the Poisson statistics, and which thus contains the information on the system parameters, such as the subcritical reactivity. However, by elaborating a simple and powerful model of the detection process of random detections [163], it was possible to extend the method for detections of neutrons in a multiplying system, where detections of neutrons from the same chain are time-correlated, leading to the deviation from the Poisson statistics [164].

This way the possibility opened up that one could unfold the statistics of the underlying point processes from measuring continuous signals. This has the advantage that by handling the continuous signals, in which the pulses from different detections overlap, one avoids the problem of the dead time, which one encounters at high count rates. This is a clear advantage, which allows the extension of the method to higher power levels (which entail higher count rates).

It is obvious that this extension is not specifically designed or meant exclusively for next generation systems. Nevertheless, there is reason for including the time-resolved method into this book. Namely, in next generation reactors, due to the faster spectrum and/or smaller volume (SMRs), higher count rates can be expected. In addition, the method, representing a new paradigm, will be more likely to gain larger applications in next generation of reactors. This is corroborated by the circumstance that the method of continuous signals can be advantageously used with fission chambers; in fact, the Campbelling techniques were developed for fission chambers. At the same time, a study investigating three detector types for sodium-cooled fast reactors (SFRs) [165], found that fission chambers are suited the best for SFRs. For these reasons, it was found worth describing this method in addition to others, which are more specifically designed for next generation reactors.

6.7.1   The model of the detector signal

In the following, the principles of the time-resolved method will be summarised in a simple form, and the Rossi- and Feynman-alpha formulae will be derived for continuous detector signals. For simplicity, similarly to the treatment of the gamma-detection-based reactivity measurement methods, all derivations will be made by assuming prompt neutrons only, and an extraneous neutron source with a simple Poisson statistics, as in [164]. The methodology was subsequently extended to including delayed neutrons, and an extraneous source with compound Poisson statistics. With the extensions, we refer to the literature [166–169].

The basic building stone of the model of the stochastic signal of a fission chamber is to assume that each arrival of a neutron to the detector induces a pulse which is allowed to have a random element, i.e. the pulses are not identical. This is represented by writing the pulse (temporal evolution of the detector current) induced by a detection at time $t = 0$ as $\varphi(x, t)$, where x is the realisation of a random variable with a density function $w(x)$. Then the probability density $h(y, t)$ of the detector signal y at time t can be written as

$$h(y, t) = \int_{-\infty}^{+\infty} \delta\left[y - \varphi(x, t)\right] w(x) \, dx, \qquad\qquad y \geq 0 \qquad\qquad (6.69)$$

Since each detection induces a pulse, the continuously arriving neutrons generate the detector current as the aggregate of such current signals, each related to different realisations x.

It can already be anticipated at this point that the final expressions for the moments or for the Feynman- and Rossi-alpha formulae will also depend on the characteristics of the detector pulse shape, In other words, unlike in the case of pulse counting, the measured quantities will not depend only on the properties of the multiplying medium, but also on the properties of the detector. However, as it will be seen, this will not lead to severe difficulties.

In further calculations we need the characteristic function $\widetilde{h}(\omega, t)$ of $h(y, t)$,

$$\widetilde{h}(\omega, t) = \int_{-\infty}^{+\infty} e^{\imath \omega y} \, h(y, t) \, \mathrm{d}y \tag{6.70}$$

which is obtained from the aforementioned as

$$
\begin{aligned}
\widetilde{h}(\omega, t) &= \int_{-\infty}^{+\infty} e^{\imath \omega y} \left(\int_{-\infty}^{+\infty} \delta \left[y - \varphi(x, t)\right] w(x) \, \mathrm{d}x \right) \, \mathrm{d}y \\
&= \int_{-\infty}^{+\infty} e^{\imath \omega \varphi(x,t)} \, w(x) \, \mathrm{d}x
\end{aligned} \tag{6.71}
$$

6.7.2   One-point densities

The next step is to derive an equation for the density $p(y, t \,|\, 1)$, where

$$p(y, t \,|\, 1) \, \mathrm{d}y \tag{6.72}$$

is the probability that in a subcritical system without a source, at the time instant $t \geq 0$, the *total* detector current $y(t)$, which is due to the aggregate of the single pulses from all individual detections from a chain of neutrons induced by *one source neutron*, lies within the interval $(y, y + \mathrm{d}y]$, provided that at time $t = 0$ the detector current was zero, while the number of the neutrons was equal to unity. We call $p(y, t \,|\, n(0) = 1)$ the single-particle induced distribution, in contrast to the one which is due to the continuous injection of source neutrons. To derive a backward-type master equation for this quantity, we need to account for the neutron multiplication in the system. This will be made by using the same variables and parameters as in the previous sections, namely by the use of the intensities of caption, detection and fission λ_c, λ_d and λ_f, respectively, as well as the number distribution $p_f(k)$ of prompt fission neutrons. The total reaction intensity λ_r is, as usual,

$$\lambda_r = \lambda_c + \lambda_d + \lambda_f \tag{6.73}$$

By applying the backward approach, by adding up the probabilities of the four mutually exclusive events of having any of the three possible reactions and no reaction, respectively, one obtains the following integral equation:

$$
\begin{aligned}
p(y, t \,|\, 1) = {} & e^{-\lambda_r t} \, \delta(y) + \lambda_d \int_0^t e^{-\lambda_r (t - t')} \, h(y, t') \, \mathrm{d}t' \\
& + \lambda_c \, \delta(y) \int_0^t e^{-\lambda_r (t - t')} \, \mathrm{d}t' \\
& + \lambda_f \int_0^t e^{-\lambda_r (t - t')} \sum_k p_f(k) \int \cdots \int_{y_1 + \cdots + y_k = y} \prod_{j=1}^k p(y_j, t' \,|\, 1) \, \mathrm{d}y_j \quad (6.74)
\end{aligned}
$$

Defining the characteristic function of $p(y, t \,|\, 1)$ as

$$g(\omega, t \,|\, 1) = \int_{-\infty}^{+\infty} e^{\imath \omega y} \, p(y, t \,|\, 1) \, \mathrm{d}y \tag{6.75}$$

we obtain from (6.74) the equation

$$g(\omega, t \,|\, 1) = e^{-\lambda_r t} + \lambda_d \int_0^t e^{-\lambda_r (t-t')} \widetilde{h}(\omega, t') \, dt' + \lambda_c \int_0^t e^{-\lambda_r (t-t')} \, dt'$$

$$+ \lambda_f \int_0^t e^{-\lambda_r (t-t')} q_f \left[g(\omega, t' \,|\, 1) \right] \, dt' \qquad (6.76)$$

where

$$q_f(z) = \sum_{k=0}^{\infty} p_f(k) \, z^k \qquad (6.77)$$

is the generating function of the number distribution $p_f(k)$ of the fission neutrons, and the sought quantity $g(\omega, t' \,|\, 1)$ in the square brackets is its argument. It is seen explicitly here how the detector properties are included into the expression, through the term containing the characteristic function $\widetilde{h}(\omega, t')$ of the random pulse shape.

From this equation for the characteristic function of the single-particle induced distribution, we need to connect to the source-induced distribution, or rather its generating function, which is the distribution of the detector signal in a subcritical system where a continuous injection of neutrons from an extraneous source takes place. Thus, denote by

$$P(y, t \,|\, 0) \, dy \qquad (6.78)$$

the probability that in a subcritical system which is driven by a neutron source with simple Poisson statistics and constant intensity S, at the time moment $t \geq 0$, the detector current is found in the interval $(y, y + dy]$, provided that at the time instant $t = 0$ the detector current and the numbers of neutrons and precursors were zero. We also need the characteristic function of $P(y, t \,|\, 0)$:

$$G(\omega, t) = \int_{-\infty}^{+\infty} e^{i\omega y} P(y, t \,|\, 0) \, dy \qquad (6.79)$$

An integral backward master equation for the source-induced distribution $P(y, t \,|\, 0)$, in terms of the single-particle induced distribution $p(y, t \,|\, 1)$ can be obtained as follows.[‡] Adding up for the mutually exclusive events that there will be or will not be a source emission between $[0, t]$, and applying the convolution theorem for the former case, one obtains the following backward equation for the source-induced distribution $P(y, t \,|\, 0)$:

$$P(y, t \,|\, 0) = e^{-St} \delta(y)$$

$$+ S \int_0^t e^{-S(t-t')} \iint_{y_1 + y_2 = y} p(y_1, t' \,|\, 1) P(y_2, t' \,|\, 0) \, dy_1 \, dy_2 \qquad (6.80)$$

[‡]Following standard praxis [6], single-particle induced distributions and their generating functions will be denoted by low case letters, whereas source-induced ones by capital letters.

From (6.80), using the convolution theorem, one obtains the following equation for the characteristic function:

$$G(\omega, t) = e^{-St} + S \int_0^t e^{-S(t-t')} g(\omega, t' \,|1) \, G(\omega, t') \, dt' \tag{6.81}$$

It is easy to show that the solution of the integral equation (6.81) is given as

$$G(\omega, t) = \exp \left\{ S \left[\int_0^t g(\omega, t' \,|n(0) = 1) - 1 \right] dt' \right\} \tag{6.82}$$

From (6.82), equations for the moments of the detector current can be obtained. It is convenient to calculate the cumulants, out of which we only need here the first two, which is the expectation and the variance. The cumulants $\kappa_n(t)$ can be obtained from the well-known relation

$$\kappa_n(t) = \left(\frac{1}{i} \right)^n \left[\frac{\partial^n K(\omega, t)}{\partial \omega^n} \right]_{\omega=0} \tag{6.83}$$

where

$$K(\omega, t) = \ln G(\omega, t) = S \left[\int_0^t g(\omega, t \,|1) - 1 \right] dt' \tag{6.84}$$

The stationary values of the moments, such as the expectation and the variance, are obtained as the asymptotic values for $t \to \infty$. Equations for these can be obtained from (6.83) and (6.84). For an arbitrary detector pulse shape, these can be solved symbolically; once the detector pulse shape distribution is fixed, concrete solutions can be obtained.

6.7.3   Two-point densities and the auto-covariance

Out of the first two cumulants, the expectation (first moment) is interesting for our purposes as it is. However, regarding the second moment, the variance of the signal in itself is not sufficient to derive the time resolved equivalents of the Feynman- or Rossi alpha methods. As mentioned earlier, the information about the subcritical reactivity is embedded into the time correlations between the detections at different times. In the case of the continuous signals, an equation for the equivalent of the Feynman-alpha method is not possible to derive. What is possible to do is to determine the stationary auto-covariance of the detector signal for two different times a time lag τ apart, which is the analogue of the Rossi-alpha method of the discrete pulse counting method. Alternatively, the cross-covariance between two detector signals as functions of the time separation τ could be considered. This would have the same advantages (suppressing the effect of background noise) as the two-detector Feynman-alpha method of pulse counting, which was treated in Section 6.5. The cross-covariance time-resolved method was tested in measurements, but its theory will not be described here.

From the auto-covariance, the time-resolved analogue of the Feynman-alpha method can be derived, with a relationship, similar to (2.10). Hence in the following we first outline the determination of the auto-covariance (Rossi-alpha), after which the time-resolved Feynman alpha formula will also be derived.

The auto-covariance is defined as

$$\lim_{t \to \infty} \mathbf{Cov}\{y(t-\tau)\,y(t)\} = \lim_{t \to \infty} \langle y(t-\tau)\,y(t)\rangle - \langle y\rangle^2 \equiv \mathbf{Cov}(\tau) \qquad (6.85)$$

For the calculation of this quantity, one needs the two-point (in time) equivalents of the one-point densities single particle induced and source-induced densities $p(y, t|1)$ and $P(y, t|0)$, respectively, i.e.

$$p(y_1, y_2, t_1, t_2|1) \qquad \text{and} \qquad P(y_1, y_2, t_1, t_2|0) \qquad (6.86)$$

or rather their characteristic functions

$$g(\omega_1, \omega_2, t_1, t_2) \qquad \text{and} \qquad G(\omega_1, \omega_2, t_1, t_2) \qquad (6.87)$$

Here,

$$p(y_1, y_2, t_1, t_2|1)\,\mathrm{d}t_1\,\mathrm{d}t_2 \qquad (6.88)$$

is the probability of the event that at time $t_1 = t - \tau$ the detector current is in the interval $(y_1, y_1 + dy_1)$ and that at $t_2 = t$ the detector current is in the interval $(y_2, y_2 + dy_2)$ in a subcritical system in which one single neutron was injected at time $t = 0$, when the detector signal was zero. $P(y_1, y_2, t_1, t_2|0)$ is the same, but for the case when at time $t = 0$ there were no neutrons in the system, the detector signal was zero, and the extraneous source with Poisson statistics and intensity S was switched on.

Neglecting the details, we only write down the equations for the characteristic functions which read as follows:

$$g(\omega_1, \omega_2, t - \tau, t) = e^{-\lambda_r t}$$
$$+ \lambda_d \int_0^t e^{-\lambda_r(t-t')} \left[\Delta(t' - \tau)\,\tilde{h}(\omega_1, t' - \tau)\,\tilde{h}(\omega_2, t')\right.$$
$$\left. + \Delta(\tau - t')\,\tilde{h}(\omega_2, t')\right]\,\mathrm{d}t' + \lambda_c \int_0^t e^{-\lambda_r(t-t')}\,\mathrm{d}t'$$
$$+ \lambda_f \int_0^t e^{-\lambda_r(t-t')}\,\{\Delta(t' - \tau)\,q_f[g(\omega_1, \omega_2, t' - \tau, t')]$$
$$+ \Delta(\tau - t')\,q_f[g(\omega_2, t')]\}\,\mathrm{d}t' \qquad (6.89)$$

and

$$G(\omega_1, \omega_2, t - \tau, t) = e^{-S_0 t}$$
$$+ S \int_0^t e^{-S(t-t')}\,[\Delta(t' - \tau)\,g(\omega_1, \omega_2, t' - \tau, t')$$
$$G(\omega_1, \omega_2, t' - \tau, t') + \Delta(\tau - t')\,g(\omega_2, t')\,G(\omega_2, t')]\,\mathrm{d}t' \qquad (6.90)$$

The one-point characteristic functions $g(\omega, t)$ and $G(\omega, t)$, as well as their defining equations were already given earlier.

From the definition (6.85) and the properties of the characteristic functions, the auto-covariance can be calculated as

$$
\mathrm{Cov}(\tau) = \lim_{t \to \infty} \left\{ \frac{1}{i^2} \left[\frac{\partial^2 G(\omega_1, \omega_2, t - \tau, t)}{\partial \omega_1 \partial \omega_2} \right]_{\substack{\omega_1 = 0 \\ \omega_2 = 0}} \right\}
$$

$$
- \lim_{t \to \infty} \left\{ \left[\frac{1}{i} \frac{\partial G(\omega_1, t - \tau)}{\partial \omega_1} \right]_{\omega_1 = 0} \left[\frac{1}{i} \frac{\partial G(\omega_2, t)}{\partial \omega_2} \right]_{\omega_2 = 0} \right\}
$$

(6.91)

With the help of (6.89) and (6.90) and their solutions, the auto-covariance can be calculated. The results remain partly symbolic as long as the characteristic function $\tilde{h}(\omega, t)$ of the random detector pulse shape is not defined. The derivation of the auto-covariance is rather laborious and lengthy and will not be given here, it can be found in [164] and [168]. Instead, first the final result for the covariance will be given with a selected concrete pulse shape, after which the time-resolved variance to mean formula will be given, as calculated from the auto-covariance.

6.7.4   Rossi-alpha formula for a given detector pulse shape

Following the aforementioned publications, a concrete pulse shape $\varphi(x, t)$ and its amplitude probability distribution $w(x)$ will be selected as follows. The pulse type will be defined as

$$
\varphi(x, t) = Q x^2 t e^{-xt} \quad \text{and} \quad w(x) = \delta(x - \alpha_e)
$$

(6.92)

This is a deterministic pulse type with a pulse shape $f(t)$ being equal to

$$
f(t) = \int_{-\infty}^{+\infty} \varphi(x, t) w(x) \, \mathrm{d}x = Q \alpha_e^2 t e^{-\alpha_e t}
$$

(6.93)

Here, α_e is the time constant of the detector electronics, whereas Q stands for the collected charge, equal to the integral of $f(t)$.

With the aforementioned pulse shape, and defining the reactivity ρ and the prompt neutron generation time Λ in terms of the reaction intensities and the expectation ν of the number of prompt neutrons per fission as in the previous chapters, after a lengthy algebra one finds the following for the covariance. To simplify the formulae, we introduce the following notations:

$$
\Phi = \frac{\lambda_d \lambda_f Q \langle \nu(\nu - 1) \rangle \alpha_e^4}{\alpha^2 (\alpha_e^2 - \alpha^2)^2}
$$

(6.94)

$$
\Psi_1 = -\frac{\alpha^2 (3\alpha_e^2 - \alpha^2)}{2\alpha_e^4} \Phi + \frac{Q}{2}
$$

(6.95)

$$
\Psi_2 = -\frac{\alpha^2 (\alpha_e^2 - \alpha^2)}{2\alpha_e^4} \Phi + \frac{Q}{2}
$$

(6.96)

These are analogues of the Y_∞ factors of traditional Feynman- and Rossi-alpha formulae.

In terms of these notations, the auto-covariance function $\text{Cov}(\tau)$ is given as

$$\text{Cov}(\tau) = \frac{1}{2} \langle y \rangle \left[\alpha\, \Phi e^{-\alpha|\tau|} + \alpha_e\, \Psi_1 e^{-\alpha_e|\tau|} + \alpha_e \Psi_2 \alpha_e\, |\tau|\, e^{-\alpha_e|\tau|} \right]. \quad (6.97)$$

Here, $\langle y \rangle$ is the expectation of the stationary detector signal, which in the present case is equal to

$$\langle y \rangle = \frac{\lambda_d S Q}{\alpha} \quad (6.98)$$

As it was shown in (2.6), in the traditional pulse counting case, the Rossi-alpha formula $R(\tau)$ is defined as the normalised (with the expectation of the intensity of the detections) autocovariance function [6]. Therefore, it is logical to define the Rossi-alpha formula for the case of the continuous signals as the auto-covariance, normalised with the expectation of the stationary detector signal. One thus obtains

$$R(\tau) = \frac{1}{2} \left[\alpha\, \Phi e^{-\alpha|\tau|} + \alpha_e\, \Psi_1 e^{-\alpha_e|\tau|} + \alpha_e \Psi_2 \alpha_e\, |\tau|\, e^{-\alpha_e|\tau|} \right] \quad (6.99)$$

This formula shows considerable resemblance to the Rossi-alpha formula, (2.7), although there are several differences. One is that the coefficients Φ, Ψ_1 and Ψ_2 are much more complicated than the coefficients A_1 and A_2. A more significant difference is that in the time-dependent parts, besides the exponential term containing the prompt neutron decay constant α, exponentials containing the decay constant of the detector pulse shape appear, and the last term even has a multiplying factor $|\tau|$.

Because of these circumstances, the unfolding of the prompt neutron decay constant from the measured data is not trivial. However, similarly to the traditional case where the prompt neutron decay constant can easily be extracted due to the fact that α and α_d are sufficiently separated, the prompt neutron decay constant α can be extracted from the continuous-signal-based Rossi-alpha measurement, if α and α_e are sufficiently separated in their magnitude. In general, one may expect that

$$\alpha_e \gg \alpha \quad (6.100)$$

in which case one might assume at the evaluation of the measurement that the second and the third terms on the r.h.s. (6.99) decay within the time resolution of the sampling of the continuous signal, and it is sufficient to fit only one exponential, the prompt neutron decay constant, to the measured data.

The evaluation of a simulated data set produced by Szieberth et. al [170] applying realistic pulse shape of a fission chamber can be seen in Figures 6.1 and 6.2, with lin-lin and log-log scale, respectively. The log-log scale allows to observe the contribution of the term related to the detector pulse shape in the shorter time range denoted by red. The contribution related the prompt decay constant is well separated in this case and a fit excluding the contribution from the pulse shape can be performed. The fitted α value shows a satisfactory agreement with the $\alpha = 213$ 1/s assumed in the simulation considering the variance of the results and the uncertainty of the non-linear fitting procedure.

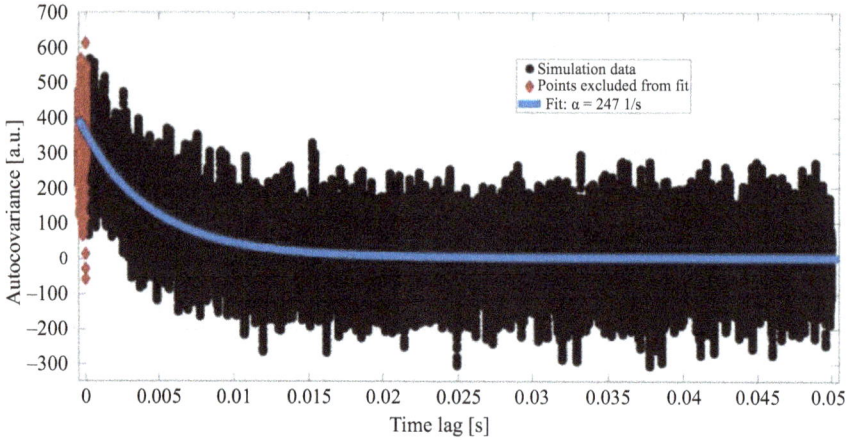

Figure 6.1   Evaluation of a Rossi-alpha measurement from a simulated data set of continuous current of a fission chamber, using a realistic pulse shape [170]

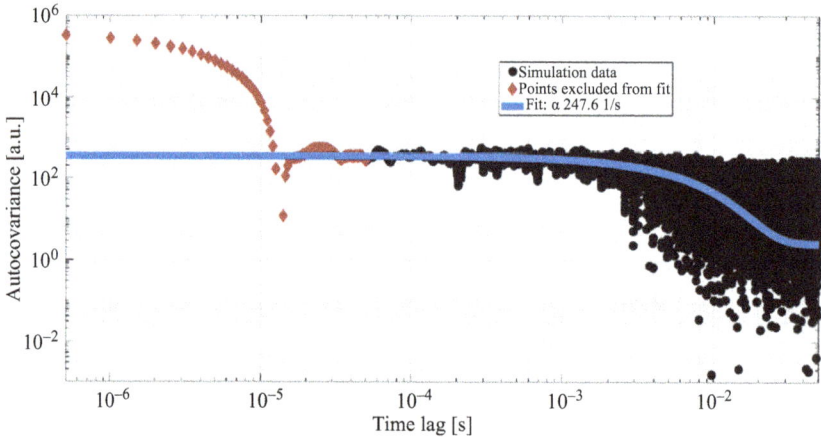

Figure 6.2   The same as Figure 6.1, but with log-log scale

Proof-of-concept measurements were also performed at the Kyoto University Critical Assembly by Szieberth *et al.* [170,171]. In these measurements the count rate was sufficiently low such that overlapping of the pulses was negligible, and no dead-time problem existed for the pulse counting method. Due to the low count rate, the measurement could be evaluated both with the pulse counting and the time-resolved method, for a comparison. Figure 6.3 shows the Rossi-α evaluation of a measurement performed in critical condition with the traditional pulse counting methods (top figure) and with the time-resolved method (bottom figure), respectively.

The α values from the two methods are close to each other, the continuous signal-based method yielding a somewhat larger value. It is seen that the relative scatter of

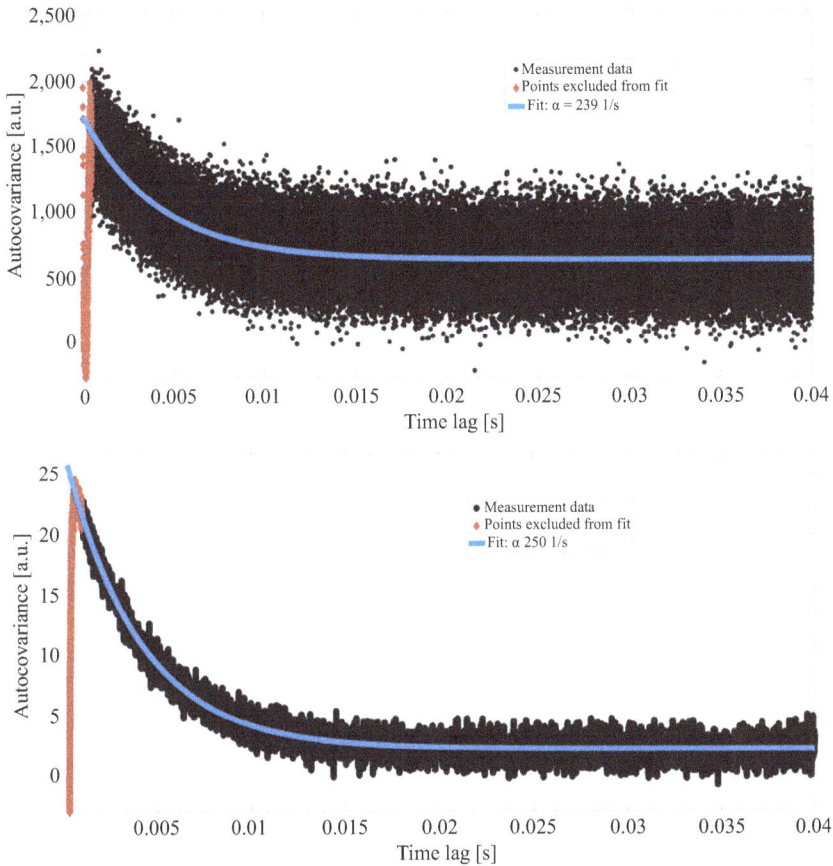

*Figure 6.3   Evaluation of a measurement, made at the Kyoto University Critical
Assembly (KUCA) a) with the pulse counting method (top figure) and b)
with the method based on continuous signals (bottom figure) [170,171]*

the data is significantly smaller for the continuous signal-based method than with
the traditional pulse-counting method. This is due to the much higher number of data
points obtained from the high-frequency sampling of the continuous signal. However,
these data points are strongly correlated. Therefore, the uncertainty of the obtained
α-values is not expected to be lower.

6.7.5   *Variance to mean formula for the time-resolved method*

In the pulse counting mode, the Feynman-alpha or variance to mean formula is given
as the dependence of the ratio of the variance and the mean of the number of counts
on the measurement time during which the count are counted. The alternative of this
for the time-resolved method is the variance to the mean of the integral of the signal

for a given time window T, as the function of the integration time. Thus, define the random function $Z(T)$ as the integral of the random function $y(t)$, i.e.

$$Z(t) = \int_0^T y(t)\, dt \tag{6.101}$$

Then, for the variance to mean method we need the mean and the variance of this quantity. The expectation is given as

$$\langle Z(T) \rangle = \int_0^\infty \langle y(t) \rangle\, dt = \frac{\lambda_d S Q}{\alpha} T \tag{6.102}$$

The variance of $Z(t)$ can be calculated with the help of the auto-covariance function as [172]

$$\sigma_Z^2(T) = \int_0^T dt_1 \int_0^T dt_2 \,\mathrm{Cov}\,(t_2 - t_1) = 2 \int_0^T dt \int_0^t d\tau \,\mathrm{Cov}(\tau) \tag{6.103}$$

Note that (6.103) differs from (2.10) in that in the latter, the expectation is added to the integral over the covariance, whereas in the former only the integral of the covariance appears. As is was discussed in [168], the reason for the difference is that (2.10) concerns a continuous parametric discrete random process (here the continuous parameter being the time during which the discrete number of pulses is counted), whereas (6.103) concerns a continuous random variable (i.e. the continuous random process $y(t)$).

Substituting (6.97) into (6.103), performing the integrations and dividing by the expectation (6.102) of $y(t)$ one arrives to the result

$$\frac{\sigma_Z^2(T)}{\langle Z(t) \rangle} \equiv \mathrm{vtm}\,(T) = \Phi f_1\,(\alpha T) + \Psi_1 f_1\,(\alpha_e T) + \Psi_2 f_2\,(\alpha_e T) \tag{6.104}$$

where the functions $f_1(X)$ and $f_2(X)$ are defined as

$$f_1(X) = 1 - \frac{1 - e^{-X}}{X} \tag{6.105}$$

and

$$f_2(X) = 1 + e^{-X} - 2\frac{1 - e^{-X}}{X} \tag{6.106}$$

As is seen from (6.104), and as it could be expected from (6.103), the variance to mean of the time-resolved method starts from zero for $T = 0$, i.e.

$$\mathrm{vtm}\,(T = 0) = 0 \tag{6.107}$$

unlike in the case of the traditional count-based methods, where it is equal to unity for $T = 0$. In principle this could be advantageous, because in the traditional method the information is in the part of the variance to mean that exceeds unity, whereas here there is no extra constant that should be subtracted from the evaluated result.

On the other hand, the same circumstances prevail here as with the Rossi-alpha method, namely that the variance to mean formula contains exponential terms containing the detector pulse decay constant, which makes the extraction of the prompt neutron decay constant more difficult. Again, if the magnitude of the detector decay constant is sufficiently separated from that of the prompt neutron decay constant, the extraction of the latter from the measurements is possible. If the detector pulse decay time is short compared by the sampling time of the signal in the measurement, then one can assume that the second and third terms on the r.h.s. of (6.104) have already reached their saturation value. In that case one has

$$\text{vtm}\,(T) = \Phi f_1\,(\alpha T) + \Psi_1 + \Psi_2 \tag{6.108}$$

which can be rewritten in the more standard Feynman-alpha formula as

$$\text{vtm}\,(T) = \Psi_1 + \Psi_2 + \Phi \cdot \left(1 - \frac{1 - e^{\alpha T}}{\alpha T} \right) \tag{6.109}$$

This formula shows that due to the influence of the detector pulse shape and a pulse width small compared to the sampling time of the signal, the time-resolved variance to mean method has a form similar to the traditional formula in that its time dependence is determined by the prompt neutron decay constant, but the initial value deviates from zero. Since both $\Psi_1 + \Psi_2$ and Φ, as well as α are unknowns, the fit needs to be made for three unknowns to extract the prompt neutron decay constant.

One can also note that this "jump" of the variance to mean at $t = 0$, due to the component of the function which already reached its asymptotic value, is an analogue of the similar jump in the gamma counting-based Feynman-alpha method, (6.42). There the extra term $\varepsilon_\eta \eta D_\eta$ corresponds to the asymptotic value of the component of the Feynman-alpha expression which is related to the decay constant due to the gamma lifetime. This lifetime is set to zero in the model applied in Section 6.4, hence the asymptotic value of this component is added at $t = 0$.

As an illustration, some results from simulations and measurements are shown in Figure 6.4 [170,171]. The figure shows the evaluation of a VTM simulation for

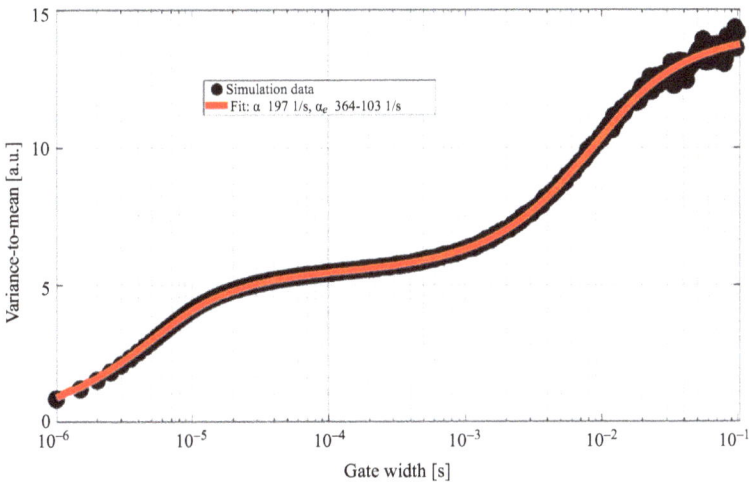

Figure 6.4   Evaluation of a Feynman-alpha measurement from a simulated data set of continuous current of a fission chamber, using a realistic pulse shape [170]

continuous signals, where the input was generated by assuming the same realistic pulse shape of a fission chamber as in the case of the Rossi-α simulations in Section 6.7.4. The logarithmic time scale allows for the observation of the contribution of both the detector pulse and the prompt decay constant. The two inflection points in the curve correspond to the detector decay constant α_e and the prompt neutron decay constant α, respectively. The detector decay constant was found to be $3.64 \cdot 10^5$ 1/s, whereas the prompt neutron decay constant was obtained as $\alpha = 197$ 1/s. These are in good agreement with the input data to the simulation.

Results from a measurement performed in KUCA in the same core configuration, evaluated by both the pulse counting and the time-resolved methods, are shown on Figure 6.5 [170,171]. These measurements were made in the critical state, and they were taken sufficiently long such that both the prompt neutron decay constant α, as

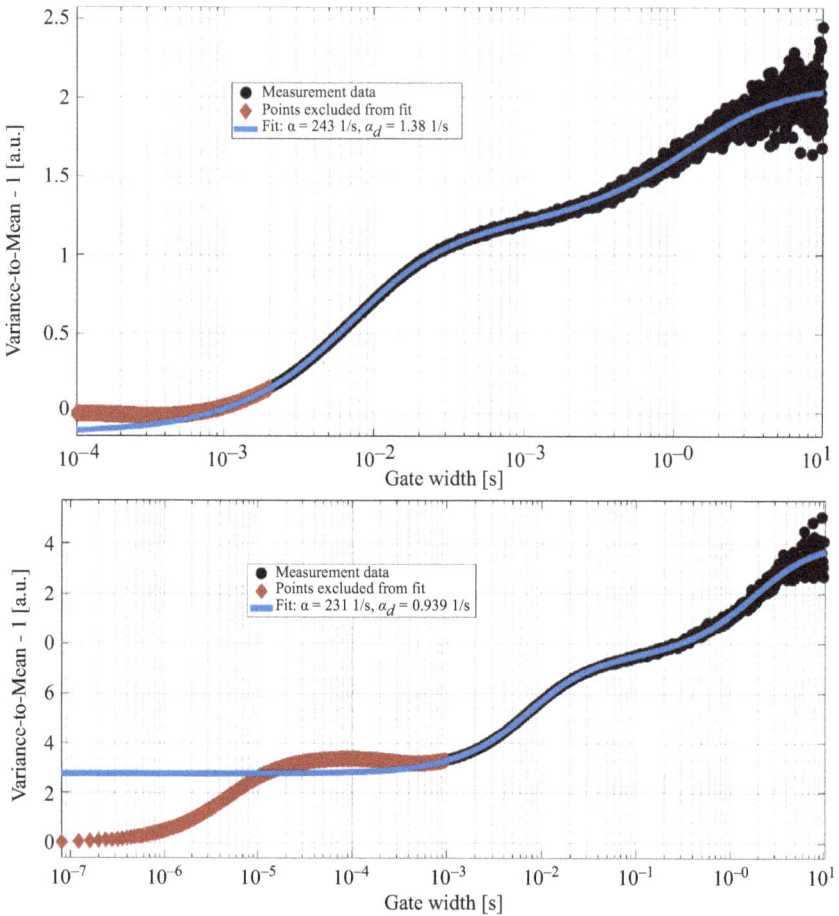

Figure 6.5 *Evaluation of a measurement performed in KUCA, a) with the pulse counting method (top figure) and b) with the method based on continuous signals (bottom figure) [170,171]*

well as the delayed neutron Feynman time constant α_d could be fitted. The latter is obtained with rather high uncertainty, as in critical condition the delayed term of the Feynman-α formula diverges. The red parts of the curves were excluded from the fit since they deviate from the theoretical curve. In the case of the pulse counting method, this is due to the dead-time effect, while in the continuous signal-based case, the effect of the pulse shape deviates from the ideal case assumed in the theory. Again, the fitted data from the two measurements are reasonably close to each other. It is seen that for the VTM, the time resolved method yields lower values, in contrast to the Rossi-alpha case.

The aforementioned results are rather encouraging, and demonstrate the potentials of the time-resolved method [173]. Work is going on to test the method with high count rates, to demonstrate its applicability in cases where the pulse counting method breaks down due to the dead time problem. It is envisaged that the method will be used widely with next generation nuclear systems, as well as in nuclear safeguards.

Chapter 7

Instrumentation and control systems diagnostics for next generation of nuclear reactors

Hash M. Hashemian[1]

This chapter describes experience with implementing in-situ or online monitoring techniques to verify the performance of the instrumentation and control (I&C) systems and the components of nuclear facilities. These techniques depend on signals from temperature sensors, pressure, level and flow transmitters, and neutron flux detectors. Almost all materials included here are based on hands-on implementation work in nuclear facilities worldwide.

Although the material in this chapter is based on testing of instrumentation and control (I&C) systems and other components of the existing fleet of nuclear power plants, they apply also to the next generation of nuclear reactors currently under design, development or construction. In fact, the motivation for writing this chapter was to entice the designers of next generation of reactors to take advantage of this material in the design and development of new plants. For example, new plants should allow for plenty of process sensors, diagnostics sensors, wireless sensors and fast data acquisition systems to allow the type of diagnostics that are articulated in this chapter.

One word on the instrumentation for next generation reactors is in order. The availability and the spacing of sensors and sampling rates of sensor data plays an important role in the possibilities and performance of diagnostics. Since once the construction is fixed and the reactor is built, there is no possibility to complement the instrumentation later on, one task of overwhelming importance in the planning of next generation reactors is to make sure that the design includes a sufficiently high level of instrumentation. This is all the more important since due to the lack of operating experience, one does not yet know all the possible perturbation types that might need to be diagnosed. Therefore, it is advisable to plan a certain redundancy in the instrumentation.

7.1 Introduction

Over the first two decades of nuclear plant operations (1960s–1970s), troubleshooting and maintenance of nuclear power plant equipment was done primarily by

[1] Analysis and Measurement Services, Knoxville, Tennessee, USA

hands-on procedures. Then, starting in the early 1980s, hands-off procedures emerged as computer technologies made it possible to acquire test data on-site in nuclear power plants. For example, a method called the loop current step response (LCSR) test was developed in the late 1970s and implemented in nuclear power plants in the early 1980s to measure the response time of temperature sensors remotely from the control room area while the sensors remained installed in the plant during operation [174–176]. The LCSR test is performed by sending a current signal to the sensor to cause heating in the sensing element. This heating produces a temperature transient in the sensor that manifests itself at the output of the sensor and can be sampled and analysed to yield the response time of the temperature sensor. It applies to resistance temperature detectors (RTDs) and thermocouples (TCs). The principle of the LCSR test for RTDs and TCs is illustrated in Figures 7.1 and 7.2, respectively. Today, almost all applications of the LCSR method in nuclear facilities have been on RTDs.

Figure 7.3 shows LCSR transients for RTDs in nuclear power plants. Two transients are shown: one for a direct immersion RTD and another from a thermowell-mounted RTD. It Is apparent that the direct immersion RTD is much faster than the thermowell-mounted RTD because thermowell adds mass and thermal resistance which slows the dynamic response of the sensor to a sudden change in temperature.

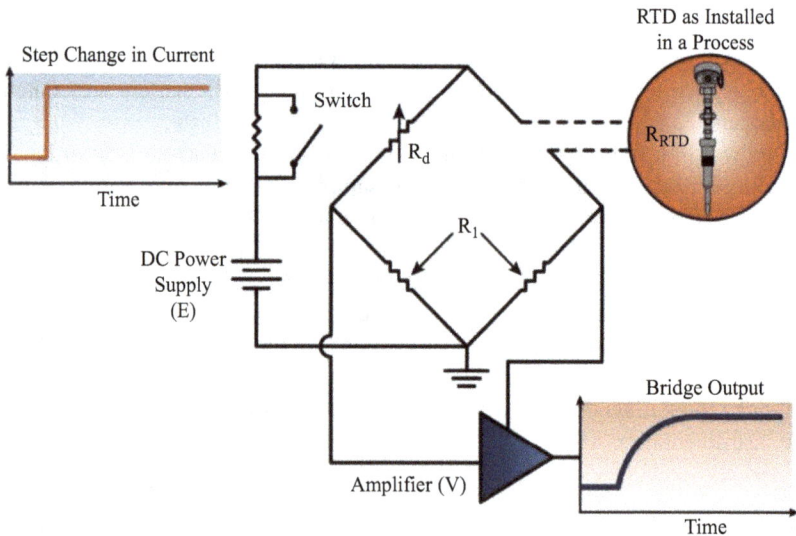

Figure 7.1   Loop current step response (LCSR) for resistance temperature detectors (RTDs)

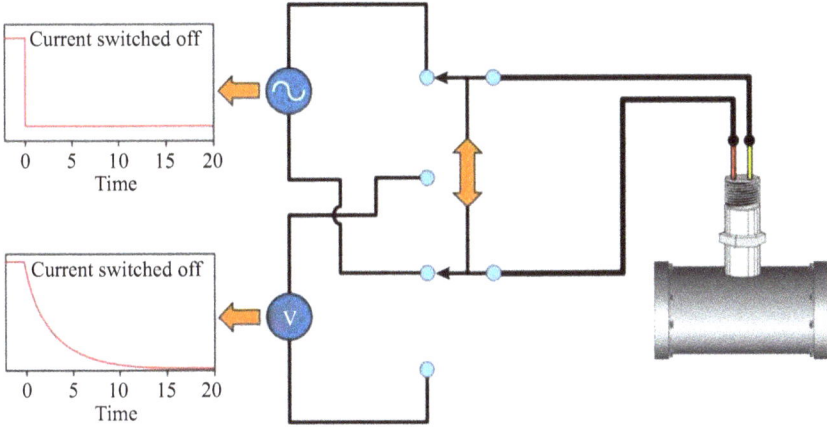

Figure 7.2   Loop current step response (LCSR) for thermocouples (TCs)

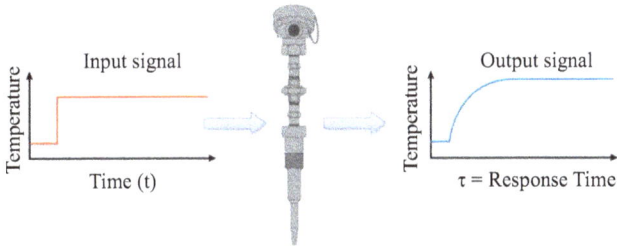

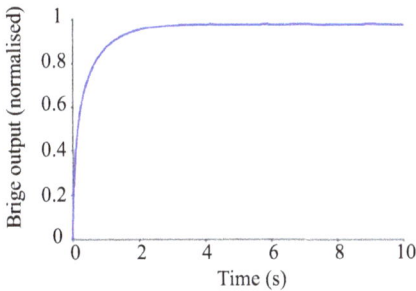

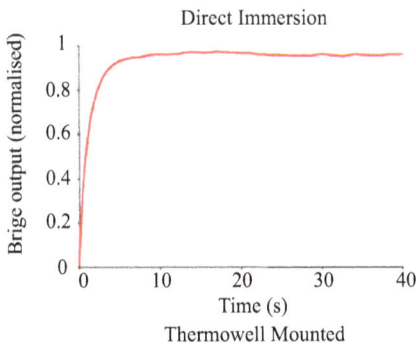

Figure 7.3   Principle of LCSR test to measure the response time of RTDs

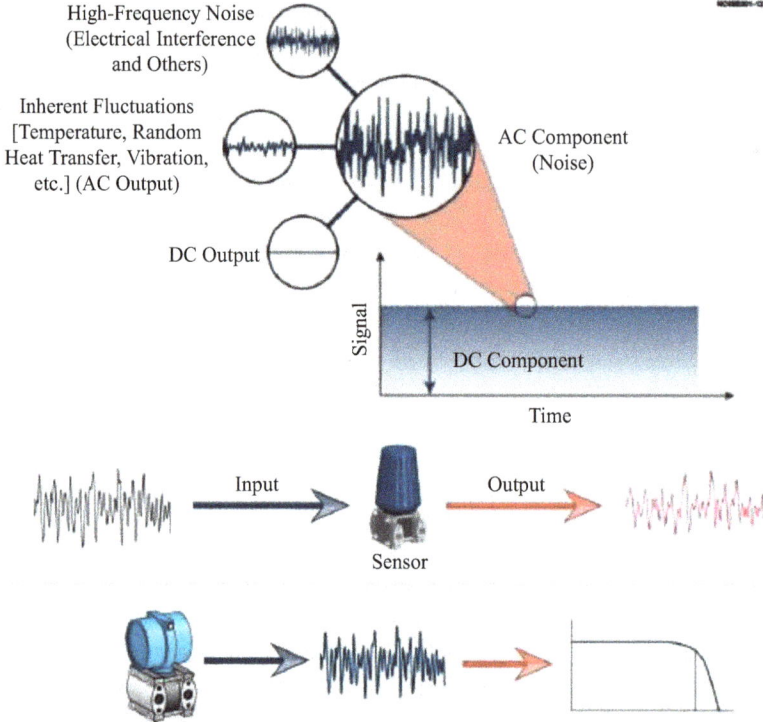

Figure 7.4   Principle of noise analysis technique for sensor response time testing

For pressure, level and flow transmitters, the noise analysis technique was adapted for response time measurements. Figure 7.4 demonstrates the principle of the noise analysis method. This is followed by Figure 7.5 illustrating the noise data acquisition process.

The noise analysis technique is a passive test that depends on natural fluctuations (noise) that exist at the output of most process sensors while the plant is operating, and it has a variety of applications in nuclear power plants besides transmitter response time testing. For example, as mentioned in Chapter 2, noise analysis can be used to measure the vibration of reactor internals using existing neutron detectors that are located outside the reactor vessel in pressurised water reactors (PWRs). Figure 7.6 shows the arrangement of the four ex-core neutron detectors in a PWR plant, each of which may be used to measure the vibration of the reactor internal components. Figure 7.7 shows the results of noise analysis given in terms of power spectral density (PSD) of the noise signal from an ex-core neutron detector in a PWR plant. The PSD is arrived at through fast Fourier transform (FFT) of the noise signal. Each peak in the PSD is indicative of vibration level of a reactor internal component.

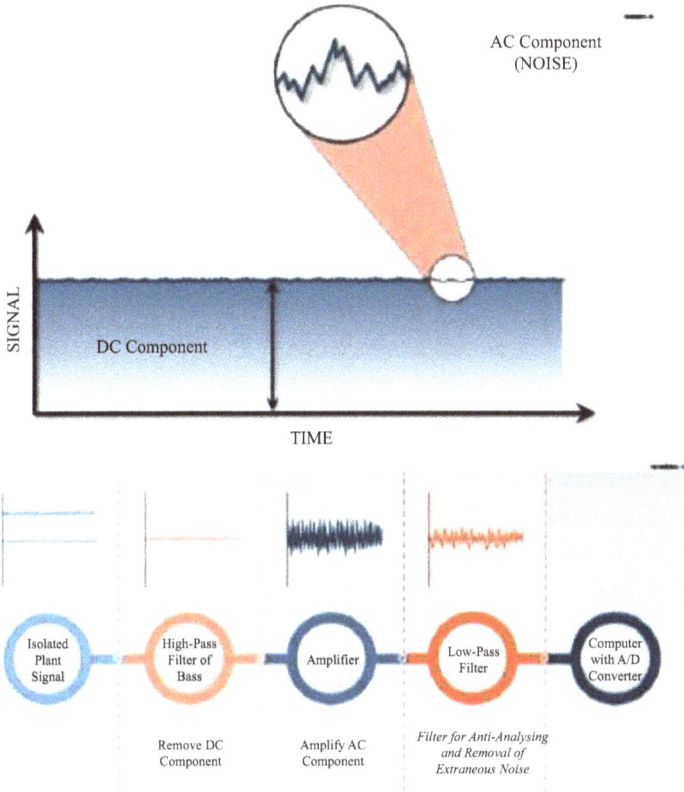

Figure 7.5   Noise data acquisition process

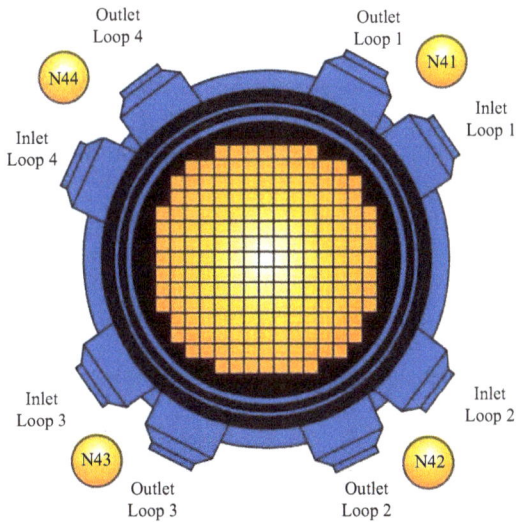

Figure 7.6   Ex-core neutron detection in a PWR plant

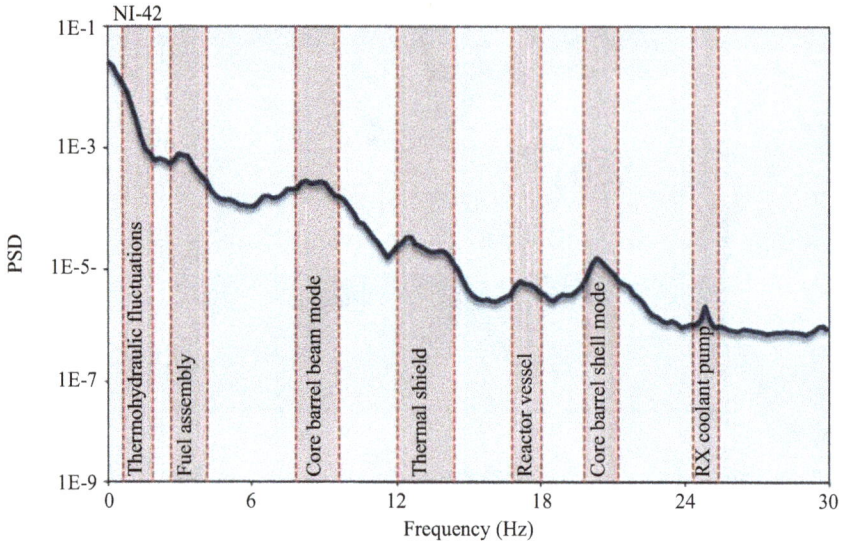

*Figure 7.7   PSD of neutron noise signals from an ex-core neutron detector in a
PWR plant*

7.2   Online flow monitoring

If noise signals from neutron detectors are cross correlated with the output of core
exit TCs in a PWR plant, the flow through the reactor core can be estimated and
monitored. Figure 7.8 shows the arrangement of neutron detectors in a PWR plant.
These detectors are at a lower elevation than core exit TCs located on top of the
reactor. Signals from the two types of sensors being a distance apart can be used
to monitor flow through the reactor core for diagnostic purposes such as detection of
blockages within the reactor coolant system (RCS). Figure 7.9 illustrates the principle
of cross-correlation techniques. The cross-correlation result is presented in terms of
a peak in the cross-correlation plot that corresponds to the time that it takes for the
fluid to travel between any pair of sensors such as an ex-core neutron detector and a
core exit TC. The signals from the two sensors are simply multiplied (after shifting
one of the signals by a time period) to arrive at the cross-correlation results.

7.3   Online detection of core flow anomalies

The author has used the noise analysis method in a number of nuclear power plants to
measure the dynamic performance of I&C systems, perform diagnostics and identify
root cause of anomalies. For example, at the Diablo Canyon nuclear power plant in
the US, the noise analysis technique was used to identify the root cause of an increasing amplitude of neutron signals towards the end of the plant operating cycle causing

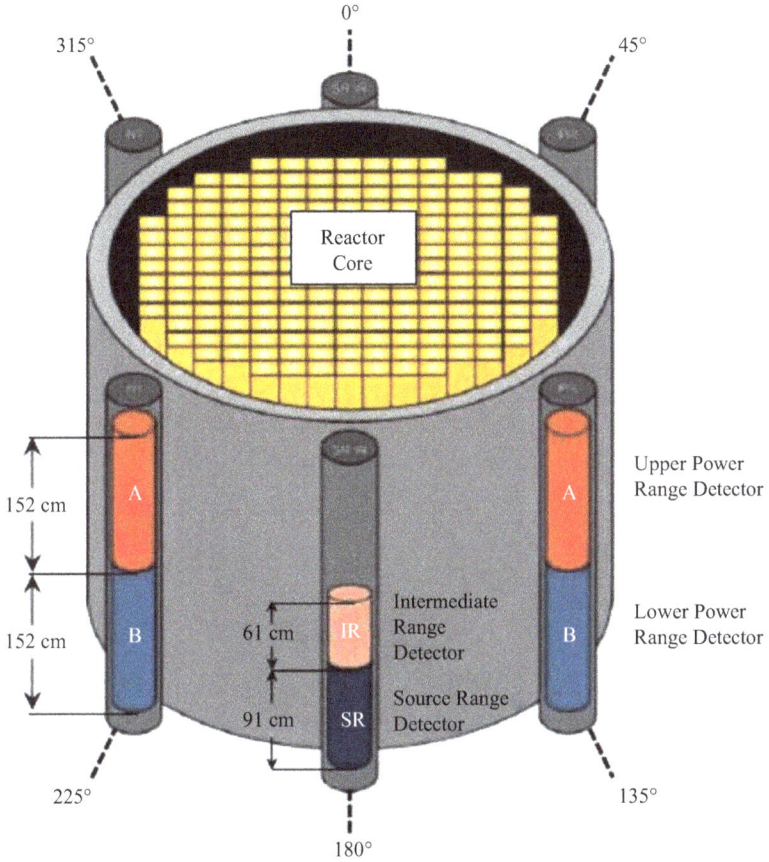

Figure 7.8   Neutron detector arrangement around reactor vessel in a PWR plant

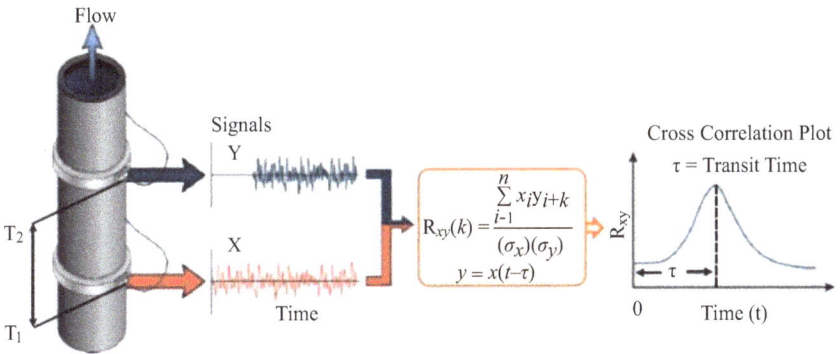

Figure 7.9   Illustration of cross-correlation technique

RX018-02

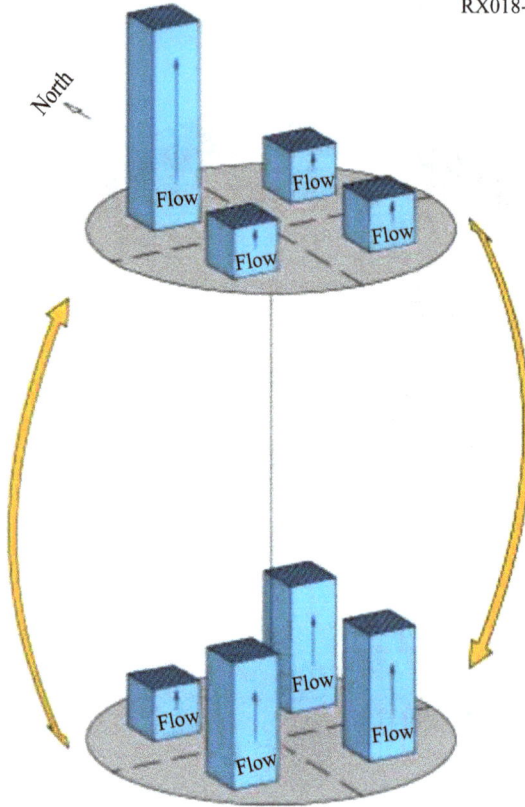

Figure 7.10   Illustration of flow fluctuations within the core of a PWR plant

neutron signal alarms. The result relieved the plant operators and the nuclear regulators who were concerned about excessive core barrel vibration or anomalous fluid flow oscillations resulting in fuel damage. The problem is illustrated in Figure 7.10 as flow fluctuations occurring in a manner in which 3/4 of the reactor coolant flow will decrease while 1/4 would increase and vice versa. This behaviour caused ex-core neutron detectors to alarm leading the plant operators to postulate that the core barrel was vibrating with a higher amplitude than in the past. Analysis and Measurement Services (AMS) ruled out this possibility using the noise analysis technique to measure the amplitude of core barrel vibration and compare it with the other reactor unit on the same site and with historical data from other plants. The results proved that the core barrel vibration level was nominal leaving the operators with the question as to the root cause of neutron signal alarms. Subsequently, AMS demonstrated that the reason for neutron signal alarms was core flow anomalies as illustrated in Figure 7.10. This conclusion was arrived at by cross-correlation of all ex-core neutron detectors with all core exit TCs. The flow oscillation shown in Figure 7.10 illustrates that the flow through 3/4 of the reactor core could decrease enough to concern the plant operators and nuclear regulators about the possibility of fuel damage. This possibility was ruled out by measuring the frequency of the coolant flow oscillations to verify that these oscillations were faster than fuel-to-coolant time constant leaving no opportunity for the flow to decrease enough for a long enough period to cause fuel damage.

7.4   Online detection of sensing line blockage and voids

The noise analysis technique has also been used to identify blockages in sensing lines that bring pressure or differential pressure information from the process to transmitters that measure pressure, level, or flow. Figure 7.11 shows the configuration of a sensing line leading from a process tank to a transmitter. It also shows a photograph of a partially blocked sensing line and two PSD plots that show the effect of sensing line blockages on the dynamic response of the affected transmitters. The photo is from an actual sensing line blockage in a PWR plant and the two PSDs are from noise data collected from two different PWR plants, one in the USA and another in Europe. It is clear from the PSD results that any blockage in a pressure sensing line can be identified by noise analysis. It is apparent in these PSDs that the blockages cause significant delay in the dynamic response of the transmitters. For example, the blockage in the sensing line of the USA plant has caused the transmitter response to become slow by about an order of magnitude.

The noise analysis technique can also be used to detect the presence of air/void in pressure sensing lines as shown in Figure 7.12. It is apparent that any significant air/void in the sensing line will not only reduce the dynamic response of the transmitter but also cause the pressure signal to fluctuate at a resonance frequency, which results in disturbance to plant operation. It should be pointed out that the data in Figure 7.12 is from laboratory experiments performed at AMS to demonstrate the capabilities of noise analysis techniques to detect air/void in pressure sensing systems.

7.5   Online detection of sensing waves

In addition to air/void in sensing lines that cause resonance at the output of pressure transmitters, standing waves in process piping can cause resonances that can disturb the plant operations. This problem can also be diagnosed using the noise analysis technique as shown in Figure 7.13. The nuclear plant affected by this problem experienced excessive flow fluctuation in the RCS causing the plant to reduce power. The PSD of this fluctuation showed a peak at 10.7 Hz while the plant was operating at full power with reactor coolant temperature at 565°F (about 300°C). This peak shifted to 12 Hz when the reactor coolant temperature decreased to 520°F (about 270°C). The following calculations starting with the standing wave equation prove that the root of the fluctuation problems was standing waves in the plant piping leading to the pressure transmitter. Standing waves are a natural phenomenon that must be taken into account in mechanical design of fluid systems.

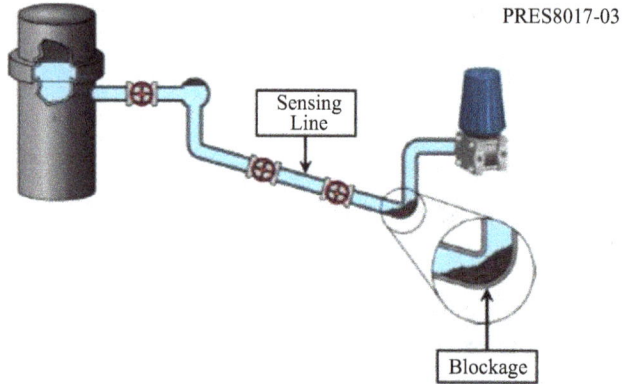

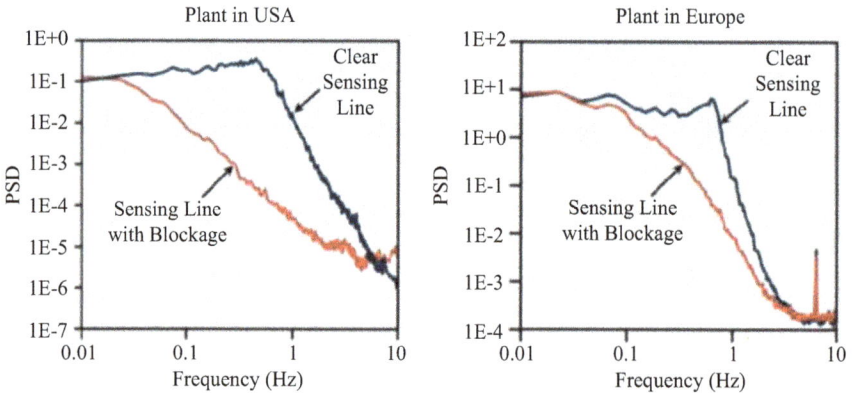

Figure 7.11   Illustration of how sensing line blockages are detected using the noise analysis technique

The frequency $F(T)$ of the standing wave, as a function of the temperature T is given as

$$F(T) = \frac{C(T)}{2L} \tag{7.1}$$

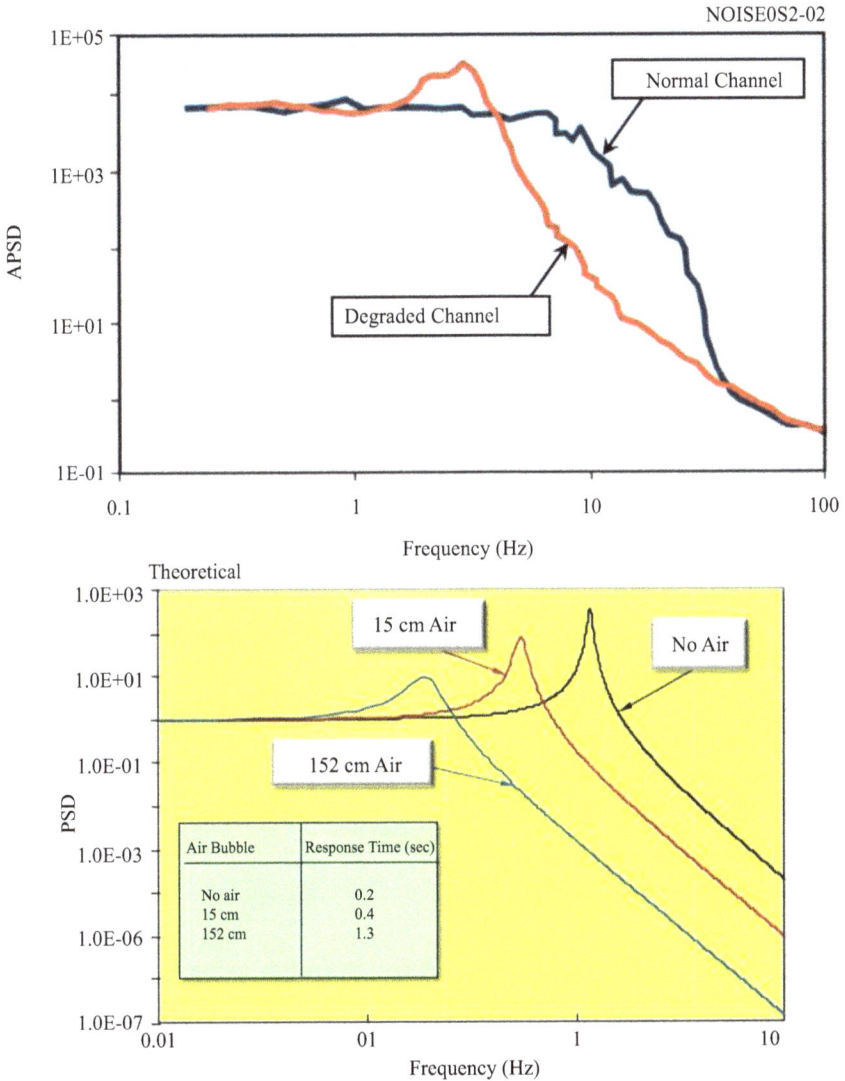

Figure 7.12   Air in sensing lines can impede pressure transmitters

where

C(T):   Speed of sound
C (565°F) = 3 100 ft/s
C (520°F) = 3 480 ft/s
L   = 145 Feet
F (565°F) = 12.0 Hz
F (520°F) = 10.7 Hz.

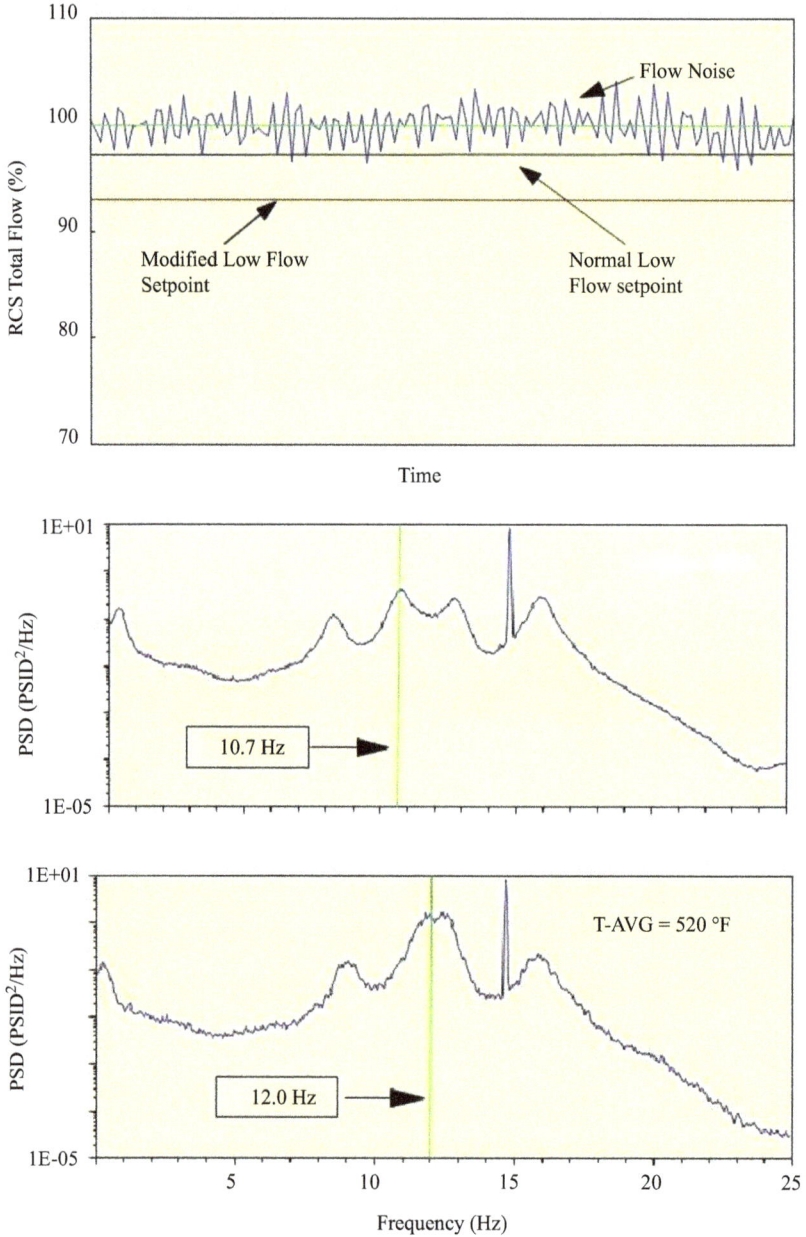

Figure 7.13   Standing waves detected using noise analysis

7.6   Online monitoring pioneers

Almost all the noise analysis work for the aforementioned online monitoring (OLM) applications was derived from the work of the late Dr Joseph A. Thie (1943–2023), who was known as the father of the nuclear noise analysis and a renowned expert in nuclear reactor surveillance and diagnostics [1,177,178]. Although he was one of the first nuclear engineers and scientists who introduced the use of noise analysis to the nuclear industry, many others worked in this area in the United States, Europe, and elsewhere [179–181]. In particular, the lead editor of the present book contributed substantially to this field and published extensively [6,7]. The author of this chapter worked with both Dr Thie and Professor Pázsit in different capacities. In fact, Dr Thie and Dr Hashemian worked together at AMS for nearly 40 years and Professor Pázsit served as the doctoral advisor of Dr Hashemian at Chalmers University of Technology in Sweden, where Dr Hashemian earned his PhD degree in nuclear engineering in 2009.

In the 1990s, AMS began to apply the noise analysis technique in nuclear power plants, commercialising the work of Dr Thie and Dr Pázsit. Subsequently, the International Atomic Energy Agency (IAEA) and the International Electrotechnical Commission (IEC) developed new standards and guidelines for the use of noise analysis techniques in nuclear power plants [182–184]. In doing this, Dr Oszvald Glöckler and Dr Gabor Por, both from Hungary but working internationally, helped with IAEA activities in this area [185,186].

7.7   Online calibration monitoring

Encouraged by the success of the noise analysis technique for online measurement of dynamic performance of sensors and systems in nuclear plants, in the 1990s, the nuclear industry began to explore the use of OLM for tracking the static behaviour of sensors and systems [187,188]. This endeavour involved the Electric Power Research Institute (EPRI), AMS, universities, and national laboratories in the United States and similar research and development (R&D) organisations in other countries. After more than two decades of R&D, OLM found its way in the Sizewell B nuclear power plant for verifying the calibration of pressure, level, and flow transmitters. The Sizewell B plant in the United Kingdom is a Westinghouse PWR near London, England and the first western nuclear plant to have a digital plant protection system I&C with a complete analogue backup I&C system. As a result, Sizewell B has more pressure, level, and flow sensors than other PWR plants and was ripe for OLM implementation [189,190].

Figure 7.14 shows the fundamentals of the OLM technique for monitoring the calibration of sensors. The advantages of OLM to monitor sensor calibrations are: 1) OLM data are already available from plant computers and can be easily retrieved from plant data historians or other existing means in the plant, and 2) there is no need for any modification of plant equipment to implement OLM; all that is needed is a plot of data over the duration of the fuel cycle to identify drift. If drift is identified, then the affected transmitter is calibrated. Otherwise, the transmitter is allowed to

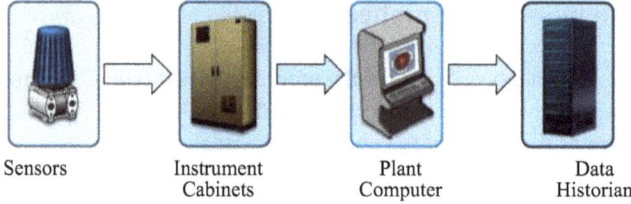

| Sensors | Instrument Cabinets | Plant Computer | Data Historian |

Figure 7.14   OLM data retrieval process

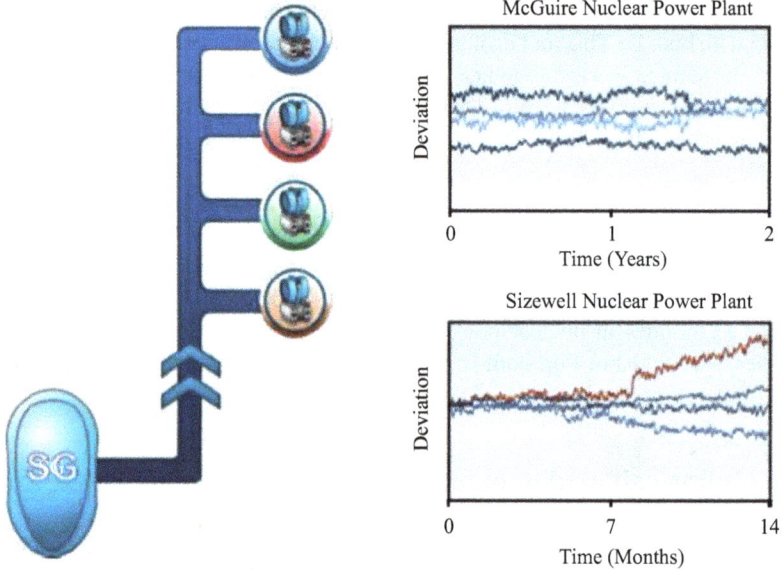

Figure 7.15   OLM data for transmitter calibration verification

continue its service for another fuel cycle. A typical fuel cycle is 14 to 24 months depending on the plant. Figure 7.15 shows results of OLM implementation to verify transmitter calibration in two different nuclear power plants: one in the United States by the name of McGuire Nuclear Power Station and the other being the Sizewell B Nuclear Plant. It is apparent that none of the four transmitters at McGuire plant drifted during the two years of operation and do not therefore need to be calibrated. In contrast, one of the transmitters at Sizewell B plant began to drift during the 14 months of monitoring. Obviously, this transmitter had to be calibrated and the other three left to operate for another 14 month fuel cycle.

7.8   OLM implementation process

To implement OLM for transmitter calibration verification, the output of pressure transmitters is monitored as the plant is operating to look for drift. If the sensor has drifted, then it is scheduled for calibration. Otherwise, the sensor is not touched until after the end of next cycle. To verify the calibration of a sensor over its entire operating range, OLM data is used from periods of startup, shutdown, and normal operation (Figure 7.16). The reference for calibration monitoring is the average of redundant sensors. Therefore, the method requires at least two-way redundancy to allow for the averaging of redundant sensors to provide a reference for calibration monitoring. Fortunately, most safety-related sensors in nuclear power plants are two or more ways redundant; therefore, OLM applies to nearly 100% of safety-related sensors in nuclear power plants. For non-safety related sensors that are usually on the secondary side of the plant and are not redundant, OLM depends on process modelling to establish the reference for calibration monitoring [191,192]. Today, most of the attention on OLM is on its application for safety-related sensors that are in the reactor containment or otherwise difficult to reach for calibration.

With OLM, plants identify drifting transmitters and calibrate only those that have drifted beyond acceptable limits as determined using plant set point data. Based on experience with OLM implementation at Sizewell B and the history of transmitter calibration drift, it has been determined that less than 10% of nuclear-grade pressure, level and flow transmitters need calibration over a single operating cycle of 14 to 24 months. In fact, it has been determined that some transmitters can operate for over 20 years without drifting enough to need a calibration. The 90% in calibration savings that nuclear plants gain from OLM can amount to over a million dollars per each two-year cycle for a single plant unit. This by itself is huge, but there are more indirect benefits when unnecessary calibrations are avoided in terms of inadvertent plant trips, damage to plant equipment, and reduction in radiation exposure to plant personnel during calibrations, not to mention human error and their consequences.

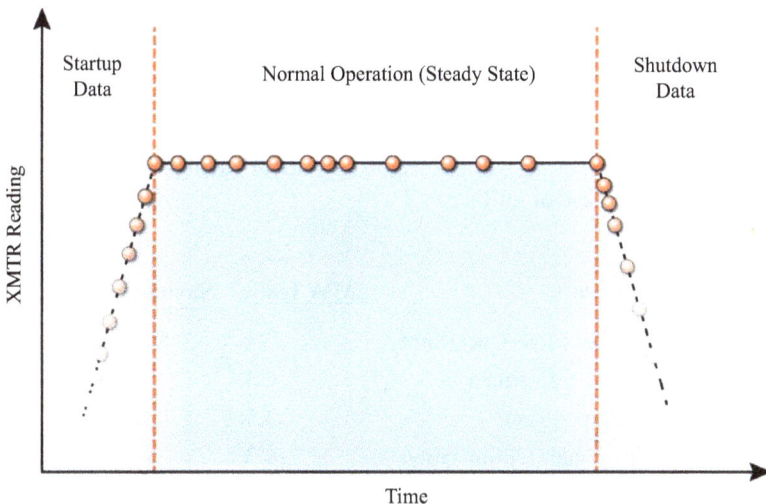

Figure 7.16   OLM data collected over all modes of plant operation

7.9   Regulatory approval of OLM

As a result of its success at Sizewell B, nuclear executives have stated that OLM contributes to "nuclear safety", which has incentivised nuclear regulators to approve OLM for the nuclear industry. In fact, British nuclear regulators approved OLM in 2006 for the Sizewell B plant in the United Kingdom, and the US Nuclear Regulatory Commission (NRC) approved OLM in 2021 for use in the US nuclear fleet. The NRC approval of OLM resulted from a review of a topical report (TR) that AMS wrote with funding from U.S. Department of Energy (DoE) [193]. DoE was enticed to fund this work for its potential to save costs for the nuclear industry while contributing to the safety, reliability and efficiency of nuclear power plants. With the NRC approval of the TR, nuclear plants are allowed to switch from time-based calibration of transmitters to condition-based calculations. Today, Plant Vogtle in the US has fully implemented OLM and has already declared substantial savings from the first use of OLM. Plant Vogtle is a Westinghouse PWR located in the state of Georgia in the United States.

In addition to an NRC approved TR, plants implementing OLM must each receive NRC approval to switch from time-based calibration practice to a condition-based calibration strategy. This approval is needed because the plant technical specifications must be modified to migrate from traditional time-based calibrations to condition-based calibrations using OLM. In the US, each plant must submit a License Amendment Request (LAR) to the NRC for approval to implement OLM. Today, the NRC has approved a LAR for Vogtle, and applications are under preparation for almost 50 other LAR approvals for more plants to embark on OLM implementation.

7.10   OLM benefits

OLM has many applications beyond sensor calibration monitoring. However, these applications are pending and are expected to take hold after OLM implementation projects for a few plants are in full use for transmitter calibration monitoring. For example, Sizewell B engineers produced the information in the following table to show the megawatt recovery and corresponding dollar savings that can potentially result from expanded use of OLM.

Applications	MW Gain	Saving/Year
Feed Water Flow Uncertainty	24	$14.0 M
Secondary Calorimetric	4	$2.3 M
Feed Temperature	3.5	$2.0 M
Cooling Water Optimisation	3	$1.8 M
Optimise Steam Drainage	2	$1.2 M
Potential Savings per Annum		**$21.3 M**

7.11   OLM for generation IV reactors

In addition to the existing generation of nuclear plants, OLM is slated to be built in the design of the next generation of reactors including water-cooled SMRs and non-water-cooled advanced modular reactors (AMRs) [194–196]. These reactors are expected to rely on autonomous operation and automated maintenance making OLM a necessity in their design.

7.12   Conclusion

This chapter reviewed the application of in-situ and OLM techniques for sensor response time testing, sensor calibration verification and anomaly detection in nuclear power plants.

Chapter 8

Worldwide activities in design and planned diagnostics of SMRs

Belle Upadhyaya[1]

8.1 Introduction

The primary objective of this chapter is to outline the development of small modular reactors (SMRs) with a focus on instrumentation and monitoring approaches being developed for efficient operation. The development and deployment of nuclear reactors are driven by the following factors as identified by Goldberg and Rosner [197]: *cost effectiveness, safety, security and nonproliferation, grid appropriateness, commercialisation roadmap, and the fuel cycle.* The design and development of SMRs must incorporate these features.

The following topics are presented in this chapter:

- An introduction to SMRs. Various stages of nuclear plant development, leading to SMRs as Generation IV (Gen-IV) plants. Types of SMR designs and vendors.
- SMR systems. Light water, gas-cooled reactors (GCRs), molten salt reactors MSRs), microreactors.
- Description of a typical SMR.
- Challenges in SMR instrumentation, built-in devices during manufacturing.
- General approaches for SMR monitoring. Use of edge computing and autonomous control design.

8.2 Phases of nuclear reactor development

Gen-I reactors were built in the 1950s and early 1960s. The focus of design and construction was on pressurised water reactors (PWRs). Large commercial reactors built in the U.S. and other countries in the 1960s and 1990s are referred to as Gen-II reactors. These include PWRs, boiling water reactors (BWRs), pressurised heavy water reactors (CANDU), GCRs, and SFRs.

The next generation reactors incorporate passive safety, standardised design, longer refuelling intervals, increased fuel enrichment, use of MOX fuels, and reduced construction time. These are the Gen-III (and Gen-III+) reactors. Examples include

[1]Department of Nuclear Engineering, The University of Tennessee, USA

advanced BWR (Japan) and AP-1000 (Westinghouse Electric Company), advanced CANDU reactor, APR-1000 (Korean advanced PWR), VVER-1200 (Russia) and the 1650 MWe European power reactor (EPR).

Gen-IV reactors [198] have new fuel cycles and operate at high temperatures. High-temperature GCRs, MSRs, and sodium fast reactors (SFRs) are some of the examples of Gen-IV design and development. High temperature operation enables higher efficiency and can be used for hydrogen production for industrial applications. SMRs cut across Gen-III and Gen-IV reactor systems with the features of semi-autonomous operation and walk-away safety systems.

8.3   SMR designs and reactor vendors

As mentioned in the last section, SMR designs include light water reactors (LWRs), GCRs, MSRs and other design types. Some of the major designs and reactor vendors are the following:

- NuScale Power: This has a basic 77 MWe natural circulation PWR module. The plant is marketed as 4-, 6-, or 12-module system.
- Holtec International design is a 160 MWe SMR.
- X-energy: Xe-100 is an 80 MWe modular high-temperature gas-cooled reactor (HTGR) planned for construction at Dow Chemical site, Texas.
- Kairos Power design consists of two 35 MWth fluoride salt-cooled pebble bed reactors. This Hermes 2 system (KP-FHR) has been approved for construction by the U.S. NRC. A 35-MWth Hermes 1 test reactor, a nonnuclear engineering test unit, has been approved for construction in Oak Ridge, Tennessee. This is the first Gen-IV design approved for construction in the US. The two-module Hermes 2 design will share a single Rankine cycle for steam generation [199].
- GE Hitachi: BWRX-300 is a 300 MWe BWR.
- The HTR-PM is a Gen-IV HTGR pebble-bed reactor, the world's first commercial-size power plant of this kind designed and built in China. It began commercial operation in December 2023. The two-module system has a single steam turbine – generator system, each module with a capacity of 250 MWth.
- OKBM-Russia: RITM-200M is a 50 MWe marine-based PWR.

"Project Pele Microreactor is a high temperature gas cooled reactor (HTGR) that can produce 1–5 MWe of power. The reactor is a collaboration between the U.S. Department of Defense and several other government agencies, including the Department of Energy, NRC, and NASA. The goal of the project is to create a small, transportable nuclear reactor that can be used in remote military locations, for civilian applications, or even in space exploration. It is designed for operation at full power for a minimum of three years. The reactor will be designed and assembled by BWXT" [200].

Some unique features of this microreactor would include remote and near-autonomous operation, "walk-away" safety, and safeguard from external threats.

8.4   Gen-IV reactor development

Gen-IV reactors use advanced fuel types with uranium enrichment just below 20%. This fuel type is referred to as high assay low enrichment uranium (HALEU) which would increase the power production capacity per unit volume and help extend the refuelling interval. One of the newer fuels is called the TRistructural ISOtropic (TRISO) fuel consisting of fuel kernels less than 1 mm diameter. Each kernel acts like its own containment system. Thousands of these fuel particles are packed together either in a cylindrical form or a pebble of approximately 60 mm in diameter (similar to a billiard ball). Several companies are involved in developing and fabricating TRISO fuel for use in advanced reactors.

HTGRs, liquid metal reactors and MSRs fall in the Gen-IV reactor category. The high temperature of the reactor coolant increases the overall plant efficiency and is used for industrial hydrogen production and for district heating. The typical HTGR temperature is as high as 750 °C both in microreactors and in SMRs. The molten salt temperature in a fluoride high-temperature reactor is as high as 750°C. The reactor pressure is almost atmospheric, thus avoiding the need for high-pressure containment. Liquid metal reactors such as the SFR and the lead fast reactor (LFR) have coolant exit temperatures around 550 °C and the system pressure is low with a cover gas. TerraPower is one of the companies involved in the development of an SFR.

8.5   Challenges to instrumentation and monitoring of SMRs

Since the overall size (diameter and height) of an SMR is somewhat limited, it is a challenge to install instrumentation systems needed for control and reactor safety. These include the following:

- Sensor placement (limited in-vessel space).
- How to get the signal and control rod drive mechanism (CRDM) cables out of the reactor vessel?
- Addressing refuelling issues.
- How to measure primary coolant flow rate?
- What is the best location for ex-vessel neutron detectors? What is the field of view?
- What are the challenges for placement of in-core neutron detectors?
- Placement of core inlet and outlet temperature detectors (needed for control actions).

Figure 8.1 shows a four-loop PWR and a typical SMR, an integral reactor. Large commercial reactors have piping that carry the reactor coolant into and out of the reactor vessel. This facilitates the installation of various sensors, such as temperature, pressure, flow, level, etc., without space restriction. Also, the neutron detectors can be placed both in-core and in ex-vessel locations. As illustrated in Figure 8.1, the entire primary and secondary loops may have to fit inside the reactor vessel.

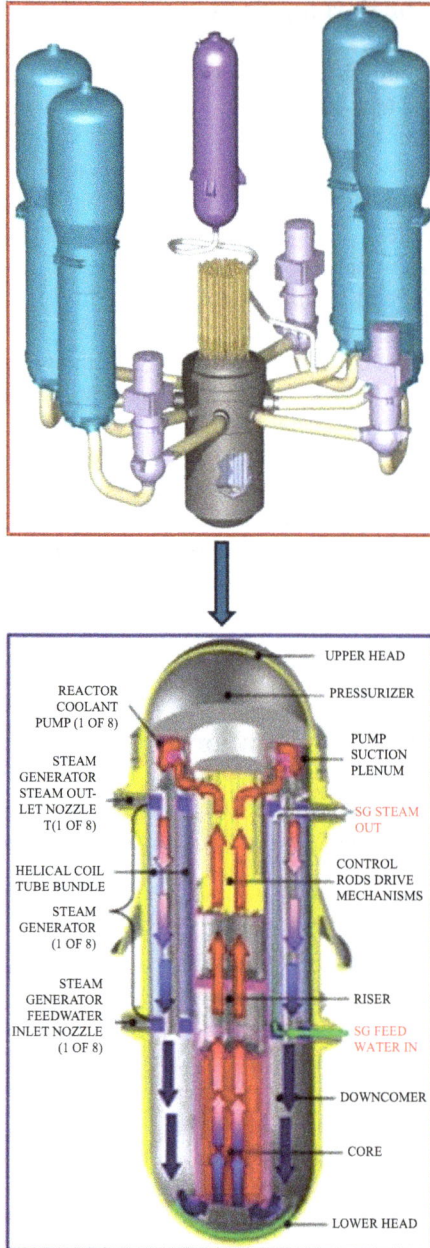

Figure 8.1   Comparison of a four-loop PWR and an integral reactor (such as an SMR). Figure 8.1a (top) is courtesy of ARIVA; Figure 8.1b (bottom) is courtesy of Westinghouse Electric Co.

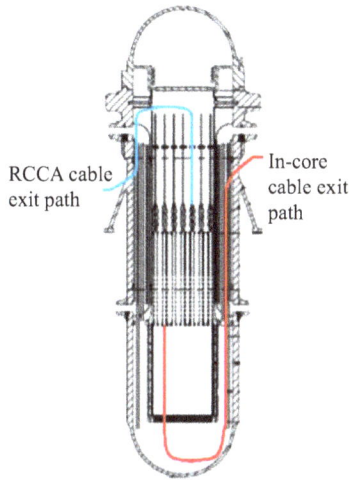

Figure 8.2 In-core detectors, CRDMs and cable routers [6]. RCCA: Reactor Control Cluster Assembly

An SMR vessel must be designed to integrate heat exchangers/steam generators, reactor coolant pumps, instrumentation and in some cases a residual heat removal system. These are some of the challenges faced in the design of SMRs. In Gen-IV reactor designs, the steam generator is external to the reactor vessel. The challenge is routing instrumentation and CRDM cables. This latter issue is illustrated in Figure 8.2 for an integral reactor system.

8.6 Development of instrumentation for SMRs

The challenges posed in the design and installation of instrument systems for SMRs are summarised in Section 8.4. Of specific interest is the design and deployment of flow sensors. Two examples of flow sensor design for light water SMRs are described.

8.6.1 Flow rate measurement using motor power analysis

Centrifugal pumps are driven by induction motors of various power ratings. As the pump flow rate changes, the load on the electric motor changes and the power drawn by the motor changes proportionately. The pump hydraulic power is given by the relationship

$$\text{Pump Hydraulic Power } (kW) = \frac{(Q.H.SG) \cdot 9.81}{3600} \tag{8.1}$$

Q is the pump flow rate and H is the developed pump head. Q and H are related, and thus the power drawn by the motor, $P_M \approx f(Q)$. This shows that a relationship between motor power and fluid flow rate exists and can be used for flow measurement.

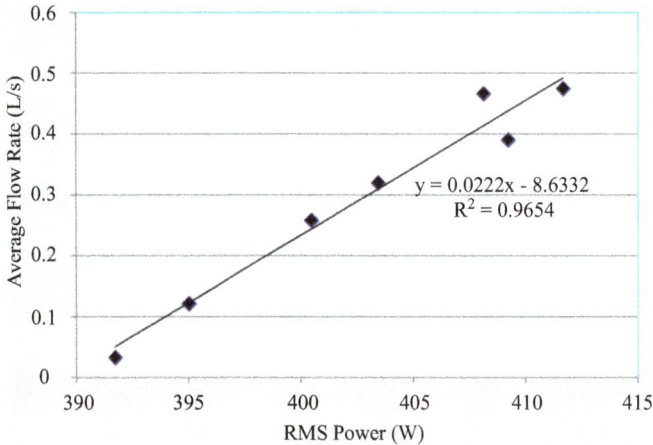

Figure 8.3   Relationship between motor power and pump discharge rate in an experimental flow control loop [201]. In general, a linear relationship may not hold

Note that this approach for measurement is an indirect technique and must be properly validated and calibrated for each pump-motor system.

A fully instrumented experimental flow control loop was used to develop relationships among motor power, flow rate, pump discharge pressure, and tank water level. This can be exploited for remote condition monitoring [201]. Results show a strong relationship between motor power- flow rate and motor power-pump discharge pressure. An example of this relationship is shown in Figure 8.3. This is a plot of the pump discharge rate as a function of the RMS motor power. The plot in Figure 8.3 indicates a systematic relationship between pump flow rate and the measured electric signature (motor power). A linear approximation to the experimental data is shown. Note that this relationship, in general, is not linear.

8.6.2   Reflection transit time ultrasonic flow meter

A transit time ultrasonic flow meter operates by emitting ultrasonic pulses between two sets of transceivers. A transceiver is both a transmitter and a receiver of ultrasonic pulses. One transceiver sends pulses against the flow (upstream) while the other sends pulses in the direction of the flow (downstream). The flow velocity is then determined by measuring the difference between the times of flight of the two pulses. The advantages of an ultrasonic flow meter are no pipe penetrations, no obstruction to coolant flow, multiple sensors can be installed to develop a liquid flow profile in the vessel, and the device can maintain measurement accuracy despite fouling of coolant conduit [202].

Transit time ultrasonic flow meters can be very useful for this application because they do not require a fully developed flow. There is no need for penetrating the reactor vessel to implement this device.

The ultrasonic flow meters are configured vertically along the reactor vessel wall, directed toward the reactor core. In this arrangement, the transceivers generate pulses that penetrate the vessel wall, cross the flow, reflect off the core barrel, and return to the other transceiver. This is a common configuration for ultrasonic flow meters that use wave reflection.

Figure 8.4 shows the configuration of the transit time reflection ultrasonic flow meter. The two transceivers are placed on the external vessel surface at a known distance apart. An estimate of the flow velocity is given by

$$V_{flow} = \frac{L\,\Delta t}{2\sin(\theta)\,t_{down}\,t_{up}} \tag{8.2}$$

where

L :   ultrasonic path length in the coolant
t_{down}:   transit time of the downstream pulse
t_{up}:   transit time of the upstream pulse
$\Delta t = t_{up} - t_{down}$
θ: angle of the reflected pulse

Because of the attenuation of the ultrasound pulses, it is necessary to amplify the received signals. The flow velocity measurement is independent of fluid properties such as pressure, temperature, Reynolds number, etc. Because of this, the ultrasonic flow meter requires minimum calibration after the initial installation [202].

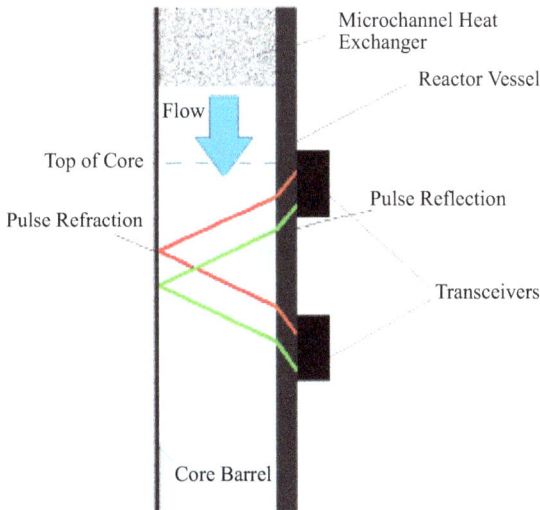

Figure 8.4   *Configuration of the reflection transit time ultrasonic flow meter. The red path indicates pulse flow in the upstream direction and the green path indicates pulse flow in the downstream direction [203]*

8.6.3   *Flow rate estimation using signal cross-correlation technique*

An indirect approach for flow rate estimation is to determine the transit time of the coolant in the flow channel. This technique was applied for flow velocity estimation in PWRs, BWRs and other situations where a direct measurement using instrumentation such as venturi meter or orifice meter is not practical. In-core neutron detectors are installed in operating PWRs and BWRs [204]. As the water flows up through the reactor core, the fluctuations in the neutron detector measurement are caused by temperature fluctuations and/or by fluid density changes. These changes are transmitted from the upstream detector location to the downstream detector location. By knowing the time of transit from one detector to the adjacent detector and the distance between the detectors, an average coolant flow rate can be estimated.

Figure 8.5 shows a configuration of two detectors placed in a fluid flow path. The direction of flow is from detector 1 to detector 2.

$X_1(t)$ and $X_2(t)$ are responses of two detectors and the measurements are assumed to be stationary over a period, T. The cross-correlation between the two measurements as a function of lag time τ, is given by

$$R_{X_1 X_2} = \frac{1}{T} \int_0^\infty X_1(t) X_2(t + \tau) \, d\tau \qquad (8.3)$$

The two detector signals are assumed to be related by a transport delay time D:

$$X_2(t) = c X_1(t - D) \qquad (8.4)$$

where c is a constant. By using (8.4), the cross-correlation between $X_1(t)$ and $X_2(t)$ is given by

$$R_{X_1 X_2} = \frac{1}{T} \int_0^\infty X_1(t) \, c X_1(t - D + \tau) \, d\tau = c R_{X_1 X_1}(\tau - D) \qquad (8.5)$$

This cross-correlation has a maximum at a lag time, D. Thus, by calculating the cross-correlation between two signals, the lag time at which the cross-correlation has a maximum value gives an estimate of the transport time D, for the fluid to flow from one detector to the next. This cross-correlation property is illustrated in Figure 8.6.

Figure 8.5   Location of two similar detectors in the fluid flow path

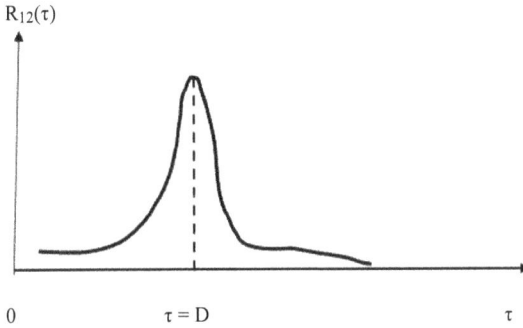

*Figure 8.6   Behaviour of cross-correlation between two signals with a dominating
transport delay-type dynamics*

If the detectors are placed a distance L apart, the average velocity v of the fluid is estimated as $v_{avg} = L/D$. The coolant mass flow rate is then calculated by knowing its density at the operating temperature and pressure. This technique was successfully applied to operating LWRs [204]. One or more of these flow monitoring techniques may be used depending on the reactor design.

8.6.4   Flowrate measurement in liquid metal reactors

Liquid metal reactor systems operate under high temperature and the coolant is chemically reactive. Liquid metal flowrate measurement in these reactors requires special design considerations. The flowmeters must be able to withstand high gamma radiation levels and thermal transients.

The following two types of flowmeters are often considered for use in commercial SFRs: (a) permanent magnet flowmeter (direct current), (b) eddy current flowmeter (alternating current). The permanent magnet flowmeters are used in some commercial SFRs, such as the BN350 to measure the primary pump flowrate [205]. These flowmeters have the disadvantage of causing damage to the magnetic components due to high temperature, radiation levels, and possible mechanical shocks.

The eddy current flowmeter has the advantage of smaller size than a magnetic flowmeter, can be implemented in a constrained space, and it is resistant to high temperature and radiation fields [205,206]. This flowmeter is an alternating current electromagnetic flow meter. The sodium flow rate is proportional to the difference between the voltages induced in the upstream and downstream coils. These are two secondary coils on either side of a primary coil that is supplied with a high frequency alternating current [206]. The coils are wound around a steel bobbin.

The eddy current flowmeter is compact in size and is more reliable in high temperature and radiation environment than the magnetic flowmeter. It is implemented in the 470 MWe Prototype Fast Breeder Reactor (PFBR) [205].

8.7   Molten salt small modular reactors

8.7.1   Types of molten salt reactors (MSRs)

The development of molten salt nuclear reactors has been in the forefront in recent days. These are generally classified as molten salt cooled reactors and fluid fuel MSRs. The following short description has been taken from Reference [207].

The molten salt reactor experiment (MSRE) was developed and operated at Oak Ridge National Laboratory (ORNL). It was operated from June 1965 through December 1969, the first of its kind, the MSRE operated at a maximum power level of 8 MWth [208]. The reactor core was composed of a matrix of rectangular graphite blocks for neutron moderation. The molten, fuel-bearing carrier salt of ^{7}LiF-BeF$_2$-ZrF$_4$-UF$_4$ at 632 °C was pumped through the core. The fission reaction increased the core-exit salt temperature to $\approx$ 655°C. The fuel salt was then circulated through an intermediate heat exchanger where the heat from the fuelled salt was transferred to a non-fuelled secondary salt (^{7}LiF-BeF$_2$) and then returned to the reactor vessel. The basic fluoride salt is often referred to as FLiBe. The thermal power generated in the reactor was calculated using the salt flow rate, difference between the outlet and inlet salt temperatures, and the specific heat capacity of salt at constant pressure.

It is important to note that the MSRE was operated with U-235 fuel (33% enrichment) and separately with U-233 fuel (91% enrichment) and achieved criticality in both cases. This suggests the use of Th-232 as the fertile material in a thermal breeder reactor for conversion to U-233. Thus, MSRs can also be designed to operate as breeder reactors using the Thorium cycle. For details of MSR dynamic modelling and molten salt breeder reactor systems review references [207,209,210].

The alternate form of the MSR is the salt-cooled reactor with solid fuel. This design is being developed by *Kairos Power* and has seen fast advancement in the commercialisation of the MSR-SMR, which is a Gen-IV reactor.

8.7.2   Hermes fluoride high-temperature reactor

The company, Kairos Power, has designed and developed a solid fuel fluoride salt-cooled high-temperature reactor (KP-FHR). Hermes 1 is a 35 MWth test reactor slated for construction and operation in Oak Ridge, U.S.A., by 2027 [211]. This is a non-nuclear system developed for testing the various components and systems of the prototype Hermes 2 reactor to be deployed as the first Gen-IV advanced reactor later in the decade. The Hermes design uses solid fuel pebbles consisting of TRISO nuclear fuel. The reactor consists of a graphite moderator and reflector. Hermes 2 is a two-unit 35 MWth system with a total electricity generation of $\approx$ 20 MWe. The coolant is ^{7}LiF-BeF$_2$ liquid salt, often called FLiBe. FLiBe remains in liquid form at high temperatures and has a heat capacity higher than that of water. The reactor operates at low pressures, less than 0.2 MPa (2 atmos.) with a cover gas of argon to capture any tritium that may be released from the molten salt coolant [212]. The core-exit coolant temperature is $\approx$ 650°C. The total coolant flow rate is $\approx$ 3000 gallons/min. The reactor is estimated to generate about 39000 spent fuel pebbles per year.

Figure 8.7 is a schematic of the Hermes reactor system, courtesy of Kairos Power [211]. Details of the reactor vessel, fuel and coolant flow are shown in Figure 8.8. The Hermes 2 reactor is a promising design and development of a Gen-IV SMR.

Figure 8.9 shows a schematic of a dual unit SMR [213]. The steam from the two modules is mixed in a steam header and then flows into single steam turbine. This approach reduces the plant equipment requirements and provides economy of operation and maintenance.

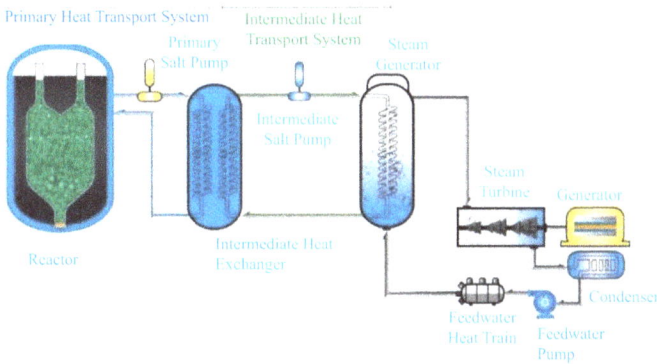

Figure 8.7   Schematic of the Kairos Power Hermes reactor [211] (courtesy of Kairos Power)

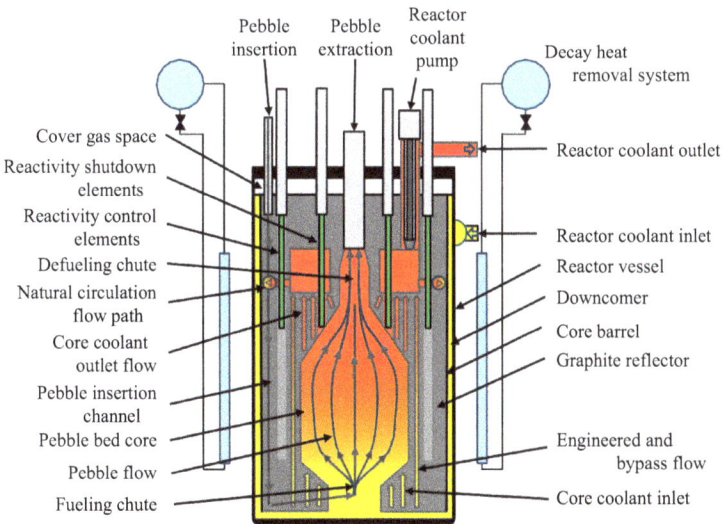

Figure 8.8   Details of Hermes reactor vessel, fuel and coolant flow systems [211] (courtesy of Kairos Power)

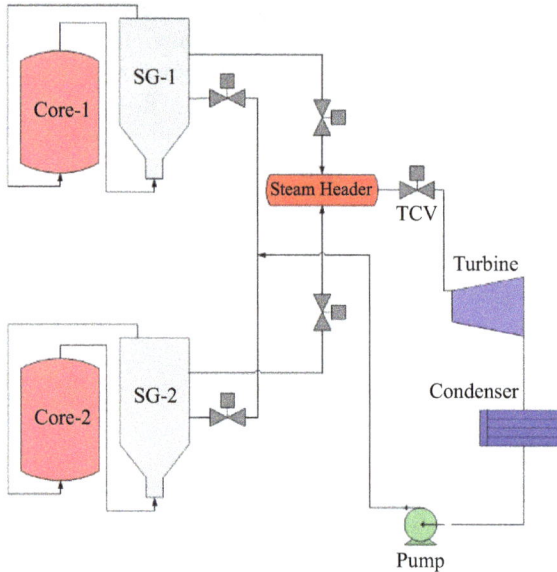

Figure 8.9   Dual module SMR power block [213]

8.8   Advances in reactor monitoring and cutting edge technologies

Reactor system instrumentation and monitoring technologies are progressing at a high rate. The following are some of the advancements that should be incorporated into new SMR designs.

8.8.1   Embedded sensors

It is prudent to consider embedded sensors for economic construction, labour reduction, and minimising retrofitting. Embedded temperature, pressure and neutron detectors can be evaluated for their effectiveness in a test reactor. Process sensors and monitoring sensors are installed in manufacturing systems and for monitoring heavy machinery such as turbines, generators, compressors, fans and others.

8.8.2   Edge computing

In a large industrial system, measurements from various devices are acquired continuously. In many cases it is necessary to make decisions about the status of the process or equipment at the site. Edge computing refers to processing the data at the equipment level and providing the necessary diagnostics information to the operator in real time. Edge computing is facilitated by sensors installed in the equipment that provide continuous measurements to edge computing devices. Edge computing is enabled by performing detailed computing in cloud-based devices and minor computing at the equipment level [214].

8.8.3   Cutting edge monitoring and diagnostics

Process and equipment monitoring technologies have advanced to the point where the operator decision-making is made easy by the intelligence embedded in these devices. An example of such a device is the industrial acoustic imaging camera developed by FLIR called the FLIR Si2. This device consists of 124 microphones and detects leakage (gas, steam, etc.), partial discharge and machinery noise caused by malfunctions. The locations of the defects in a large facility are captured visually that enables further detailed diagnostics of the cause of leak or other malfunctions. This is one such example of advanced technology that can be deployed in an SMR plant, without radiation exposure to the operator.

8.8.4   Motion amplification technology

The Motion Amplification® technology uses digital video and image processing techniques to measure vibration and movement, pixel by pixel, extracting and scaling up data to produce a visual representation of movement [215].

RDI Technologies, Knoxville, Tennessee, USA, has developed and implemented the advanced IRIS-M [215] equipment for industrial applications. A large area of manufacturing or processing plants can be monitored from a distance and movement of equipment can be tracked continuously. In addition to the visual monitoring of the equipment motion/vibration, a complete spectral analysis of the system can be made, and various faults can be identified.

Chapter 9
Conclusions and future directions

Imre Pázsit[1]

In this book, a collection of ideas was presented about the possible use of noise analysis methods in next generation nuclear systems, and whenever it was possible, some quantitative estimates were also given about the expected efficiency and accuracy of the problems and methods investigated. The task was certainly difficult, and the material presented can by no means be considered as complete and covering all aspects. One obvious reason for this is the fact that most of the next generation systems are still in the design phase, and hence the concrete data of these cores are either not available or are not public. Even more so, there has been no operational experience with these systems which, among others, means that the list of possible operational concerns or malfunctions is rather incomplete.

The only pragmatic way of handling the situation was to try to extrapolate from the experience with currently operating reactors and to check the possibilities of diagnostics and surveillance with the differing material and geometrical properties of these new systems, as well as extending the existing methods to suit the characteristic of these systems (such as developing an analytical model for fast reactors in two-group theory).

Some of the recently developed new methods are not specifically for next generation systems. Examples are the ICFM-quality noise simulators described in Chapter 5; the increasing use of machine learning methods in solving the inverse task of noise source unfolding; or the use of time-resolved detector current instead of pulse counting for zero-power reactor diagnostics (reactivity measurements). The inclusion of these into the book felt completely logical. First, these will be used in next generation systems from the beginning, whereas these are still very moderately used in the current reactors, if at all. Second, the description of these is not yet found in existing monographs. Conversely, all the extended methods described here can also be applied to recent Gen-II and Gen-III reactors, which can be considered as a subset of all reactor types. Therefore, the book should also be of some use not only for those interested in next generation reactors but also for practitioners of current systems.

The conceptual study of the dynamic properties of the various representative next generation systems was performed in this book with simple analytical tools; only

[1]Division of Subatomic, High Energy and Plasma Physics, Department of Physics, Chalmers University of Technology, Sweden

the group constants of the two-group model were based on full Serpent models and homogenisation methods. The principles of noise simulators, i.e. system codes capable of treating real inhomogeneous cores of both thermal and fast systems, described in Chapter 5, were not used yet for a similar analysis. Here lies one of the envisaged lines of further development. Most of the noise simulators so far were developed for specifically thermal light water systems and solid fuel, using traditional two-group diffusion theory. Further development of these codes for fast systems, and both thermal and fast spectrum MSRs is expected, especially as the design of next generation systems becomes more detailed and publicly available. This refers also to the instrumentation of the next generation systems, which is largely undetermined yet.

Another area where more efforts need to be spent is the representation of propagating perturbations in terms of cross-section fluctuations. As long as only one of the cross sections is affected by the perturbation, such as vibrations of an absorber or fuel rod, the situation is simple, one only needs a model of the spatial and temporal variation of the geometry of the perturbation. For the propagating perturbations, a temperature or density variation in the coolant of liquid metal-cooled cores or MSRs affects several cross sections simultaneously. The correct representation of the ratio of the cross-section perturbations needs to be known for a correct estimation of the induced neutron noise. In this book, only qualified guesses were made based on the character of the perturbation and the specific properties of the cores based on the ratios of the static cross sections of the homogenised model. This procedure should be refined by using high-precision static assembly-level transport codes, the same ones that are used to generate the group constants, to also generate the perturbation of the group constants, such as was done in [33] for traditional cores.

The usefulness and limits of the validity of the two-group theory, suggested in this book for fast systems and which was used to analyse those, also needs to be investigated and confirmed. This requires the establishing of a link between the analytical two-group methods elaborated for fast cores, and the noise simulators under development for fast systems, which use a multi-group approach (e.g. 33 energy groups). These codes are also capable to calculate the response of detectors with various spectral sensitivities. Such a comparative work could also find an optimum energy threshold between the two groups, which need to be used in the analytical model for best results. This work is yet to be done.

In this book, emphasis was put on the detection and identification of various possible in-core disturbances, which is one of the prime objectives of surveillance and diagnostics. However, as mentioned in Section 2.3.8, subsections 2.3.8.1 and 2.3.8.2, noise analysis is equally useful in determining and monitoring stability properties and reactivity coefficients with non-intrusive methods during operation. Prominent examples of these in the current reactor are the stability issues of BWRs and the determination of the moderator temperature coefficient (MTC) of pressurised water reactors.

Stability issues might also occur in both water-cooled and lead-cooled Gen-IV small modular reactors (SMRs), which are planned to use natural circulation. While maintaining coolant circulations without pumps is advantageous in handling emergency situations, such as in the case of station blackout and dropout of external power supply, stability properties are as a rule worse in the case of natural circulation, especially in the case of reduced flow, than with forced circulation. Some studies along these lines will be useful.

Reactivity coefficients of various types might also be interesting to monitor in Gen-IV systems and SMRs, such as the void coefficient in SFRs. As the experience with the MTC showed, the possibility of accurately determining the reactivity coefficient depends on both the transfer properties of the system for the variations of the parameter in question, as well as on the spatial and temporal correlation properties of the parameter whose reactivity coefficient needs to be determined and the possibility of measuring these latter. The performance of such noise-based methods in determining the relevant reactivity coefficients in concrete designs can be effectively investigated by noise simulators. Work is expected to be done in this direction in the future.

One option of further development concerns the inclusion of the calculation of gamma noise in the noise simulators. Power reactor diagnostics with gamma measurements is expected to gain importance in next generation of nuclear reactors. It would be therefore useful to extend the capability of the currently developed noise simulators to account also for the transfer between the in-core perturbations and the induced gamma noise. For different reasons and with a different underlying physics, calculations of the gamma flux are already included in the static ICFM codes. The principles of the gamma noise induced by cross-section fluctuations, as described in Section 2.3.7, are clear, and the implementation should be straightforward.

The application of machine learning methods in nuclear engineering applications is predicted to grow steadily. Integration of deep learning and convolutional artificial neural networks into online monitoring and surveillance systems will certainly be part of future development.

In addition to the lack of operational experience with the new systems and that of the details of the core design, one further challenge is the lack of knowledge of the planned instrumentation. The type of analysis made in this book, and in particular the foreseen ones with the use of noise simulators, could yield valuable added information to the design of instrumentation of the planned systems. The efficiency of diagnostics of future systems could be largely enhanced if designers utilised the capabilities of the methods described here for planning the instrumentation, such as optimising the location of the various detectors within the existing constraints.

To conclude, advanced noise analysis is expected to play an important role for next generation nuclear systems. The recently developed new powerful methods, some of them originally designed for current reactors but adopted for Gen-IV systems and SMRs, will hopefully find their place in the surveillance and diagnostics of next generation nuclear systems as they become deployed.

Appendix A

A.1 Detailed description and user manual to the Notebook "Two-group power reactor noise calculations" [10]

A.1.1 Summary

All calculations and plots in Chapters 3 and 4, regarding the Gen-IV reactors and SMRs, respectively, were made by a Wolfram Mathematica Notebook [10]. A streamlined and user friendly version of the notebook is uploaded to the Wolfram Notebook Archive, from where it is freely accessible and downloadable through the link

https://notebookarchive.org/2025-05-7dypmeq

Interested readers who do not have access to Mathematica, can run the downloaded Notebook with the open access software Wolfram Player,* available for free from

https://www.wolfram.com/player/

Alternatively, the Notebook can be run directly online in the Wolfram Cloud. To do this one needs to have an account, for which one can sign up for free at

https://www.wolframcloud.com/

With the notebook, all calculations reported in the book can be repeated and the plots reconstructed. Moreover, the code in the Notebook can be changed, which makes it possible to make calculations with changed input parameters, i.e, study the same systems as in the book but with some parameters changed, or make calculations for new systems, or just change the format of the plots of the existing calculations. One particular option is to change the size of the core, to check how the dynamic properties of the system transfer change if a specific full-size Gen-IV core was converted into an SMR or vice versa.

In this Appendix, partly a simple user manual is given, and partly the work flow of the Notebook is described and is related to the corresponding formulae in the book.

---

*The concept of the Player is the same as behind the Adobe pdf reader, a free software to read pdf files. The reason why it is called Wolfram Player and not Wolfram Reader is that with it, one can also run the interactive animations and manipulations of a notebook.

A.1.2 *The input data*

The calculations in the Notebook need a set of input data, which the Notebook reads from external sources, they are not stored in the Notebook. These data are the two-group diffusion coefficients and macroscopic cross sections, the fission spectrum parameters χ_1 and χ_2 for both the prompt and delayed neutrons, the group velocities, and the effective delayed neutron fractions and decay constants for one averaged delayed neutron group. In addition, the radius R and the height H of the core, as well as the type of the reactor needs to be specified. The type is basically the same as the name of the reactor, except for Allegro, for which it is GFR.

For each reactor these data are stored in a JSON (JavaScript Object Notation) file. The data format is shown next for the file "SFR.json".

```
{
"description": "SFR based on the OECD NEA benchmark,
this is the 3600MWth oxide core",
"Type": "SFR",
"H": 100.55,
"R": 256.47675,
"D1": 2.66184,
"D2": 1.24986,
"nuSig1": 0.0217533,
"nuSig2": 0.00582776,
"Siga1": 0.00809302,
"Siga2": 0.00585308,
"SigR1": 0.0484388,
"SigR2": 0.0058526,
"Sig1": 0.186146,
"Sig2": 0.31168,
"beta0": 0.00447966,
"lambda": 0.600723,
"v1": 2107130741.1410954,
"v2": 238608815.16406745,
"Chi1": 0.595249,
"Chi2": 0.404751,
"Chi1p": 0.597823,
"Chi2p": 0.402177,
"Chi1d": 0.0417677,
"Chi2d": 0.958232,
"Eg": 1.35
}
```

These files are available from the open access source Mendeley Data [12] and from a public GitHub repository. The link to the public GitHub repository is available in the Notebook. Contributions with data to further systems in the format given before here are endorsed to be uploaded to this repository by the interested readers.

The Notebook reads the .json file of the selected reactor from the GitHub repository into the file "data". The variables used in the code are then assigned from that file in the next step.

There is one peculiarity worth mentioning, namely that the Serpent calculation does not yield the traditional removal cross section Σ_R directly, rather it is embedded into the variable SigR1, which is the sum of the absorption and the removal cross sections, i.e. $\Sigma_R = \text{SigR1} - \text{Siga1}$.

These data are used in the calculations, which will be described next.

A.1.3   Executing the Notebook

Using the Notebook in its default version, i.e. executing the instructions without changing either the input data or the plot formatting options, needs only a few clicks. After opening the code, its content is shown as in Figure A.1.

The content and function of the various sections is as follows.

Abstract

Contains a short description, not executable.

Preamble

Sets the Notebook directory, the date, and defines a few new functions of the Arg command that are used in the calculations. These are needed to make sure that the correct physical phase delay is shown (i.e. that the phase is always negative), in contrast to the Arg command itself, whose results are between $-180°$ and $180°$. The

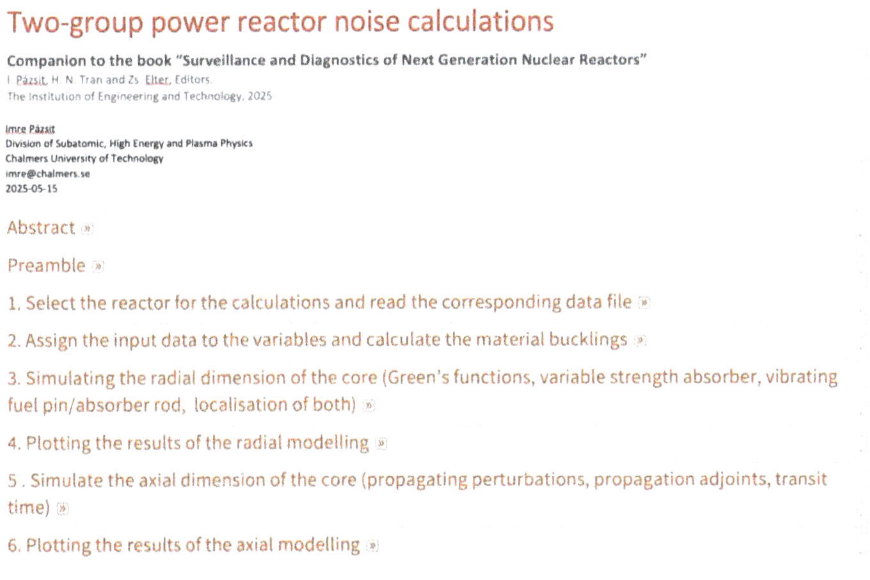

Two-group power reactor noise calculations

Companion to the book "Surveillance and Diagnostics of Next Generation Nuclear Reactors"
I. Pázsit, H. N. Tran and Zs. Elter, Editors
The Institution of Engineering and Technology, 2025

Imre Pázsit
Division of Subatomic, High Energy and Plasma Physics
Chalmers University of Technology
imre@chalmers.se
2025-05-15

Abstract »

Preamble »

1. Select the reactor for the calculations and read the corresponding data file »

2. Assign the input data to the variables and calculate the material bucklings »

3. Simulating the radial dimension of the core (Green's functions, variable strength absorber, vibrating fuel pin/absorber rod, localisation of both) »

4. Plotting the results of the radial modelling »

5. Simulate the axial dimension of the core (propagating perturbations, propagation adjoints, transit time) »

6. Plotting the results of the axial modelling »

Figure A.1   The first page of the Notebook, showing the table of contents of the code

Preamble has to be executed only once at the beginning, which can be done even without opening the corresponding section.

1. Select the reactor for the calculations and read the corresponding data file

Opening this section displays a drop-down menu, by which any of the eight reactor types (three Gen-IV and five small modular reactors) can be selected for calculations. The page is illustrated in Figure A.2. After having selected the core, the corresponding data are read from the JSON file on GitHub to the variable "data".

2. Assign the input data to the variables and calculate the material bucklings

This is a short section, where the variables used in the Notebook are assigned from the file "data", and the material bucklings, i.e. the two roots of the characteristic equation are calculated according to (2.107) or its shortened form (2.111). At this point only the global root B_0^2 is interesting, which is given by (2.112).

In addition, the value of certain parameters, are adjusted or defined, which depend of the type of the core. For the two MSRs, i.e. the MSDR and the MSFR, the parameter β is downscaled by a factor 0.4, for reasons explained in Section 2.3.6.2. Also, an index variable i is defined, depending on the type of the core. This variable is used when selecting the weight parameters α_1 and α_2 (as e.g. in (2.158)) of the propagation adjoints, (2.162) and (2.163).

This section can also be executed without opening it. On execution, it also plots the static fluxes, but these will be seen only if the section is opened.

Choose between the following : SFR, Allegro, MSDR, PWRSMR, LWR SMR, MSFR, LFRSMR or FHR
========================
SFR: Sodium fast reactor based on the OECD NEA benchmark, 3600 MWth oxide core
Allegro: A small Allegro core (but it counts as Gen-IV)
MSDR: Molten Salt Demonstrator Reactor, large thermal MSR

PWRSMR: Small PWR, from half-length Westinghouse fuel elements
LWRSMR: Advanced light water SMR, inspired by the Rolls Royce SMR
LFRSMR: Lead cooled fast reactor SMR
MSFR: Molten Salt Fast Reactor, inspired by Samofar
FHR: Fluoride Salt-cooled High-temperature Reactor (Hermes). Thermal core.

fileName SFR

Selected file: SFR.json
Data preview:
<| <<1>> |>

Figure A.2   Section No. 1 of the Notebook, showing the page for the selection of the reactor type

After this step, one can continue to two different options. One can either model the core in the horizontal dimension, which is the one to select when the noise by a variable strength absorber and a vibrating fuel pin or absorber rod is calculated (Section No. 3 of the Notebook), or model the core in the axial dimension (Section No. 5), when the effect of the propagating perturbation is calculated. Because in either case one turns to a 1D model, to maintain criticality of the original 3D system, one has to adjust the absorption cross sections, such that the 1D system remains critical with the selected radial or axial dimensions. This is made automatically by the code, no user action is required.

Since the diameter of most of the systems is different from their heights, this adjustment leads to different cross sections in the two different options. This also means that although for a given system, one can freely select whether one wants to model the radial or the axial dimension first, after having performed the calculations in one dimension and turning to the calculations in the other dimension, one has to run Section No. 2 again, to restore the original value of the 3D cross sections.

3. Simulate the radial dimension of the core

This is a long computational section, which can nevertheless be executed without opening it. Actually, in this section, no computations proper are performed, only the formulae for the various functions (components of the Green's function, the noise by variable strength absorber and of a vibrating fuel rod, etc.), are defined as functions of their parameters (space and frequency). The actual evaluations take place in the subsequent Section No. 4, where the plotting is performed. Hence in this section "calculation" means generating the appropriate formula for later evaluation.

To start with, the static absorption cross sections are adjusted to make the 1D system critical with the given radius, such that the slab thickness of the 1D model is equal to $2R$. Then the two roots of the characteristic equation of the static (critical) core are calculated. This is followed by the definition of the frequency-dependent cross sections and spectral parameters, (2.115)–(2.119), after which the two frequency-dependent complex valued roots $\mu(\omega)$ and $\nu(\omega)$ of the dynamic equation (2.114) are calculated.

Next, the step-wise continuous basic Green's functions $g_\mu(x, x_0, \omega)$ and $g_\nu(x, x_0, \omega)$ are calculated from (2.89) and (2.90), as well as their derivatives w.r.t. x_0. The formulae for the frequency-dependent coupling coefficients $c_\mu(\omega)$ and $c_\nu(\omega)$, (2.124) and (2.125), as well as the coefficients $A_{\mu 1}(\omega)$, $A_{\nu 1}(\omega)$, $A_{\mu 2}(\omega)$ and $A_{\nu 2}(\omega)$, (2.92)–(2.96) are also generated. The notations of the coefficients and the basic Green's functions $g_\mu(x, x_0, \omega)$ and $g_\nu(x, x_0, \omega)$, as well as the Green's function elements $G_{ij}(x, x_0, \omega)$ are the same in the Notebook as in the book (and as given here). The spatial derivatives w.r.t. x_0 are denoted by adding $x0$ after g and G, i.e. $gx0_\mu(x, x_0, \omega)$ and $Gx0_{11}(x, x_0, \omega)$, etc.

In possession of the components of the Green's function and its derivatives w.r.t. x_0, the noise induced by a variable strength absorber and by a vibrating fuel pin or absorber rod is calculated. The selection of whether a vibrating fuel or absorber pin is performed is chosen by the program itself, according to the type of the reactor. For

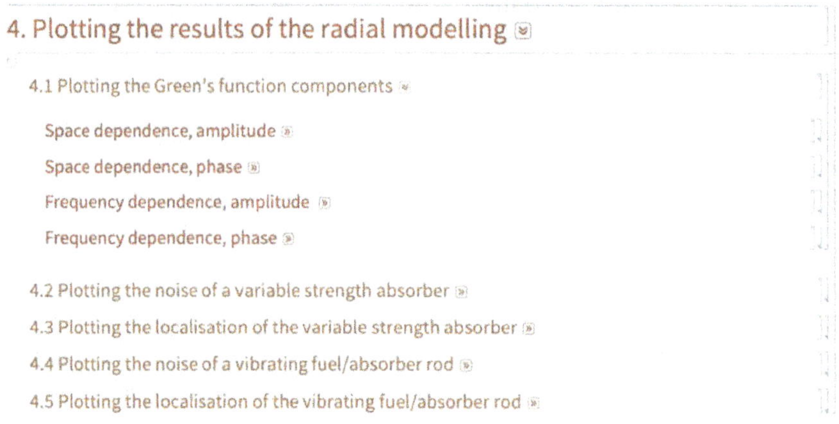

4. Plotting the results of the radial modelling ⊗

4.1 Plotting the Green's function components ⊗

Space dependence, amplitude ⊗

Space dependence, phase ⊗

Frequency dependence, amplitude ⊗

Frequency dependence, phase ⊗

4.2 Plotting the noise of a variable strength absorber ⊗

4.3 Plotting the localisation of the variable strength absorber ⊗

4.4 Plotting the noise of a vibrating fuel/absorber rod ⊗

4.5 Plotting the localisation of the vibrating fuel/absorber rod ⊗

Figure A.3   The section page of the Notebook, showing the plotting options

MSRs, a vibrating absorber is selected, because there are no fuel pins which could vibrate. For the other reactor types, the noise by a vibrating fuel pin is calculated.

Once the formulae for the noise by the variable strength absorber and the vibrating fuel/absorber pin or rod are available, one can also calculate the localisation curves $\Delta_i(x_p)$ and $\theta_i(x_p)$, (2.138) and (2.139), which is the last step in this section.

The formulae generated in Section No. 3 are used in the next section to plot the results.

4. Plotting the results of the radial modelling

In this section, the amplitude and the phase of the space and frequency dependence of the Green's functions components can be calculated for fixed values of the other parameters, space dependence of the noise, and dependence of the localisation curves on the position x_p of the perturbation can be plotted.

Opening this section, the options available are shown in Figure A.3. For illustration, option 4.1, plotting the Green's functions, is also open. The other numbered subsections also contain such options. Each option contains only one single plot command. Some plotting parameters need to be adjusted for the best plot, since these should be different for the different reactors. To help the user, these parameters are defined at the beginning of the command and are in red colour. A full list of these with explanations are given at the beginning of Section No. 4 of the Notebook.

5. Simulate the axial dimension of the core

This section starts the same way as the radial/horizontal simulation, namely finding the absorption cross sections to make the system critical with the given core height. For this, Section No. 2 needs to be rerun first, to restore the original 3D values of the cross sections, from which the new absorption cross sections, making the 1D system critical with the height of the system as the thickness of the slab. The calculation of the new values of the absorption cross sections is made at the beginning of the section.

After that all calculations leading to the Green's function components with the new cross-section values are performed in an identical way as in Section No. 3. As mentioned, this can be done also independently from the modelling of the radial dimension, i.e. one can directly jump to Section No. 5 after Section No. 2.

First the propagation adjoints in the two groups are calculated, as one of the quantities that can be plotted in Section No. 6. To calculate the neutron noise induced by the propagating perturbation, the integrals of the propagation adjoints need to be calculated. This integral, similarly to how the spatial derivatives of the Green's function components are made, boils down to the calculation of the integrals of the piecewise continuous basic functions $g_\mu(x, x_0, \omega)$ and $g_\nu(x, x_0, \omega)$ multiplied by the perturbation represented by the propagating perturbation with respect to x_0.

Since in this bare homogeneous system $\phi_2(x) = c_\mu \, \phi_1(x)$, as is also seen from (2.160) and (2.161), it suffices to calculate the integral of the piecewise continuous basic functions as

$$g_{\mu,\nu}int(z, \omega) = \int_0^H g_{\mu,\nu}(z, z_0, \omega) \, \phi_1(z_0) \, e^{-\frac{i\omega z_0}{u}} \tag{A.1}$$

Here in the notations, we have switched to the axial coordinate system, as is usual with propagating perturbations. It is also to be noted that in the notebook, in the calculations the velocity of the propagating perturbation is denoted by u, in order not to lead to conflict with the group velocities v_1 and v_2, but in the figure legends, it is designated as v.

The integrals in (A.1), which have to be taken over piecewise continuous functions, consisting of triple products of trigonometric and complex exponential function, can be performed analytically, but they lead to rather lengthy formulae. These integrals were performed off-line with Mathematica, and the resulting formulae for $g_{\mu,\nu}int(z, \omega)$ are implemented and used in the Notebook, with a similar notation.

With the aforementioned, the formulae for the calculation and plotting of the propagation adjoints, and the formulae for the neutron noise in the two groups are generated. This section ends here. The dependence of the phase between two axially displaced neutron detectors is calculated in the next section, where it also can be plotted.

6. Plotting the results of the axial modelling

This section has a structure similar to that of Section No. 4, but with considerably fewer options, only three, namely to plot either the amplitude or the phase of the propagation adjoints, or the frequency dependence of the phase of the cross-spectrum between two axially displaced detectors. The plotting of the propagation adjoints is completely in line with that of the Green's function elements, in that they can be plotted with different parameters, since the formulae are evaluated in the plot command.

Plotting the phase of the CPSD between two detectors is a considerably more complicated task. The main complication arises from the fact that from physical expectations, and in order to be able to fit a straight line to the phase to determine the transit time of propagation between the two detectors, the phase must be a monotonically decreasing function of the frequency. On the other hand, the routines in every code, including Mathematica, give the argument of a complex number between $-180°$ and $+180°$, i.e. the phase is "folded back" to $+180°$ when it decreased below $-180°$.

To prevent the code to do so is solved in a somewhat ad-hoc way in the notebook. The phase is calculated in discrete points, and if the jump of the phase between a point and two consecutive points both exceeds a threshold, $360°$ is subtracted in the continuation. This happens every time when the criteria for a too big jump are fulfilled.

Finding the proper threshold appears to require some trial and error. It depends on both the reactor type, and the propagation velocity of the coolant. The latter needs to be adjusted to the height of the core to have realistic data. In addition, in some cases the phase in the fast group starts from $-360°$, which needs to be compensated for.

Due to the discretisation of the phase and the aforementioned procedure, the phase of the cross section needs to be calculated in connection of the plotting, and some parameters need to be adjusted to get the proper plot. The following adjustable parameters can be tuned to get the proper plot:

- **u**: the velocity of the propagation
- **threshold**: the threshold, deciding when the phase need to be shifted $-360°$ to avoid the discontinuity, Its value should be between 100 and 359.
- **offset**: a parameter taking the value 0 or 1. In the latter case, $+360°$ will be added to the phase of the fast noise CPSD, to avoid that it starts at $-360°$.

There are default values in the code for all reactors for the aforementioned tunable parameters in the variables **udef**, **thresdef** and **offdef**. These are specified in Section No. 2, which needs to be run before running sections 3 and 5 of the Notebook (see Figure A.1).

During the calculation and plotting of the frequency dependence of the phase, some error message about "Range" may occur, and the plot will not be continuous, rather it will be still folded back when the phase reaches $-360°$. This should not occur with the default values defined in the Notebook. When such an error message occurs, often it is sufficient to rerun the plot command. If it does not help, then one can experiment with changing the threshold. In some cases it is advisable to quit Mathematica to reset all variables, and start the calculations anew.

Bibliography

[1]   Thie JA. *Reactor Noise*. La Grange Park, Illinois, USA: American Nuclear Society; 1963.

[2]   Uhrig RE, editor. *Noise Analysis in Nuclear Systems*. U.S. Atomic Energy Commission, Washington, DC: Division of Technical Information; 1964. AEC Symposium Series No. 4.

[3]   Uhrig RE, editor. *Neutron Noise, Waves, and Pulse Propagation*. U.S. Atomic Energy Commission, Washington, DC: Division of Technical Information; 1967. USAEC, CONF-660206.

[4]   Williams MMR. *Random Processes in Nuclear Reactors*. Oxford: Pergamon Press; 1974.

[5]   Thie JA. *Power Reactor Noise*. La Grange Park, Illinois, USA: American Nuclear Society; 1981.

[6]   Pázsit I and Pál L. *Neutron Fluctuations: A Treatise on the Physics of Branching Processes*. New York: Elsevier; 2008.

[7]   Pázsit I and Demazière C. Noise techniques in nuclear systems. In: Cacuci DG, editor. Vol. 2 of *Handbook of Nuclear Engineering*. Springer Science; 2010. pp. 1631–1737.

[8]   Pázsit I. Symmetries and asymmetries in branching processes [Electronic Article]. *Symmetry*. 2023;**15**(6):1154.

[9]   Wolfram Research, Inc. Mathematica, Version 14.2; 2024. Champaign, IL.

[10] Pázsit I. Two-group power reactor noise calculations. From the Wolfram Notebook Archive; 2025. Available from: https://notebookarchive.org/2025-05-7dypmeq.

[11] Wolfram Foundation. The Notebook Archive; 2025. Accessed: 2025-05-21. Available from: https://www.notebookarchive.org/.

[12] Elter Zs and Pázsit I. Simulated homogenized two-group cross sections of advanced nuclear reactor designs for noise analysis. Mendeley Data, V1. 2025;DOI: 10.17632/k9cm3m22vs.1. Available from: https://data.mendeley.com/datasets/k9cm3m22vs/1.

[13] Sjöstrand NG. Measurements on a subcritical reactor using a pulsed neutron source. Arkiv för Fysik. 1956;**11**(13):233–246.

[14] Williams MMR. Note on relationship between variance and correlation function; 2024. Personal communication.

[15] Pázsit I. Duality in transport theory. *Annals of Nuclear Energy*. 1987;**14**(1):25–41.

[16] Pázsit I and Chakarova R. Variance and correlations in sputtering and defect distributions. *Transport Theory and Statistical Physics*. 1997;**26**(1–2):1–25.

[17]   Pázsit I. *Ringhals Diagnostics and Monitoring: An overview of 30 years of collaboration 1993–2023.* Göteborg, Sweden: Chalmers University of Technology; 2023. CTH-NT-350/RR-27. Available from: https//research.chalmers.se/publication/537867/file/537867_Fulltext.pdf.

[18]   Saito K. Source papers in reactor noise. *Progress in Nuclear Energy.* 1979;**3**(3):157–218.

[19]   Kosály G. Noise investigations in boiling-water and pressurized-water reactors. *Progress in Nuclear Energy.* 1980;**5**:145–199.

[20]   Pázsit I and Glöckler O. On the neutron noise diagnostics of pressurized water reactor control rod vibrations III. Application at a power plant. *Nuclear Science and Engineering.* 1988;**99**(4):313–328.

[21]   Karlsson JK-H and Pázsit I. Localisation of a channel instability in the Forsmark-1 boiling water reactor. *Annals of Nuclear Energy.* 1999;**26**(13):1183–1204.

[22]   Demazière C, Vinai P, Hursin M, *et al.* Overview of the CORTEX Project. In: *Proceedings of the PHYSOR 2018.* Cancun, Mexico; 2018.

[23]   Demazière C, editor. Neutron noise-based core diagnostics and monitoring of nuclear reactors: Outcomes of the Horizon 2020 CORTEX project. *Annals of Nuclear Energy.* 2025;**212**. Special issue on the CORTEX project.

[24]   Pázsit I and Kitamura M. The role of neural networks in reactor diagnostics and control. In: Lewins J and Becker M, editors. Vol. 24 of *Advances in Nuclear Science and Technology.* Boston, MA: Springer US; 1996. pp. 95–130.

[25]   Pázsit I and Pál L. Multiplicity theory beyond the point model. *Annals of Nuclear Energy.* 2021;**154**:108119.

[26]   Pázsit I, Dykin V and Darby F. Space-dependent calculation of the multiplicity moments for shells with the inclusion of scattering. *Nuclear Science and Engineering.* 2023;**197**:2030.

[27]   Pázsit I, Garis NS and Glöckler O. On the neutron noise diagnostics of pressurized water reactor control rod vibrations – IV: Application of neural networks. *Nuclear Science and Engineering.* 1996;**124**(1):167–177.

[28]   Meem JL. *Two Group Reactor Theory.* New York: Gordon and Breach; 1964.

[29]   Pázsit I. Neutron noise theory in the P_1 approximation. *Progress in Nuclear Energy.* 2002;**40**(2):217–236.

[30]   Henry AF. The application of reactor kinetics to the analysis of experiments. *Nuclear Science and Engineering.* 1958;**3**:52–70.

[31]   Henry AF. *Nuclear Reactor Analysis.* The MIT Press; 1975.

[32]   Weinberg AM and Schweinler HC. Theory of oscillating absorber in a chain reactor. *Physical Review.* 1948;**74**(8):851–863.

[33]   Dykin V, Jonsson A and Pázsit I. Qualitative and quantitative investigation of the propagation noise in various reactor systems. *Progress in Nuclear Energy.* 2014;**70**:98–111.

[34]   van Dam H. A perturbation method for analysis of detector response to parametric fluctuations in reactors. *Atomkernenergie.* 1975;**25**:70.

[35] van Dam H. Neutron noise in boiling water reactors. *Atomkernenergie*. 1975;**27**:8.

[36] Pázsit I. Dynamic transfer function calculations for core diagnostics. *Annals of Nuclear Energy*. 1992;**19**(5):303–312.

[37] Pázsit I. Hugo van Dam and the dynamic adjoint function. *Annals of Nuclear Energy*. 2003;**30**(17):1757–1775.

[38] Behringer K, Kosály G and Pázsit I. Linear response of the neutron field to a propagating perturbation of moderator density (2-group theory of boiling water-reactor noise). *Nuclear Science and Engineering*. 1979;**72**(3): 304–321.

[39] Dykin V and Pázsit I. The molten salt reactor point-kinetic component of neutron noise in two-group diffusion theory. *Nuclear Technology*. 2016;**193**(3):404–415.

[40] Hansson PT and Foulke LR. Investigations in spatial reactor kinetics. *Nuclear Science and Engineering*. 1963;**17**(4):528–533.

[41] Bell G and Glasstone S. Nuclear Reactor Theory. New York, USA: Van Nostrand Reinhold Company; 1970.

[42] Guo H, Li W, Zhang J, *et al.* Analysis of flow-induced vibration of wire-wrapped fuel assemblies under the liquid metal axial flow in the Gen-IV nuclear reactor. *Annals of Nuclear Energy*. 2023;**188**:109811.

[43] Eswaran M, Sajish SD and Natesan K. Investigation of flow-induced vibration in core sub-assemblies of sodium-cooled fast reactors. *Progress in Nuclear Energy*. 2025;**185**:105780.

[44] Pázsit I. Investigation of the space-dependent noise induced by a vibrating absorber. *Atomkernenergie*. 1977;**30**(1):29–35.

[45] Pázsit I. 2-Group theory of noise in reflected reactors with application to vibrating absorbers. *Annals of Nuclear Energy*. 1978;**5**(5):185–196.

[46] Jonsson A, Tran HN, Dykin V, *et al.* Analytical investigation of the properties of the neutron noise induced by vibrating absorber and fuel rods. *Kerntechnik*. 2012;**77**(5):371–380.

[47] Wach D and Kosály G. Investigation of the joint effect of local and global driving sources in incore-neutron noise measurements. *Atomkernenergie*. 1974;**23**:244–250.

[48] Kosály G. Investigation of the local component of power-reactor noise via diffusion theory. Central Research Institute for Physics, Budapest; 1975. KFKI-75-27.

[49] Adorján F, Czibók T, Kiss S, *et al.* Core asymmetry evaluation using static measurements and neutron noise analysis. *Annals of Nuclear Energy*. 2000;**27**(7):649–658.

[50] Czibók T, Kiss G, Kiss S, *et al.* Regular neutron noise diagnostics measurements at the Hungarian Paks NPP. *Progress in Nuclear Energy*. 2003;**43** (1–4):67–74.

[51] Saraswat A, Fraile A, Gedupudi S, *et al.* A comprehensive review of experimental and numerical studies on liquid metal-gas two-phase flows and associated measurement challenges. *Annals of Nuclear Energy*. 2025;**213**:111104.

[52]   Nishihara H and Konishi H. A new correlation method for transit-time estimation. *Progress in Nuclear Energy*. 1977;**1**(2):219–229.

[53]   Kosály G, Maróti L and Meskó' L. A simple space dependent theory of the neutron noise in a boiling water reactor. *Annals of Nuclear Energy*. 1975;**2**(2):315–321.

[54]   Loberg J, Österlund M, Blomgren J, *et al.* Neutron detection-based void monitoring in boiling water reactors. *Nuclear Science and Engineering*. 2010;**164**(1):69–79.

[55]   Dykin V and Pázsit I. Simulation of in-core neutron noise measurements for axial void profile reconstruction in boiling water reactors. *Nuclear Technology*. 2013;**183**(3):354–366.

[56]   Pázsit I, Torres LA, Hursin M, *et al.* Development of a new method to determine the axial void velocity profile in BWRs from measurements of the in-core neutron noise. *Progress in Nuclear Energy*. 2021;**138**:103805.

[57]   Pázsit I and Glöckler O. On the neutron noise diagnostics of pressurized water-reactor control rod vibrations. II. Stochastic vibrations. *Nuclear Science and Engineering*. 1984;**88**(1):77–87.

[58]   Uhrig RE. Potential application of neural networks to the operation of nuclear power plants. *Nuclear Safety*. 1991;**32**(1):68–79.

[59]   Van Der Hagen THJJ. Artificial neural networks versus conventional methods for boiling water reactor stability monitoring. *Nuclear Technology*. 1995;**109**(2):286–305.

[60]   Garis NS, Pázsit I, Sandberg U, *et al.* Determination of PWR control rod position by core physics and neural network methods. *Nuclear Technology*. 1998; **123**(3):278–295.

[61]   Sunde C, Avdic S and Pázsit I. Classification of two-phase flow regimes via image analysis by a neuro-wavelet approach. *Applied Computational Intelligence*. 2004;236–239. https://doi.org/10.1142/9789812702661_0045

[62]   Dulla S. Models and methods in the neutronics of fluid fuel reactors; 2005. PhD thesis, Graduate School of Nuclear Engineering University of Turin.

[63]   Pázsit I and Jonsson A. Reactor kinetics, dynamic response and neutron noise in molten salt reactors (MSR). *Nuclear Science and Engineering*. 2011;**167**:61 –76.

[64]   Jonsson A and Pázsit I. Two-group theory of neutron noise in molten salt reactors. *Annals of Nuclear Energy*. 2011;**38**(6):1238–1251.

[65]   Pázsit I, Jonsson A and Pál L. Analytical solutions of the molten salt reactor equations. *Annals of Nuclear Energy*. 2012;**50**:206–214.

[66]   Wang J and Cao X. Characters of neutron noise in full-size molten salt reactor. *Annals of Nuclear Energy*. 2015;**81**:179–187.

[67]   Wang J and Cao X. Applicability of adiabatic approximation on neutron noise analysis in molten salt reactor. *Annals of Nuclear Energy*. 2016;**92**:295–303.

[68]   Carter JP and Arcilesi Jr DJ. Investigation of neutron noise in a micro-scale, natural circulation molten salt fission battery system. *Nuclear Engineering and Design*. 2021;**383**:111437.

[69]   Ozgener HA and Ozgener B. Numerical and semi-analytical solution of a transient during the start-up of a slab reactor with circulating fuel at zero power. *Annals of Nuclear Energy*. 2021;**153**:108080.

[70]   Pázsit I and Dykin V. Chapter 5 – Kinetics, dynamics, and neutron noise in stationary molten salt reactors. In: Dolan TJ, Pázsit I, Rykhlevskii A, *et al.*, editors. *Molten Salt Reactors and Thorium Energy*. Woodhead Publishing; 2024. pp. 199–261.

[71]   Bureš L. Comparison of analytical models for frequency response and linear stability of molten salt reactors. *Nuclear Science and Engineering*; 1–24. https://doi.org/10.1080/00295639.2025.2460318.

[72]   Lapenta G and Ravetto P. Basic Reactor Physics Problems in Fluid-Fuel Recirculated Reactors. *Kerntechnik*. 2000;**65**(5–6):250–253.

[73]   Gelinas RJ and Osborn RK. Reactor Noise Analysis by Photon Observation. *Nuclear Science and Engineering*. 1966;**24**(2):184–192.

[74]   Kenney ES. Noise analysis of nuclear reactors with the use of gamma radiation. In: Uhrig RE, editor. *Neutron Noise, Waves, and Pulse Propagation*. Springfield, Virginia: USAEC, CONF-660206; 1967. pp. 399–411.

[75]   Kenney ES and Schultz MA. Local in-core power measurements with out-of-core gamma detectors. In: Preprints, *Japan–United States Seminar on Nuclear Reactor Noise Analysis*. Tokyo and Kyoto, Japan; 1968. pp. 103–122.

[76]   Preprints, *Japan–United States Seminar on Nuclear Reactor Noise Analysis*; 1968.

[77]   Osborn RK. Gamma-ray fluctuation measurements versus neutron fluctuation measurements. In: *Preprints, Japan–United States Seminar on Nuclear Reactor Noise Analysis*. Tokyo and Kyoto, Japan; 1968. pp. 25–33.

[78]   Jammes C, Filliatre P, Elter Zs, *et al.* Progress in the development of the neutron flux monitoring system of the French GEN-IV SFR: Simulations and experimental validations. *Journal of Nuclear Science and Technology*. 2025;**62**(4):123–145.

[79]   Kostic L and Seifritz W. The theory of space dependent reactor noise analysis using gamma radiation. *Journal of Nuclear Energy*. 1971;**25**(12):637–655.

[80]   van Dam H and Kleiss EBJ. Response of incore γ-detectors to parametric fluctuations in a reactor core. *Annals of Nuclear Energy*. 1985;**12**(4): 201–207.

[81]   Behringer K and Nishihara H. The field of view of a γ-sensitive incore detector in BWRs. *Annals of Nuclear Energy*. 1985;**12**(1):1–7.

[82]   Thie JA. Core motion monitoring. *Nuclear Technology*. 1979;**45**(1):5–45.

[83]   Fontaine B, Prulhière G, Vasile A, *et al.* Description and preliminary results of PHENIX core flowering test. *Nuclear Engineering and Design*. 2011;**241**(10):4143–4151.

[84]   Zylbersztejn F, Tran HN, Pázsit I, *et al.* Calculation of the neutron noise induced by periodic deformations of a large sodium-cooled fast reactor core. *Nuclear Science and Engineering*. 2014;**177**(2):203–218.

[85]   Thie JA. Boiling water reactor instability. *Nucleonics*. 1958;**16**(3):102–111.

[86]   Thie JA. *Dynamic Behavior of Boiling Reactors*. Argonne National Lab., Lemont, Ill.; 1959. ANL-5849.

[87]   D'Auria F, Ambrosini W and Anegawa T. State-of-the-art report on BWR stability. OECD-CSNI Report, Paris, France; 1997. OECD/GD (97) 13.

[88]   Lefvert T. *BWR Stability Benchmark, Final Specifications*. NEA, OECD Publishing, Paris; 1994. NSC/DOC(96)22.

[89]   Lu Q and Rizwan u. Stability analysis of nuclear-coupled thermal hydraulics for a natural circulation lead-cooled fast reactor. *Annals of Nuclear Energy*. 2020;**149**:107747.

[90]   Demazière C and Pázsit I. Theoretical investigation of the MTC noise estimate in 1-D homogeneous systems. *Annals of Nuclear Energy*. 2002;**29**(1):75–100.

[91]   Demazière C. Development of a noise-based method for the determination of the moderator temperature coefficient of reactivity (MTC) in pressurized water reactors (PWRs) [PhD thesis]. Chalmers University of Technology; 2002.

[92]   Leppänen J, Valtavirta V, Rintala A, *et al.* Status of Serpent Monte Carlo code in 2024. *European Physical Journal: Nuclear Sciences & Technologies*. 2025;**11**:3.

[93]   OECD Nuclear Energy Agency. *Benchmark for Neutronic Analysis of Sodium-cooled Fast Reactor Cores with Various Fuel Types and Core Sizes*. OECD NEA; 2015. NEA/NSC/R(2015)9.

[94]   Conti A, Gerschenfeld A, Gorsse Y, *et al.* Numerical analysis of core thermal-hydraulic for sodium-cooled fast reactors. In: *NURETH 2015 – 16th International Topical Meeting on Nuclear Reactor Thermal Hydraulics*. Chicago, United States; 2015.

[95]   Bělovský L, Gadó J, Hatala B, *et al.* The ALLEGRO Experimental Gas Cooled Fast Reactor Project. International Atomic Energy Agency; 2017. IAEA-CN–245. Available from: https://inis.iaea.org/records/6tw2q-9qy93.

[96]   *Advanced Three-Dimensional Two-Group Reactor Analysis Code, The User's Manual for SIMULATE-3*. USA; 2001.

[97]   Allibert M, Delpech S, Gerardin D, *et al.* Chapter 7 – Homogeneous molten salt reactors (MSRs): The molten salt fast reactor (MSFR) concept. In: Pioro IL, editor. *Handbook of Generation IV Nuclear Reactors* (Second Edition). Woodhead Publishing; 2023. pp. 231–257.

[98]   Vu TM, Bui TH and Tran LQL. Feasibility study on the application of boron carbide for long-term reactivity control in the LOTUS small fast reactor. *Journal of Nuclear Engineering and Radiation Science*. 2024;**10**(2):021503.

[99]   Satvat N, Sarikurt F, Johnson K, *et al.* Neutronics, thermal-hydraulics, and multi-physics benchmark models for a generic pebble-bed fluoride-salt-cooled high temperature reactor (FHR). *Nuclear Engineering and Design*. 2021;**384**:111461.

[100]   Duchnowski EM, Satvat N and Brown NR. Neutronic and thermal-hydraulic calculations under steady state and transient conditions for a generic pebble

bed fluoride salt cooled high-temperature reactor. *Nuclear Engineering and Design*. 2023;**414**:112520.

[101]   Satvat N, Hernandez R, Vitullo F, *et al.* Hermes reactor demonstration, initial startup, and physics testing. *Nuclear Science and Engineering*. 2025;1–11. https://doi.org/10.1080/00295639.2025.2462892

[102]   Kairos Power. Public Release, Equilibrium core concentrations. Kairos Power; 2021. Available from: https://kairospower.com/generic-fhr-core-model.

[103]   Rohde U, Seidl M, Kliem s, *et al.* Neutron noise observations in German KWU built PWRs and analyses with the reactor dynamics code DYN3D. *Annals of Nuclear Energy*. 2018;**112**:715–734.

[104]   Viebach M, Bernt N, Lange C, *et al.* On the influence of dynamical fuel assembly deflections on the neutron noise level. *Progress in Nuclear Energy*. 2017;**104**:32–46.

[105]   Viebach M, Lange C, Bernt N, *et al.* Simulation of low-frequency PWR neutron flux fluctuations. *Progress in Nuclear Energy*. 2019;**117**:103039.

[106]   Gammicchia A, Santandrea S, Zmijarevic I, *et al.* A MOC-based neutron kinetics model for noise analysis. *Annals of Nuclear Energy*. 2020;**137**:107070.

[107]   Kolali A, Ghafari M and Vosoughi N. Power reactor noise simulation and analysis by developing time-domain neutron noise simulator: iPWR case study. *Engineering and Design*. 2025;**433**:113894.

[108]   Demazière C and Pázsit I. Numerical tools applied to power reactor noise analysis. *Progress in Nuclear Energy*. 2009;**51**(1):67–81.

[109]   Malmir H, Vosoughi N and Zahedinejad E. Development of a 2-D 2-group neutron noise simulator for hexagonal geometries. *Annals of Nuclear Energy*. 2010;**37**(8):1089–1100.

[110]   Demazière C. Core sim: A multi-purpose neutronic tool for research and education. *Annals of Nuclear Energy*. 2011;**38**(12):2698–2718.

[111]   Yamamoto T. Monte Carlo method with complex-valued weights for frequency domain analyses of neutron noise. *Annals of Nuclear Energy*. 2013;**58**:72–79.

[112]   Rouchon A, Zoia A and Sanchez R. A new Monte Carlo method for neutron noise calculations in the frequency domain. *Annals of Nuclear Energy*. 2017;**102**:465–475.

[113]   Chionis D, Dokhane A, Belblidia L, *et al.* Development and verification of a methodology for neutron noise response to fuel assembly vibrations. *Annals of Nuclear Energy*. 2020;**147**:107669.

[114]   Vidal-Ferrándiz A, Carreno A, Ginestar D, *et al.* A time and frequency domain analysis of the effect of vibrating fuel assemblies on the neutron noise. *Annals of Nuclear Energy*. 2020;**137**:107076.

[115]   Verma V, Chionis D, Dokhane A, *et al.* Studies of reactor noise response to vibrations of reactor internals and thermal-hydraulic fluctuations in PWRs. *Annals of Nuclear Energy*. 2021;**157**:108212.

[116]   Larsson V, Demazière C, Pázsit I, *et al.* Neutron noise calculations using the analytical nodal method and comparisons with analytical solutions. *Annals of Nuclear Energy*. 2011;**38**(4):808–816.

[117]   Hosseini SA and Vosoughi N. Neutron noise simulation by GFEM and unstructured triangle elements. *Nuclear Engineering and Design*. 2012;**253**:238–258.

[118]   Hosseini SA and Vosoughi N. Development of 3D neutron noise simulator based on GFEM with unstructured tetrahedron elements. *Annals of Nuclear Energy*. 2016;**97**:132–141.

[119]   Larsson V and Demazière C. Comparative study of 2-group P1 and diffusion theories for the calculation of the neutron noise in 1D 2-region systems. *Annals of Nuclear Energy*. 2009;**36**(10):1574–1587.

[120]   Yi H, Vinai P and Demazière C. On the simulation of neutron noise using a discrete ordinates method. *Annals of Nuclear Energy*. 2021;**164**:108570.

[121]   Gong H, Chen Z, Wu W, *et al.* Neutron noise calculation: A comparative study between SP3 theory and diffusion theory. *Annals of Nuclear Energy*. 2021;**156**:108184.

[122]   Bahrami M and Vosoughi N. Sn transport method for neutronic noise calculation in nuclear reactor systems: Comparative study between transport theory and diffusion theory. *Annals of Nuclear Energy*. 2018;**114**:236–244.

[123]   Yamamoto T and Sakamoto H. New findings on neutron noise propagation properties in void containing water using neutron noise transport calculations. *Progress in Nuclear Energy*. 2016;**90**:58–68.

[124]   Yamamoto T and Sakamoto H. Deterministic and stochastic methods for sensitivity analysis of neutron noise. *Progress in Nuclear Energy*. 2022;**145**:104130.

[125]   Olmo-Juan N, Demazière C, Barrachina T, *et al.* PARCS vs CORE SIM neutron noise simulations. *Progress in Nuclear Energy*. 2019;**115**:169–180.

[126]   Mylonakis A, Vinai P and Demazière C. Numerical solution of two-energy-group neutron noise diffusion problems with fine spatial meshes. *Annals of Nuclear Energy*. 2020;**140**:107093.

[127]   Rohde U, Kliem S, Grundmann U, *et al.* The reactor dynamics code DYN3D – Models, validation and applications. *Progress in Nuclear Energy*. 2016;**89**:170–190.

[128]   Kliem S, Bilodid Y, Fridman E, *et al.* The reactor dynamics code DYN3D. *Kerntechnik*. 2016;**81**(2):170–172.

[129]   Hosseini SA, Vosoughi N and Vosoughi J. Neutron noise simulation using ACNEM in the hexagonal geometry. *Annals of Nuclear Energy*. 2018;**113**:246–255.

[130]   Kolali A, Vosoughi J and Vosoughi N. Development of SD-HACNEM neutron noise simulator based on high order nodal expansion method for rectangular geometry. *Annals of Nuclear Energy*. 2021;**162**:108496.

[131]   Yamamoto T. Implementation of a frequency-domain neutron noise analysis method in a production-level continuous energy Monte Carlo code: Verification and application in a BWR. *Annals of Nuclear Energy*. 2018;**115**:494–501.

[132]   Mylonakis A, Vinai P and Demazière C. CORE SIM+: A flexible diffusion-based solver for neutron noise simulations. *Annals of Nuclear Energy*. 2021;**155**:108149.

[133] Tran HN, Zylbersztejn F, Demazière C, *et al.* A multi-group neutron noise simulator for fast reactors. *Annals of Nuclear Energy*. 2013;**63**:158–169.

[134] Rimpault G, Plisson D, Tommasi J, *et al.* The ERANOS code and data system for fast reactor neutronic analyses. *PHYSOR 2002 – International Conference on the New Frontiers of Nuclear Technology: Reactor Physics, Safety and High-Performance Computing*, October 2002, Seoul, South Korea. 2002.

[135] Gen-IV International Forum. A technology roadmap for Generation IV nuclear energy systems. Tech Rep. 2002; p. GIF–002–00.

[136] Gen-IV International Forum. Technology roadmap update for Generation IV nuclear energy systems. Tech Rep. 2014.

[137] IAEA. Technology roadmap for small modular reactor deployment. No NR-T-1.18, IAEA Nuclear Energy Series, Vienna. 2021.

[138] Tran HN. Properties of neutron noise induced by localized perturbations in an SFR. *Science and Technology of Nuclear Installations*. 2015;140979.

[139] Nikitin E and Fridman E. A coordinate transformation method to simulate non-uniform radial deformation of nuclear reactor cores. *Annals of Nuclear Energy*. 2025;**216**:111292.

[140] Hutchinson J, Nelson M, Grove T, *et al.* Validation of statistical uncertainties in subcritical benchmark measurements: Part I – Theory and simulations. *Annals of Nuclear Energy*. 2019;**125**:50–62.

[141] Hutchinson J, Bahran R, Cutler T, *et al.* Validation of statistical uncertainties in subcritical benchmark measurements: Part II – Measured data. *Annals of Nuclear Energy*. 2019;**125**:342–359.

[142] Miller CA, Peters WA, Odeh FY, *et al.* Sub-critical assembly die-away analysis with organic scintillators. *Nuclear Instruments and Methods in Physics Research Section A: Accelerators, Spectrometers, Detectors and Associated Equipment*. 2020;**959**:163598.

[143] Hutchinson J, Clark AR, Cutler T, *et al.* Subcritical neutron noise measurements for plutonium systems with varying geometry and mass. *Annals of Nuclear Energy*. 2024;**195**:110179.

[144] Caldwell JT, Kunz WE and Atencio JD. Apparatus and method for quantitative assay of generic transuranic wastes from nuclear reactors. US Patent 363,979; 1982.

[145] Menlove HO, Menlove SH and Tobin SJ. Fissile and fertile nuclear material measurements using a new differential die-away self-interrogation technique. *Nuclear Instruments and Methods in Physics Research Section A – Accelerators Spectrometers Detectors and Associated Equipment*. 2009;**602**(2):588–593.

[146] Pál L and Pázsit I. A special branching process with two particle types. *European Physical Journal Plus*. 2011;**126**(2):20.

[147] Anderson J, Pál L, Pázsit I, *et al.* Derivation and quantitative analysis of the differential self-interrogation Feynman-alpha method. *European Physical Journal Plus*. 2012;**127**(2):21.

[148]   Pál L and Pázsit I. Two-Group Theory of the Feynman-Alpha Method for Reactivity Measurement in ADS. *Science and Technology of Nuclear Installations*. 2012;**2012**:620808.

[149]   Chernikova D, Wang Z, Pázsit I, *et al*. A general analytical solution for the variance-to-mean Feynman-alpha formulas for a two-group two-point, a two-group one-point and a one-group two-point cases. *European Physical Journal Plus*. 2014;**129**(11):259.

[150]   Chernikova D, Axell K, Avdic S, *et al*. The neutron-gamma Feynman variance to mean approach: Gamma detection and total neutron-gamma detection (theory and practice). *Nuclear Instruments and Methods in Physics Research Section A: Accelerators, Spectrometers, Detectors and Associated Equipment*. 2015;**782**:47–55.

[151]   Chernikova D, Pázsit I, Favalli A, *et al*. The Inclusion of Photofission, Photonuclear, (n,xn), (n,n'xγ), and (n,xγ) Reactions in the Neutron-Gamma Feynman-Alpha Variance-to-Mean Formalism. *Nuclear Science and Engineering*. 2017;**185**(1):206–216.

[152]   Goto M, Sano T, Nakajima K, *et al*. Feynman-α analysis using BGO gamma-ray detector in a university training and research reactor. *Nuclear Science and Engineering*. 2023;**197**(8):1814–1822.

[153]   Darby FB, Pakari OV, Hua MY, *et al*. Neutron-gamma noise measurements in a zero-power reactor using organic scintillators. *IEEE Transactions on Nuclear Science*. 2024;**71**(5):1033–1040.

[154]   Pakari O, Mager T, Frajtag P, *et al*. Gamma noise to non-invasively monitor nuclear research reactors. *Scientific Reports*. 2024;**14**(1):8409. https://doi.org/10.1038/s41598-024-59127-y.

[155]   Seifritz W, Stegemann D and Väth W. Two-detector cross-correlation experiments in the fast-thermal Argonaut reactor (STARK). In: Uhrig RE, editor. *Neutron Noise, Waves, and Pulse Propagation*. Springfield, Virginia: USAEC, CONF-660206; 1967. pp. 195–216.

[156]   Nomura T, Gotoh S and Yamaki K. Reactivity measurements by the two-detector cross-correlation method and supercritical reactor noise analysis. In: Uhrig RE, editor. *Neutron Noise, Waves, and Pulse Propagation*. Springfield, Virginia: USAEC, CONF-660206; 1967. pp. 217–246.

[157]   Pál L. On the theory of stochastic processes in nuclear reactors. *II Nuovo Cimento*. 1958;**7**(Suppl. 1):25–42.

[158]   Abderrahim HA, Baeten P, De Bruyn D and Fernandez R. MYRRHA – A multi-purpose fast spectrum research reactor. *Energy Management and Conversion*. 2012;**63**:4–10.

[159]   Degweker SB. Some variants of the Feynman alpha method in critical and accelerator driven sub critical systems. *Annals of Nuclear Energy*. 2000;27(14):1245–1257.

[160]   Degweker SB. Reactor noise in accelerator driven systems. *Annals of Nuclear Energy*. 2003;**30**(2):223–243.

[161]   Degweker SB and Rana YS. Reactor noise in accelerator driven systems – II. *Annals of Nuclear Energy*. 2007;**34**(6):463–482.

[162] Pázsit I and Yamane Y. Theory of neutron fluctuations in source-driven subcritical systems. *Nuclear and Instrumentation Methods A.* 1998;**403**:431–441.

[163] Pál L, Pázsit I and Elter Zs. Comments on the stochastic characteristics of fission chamber signals. *Nuclear Instruments and Methods* 2014;**A 763(0)**:44–52.

[164] Pál L and Pázsit I. Campbelling-type theory of fission chamber signals generated by neutron chains in a multiplying medium. *Nuclear Instruments and Methods in Physics Research Section A: Accelerators, Spectrometers, Detectors and Associated Equipment.* 2015;**794**:90–101.

[165] Filliatre P, Jammes C, Geslot B, *et al.* In vessel neutron instrumentation for sodium-cooled fast reactors: Type, lifetime and location. *Annals of Nuclear Energy.* 2010;**37**(11):1435–1442.

[166] Kitamura Y and Fukushima M. Count-loss effect in subcriticality measurement by pulsed neutron source method, (II) proposal for utilization of neutron detection system operated in current mode. *Journal of Nuclear Science and Technology.* 2014;**51**(6):752–765.

[167] Pál L and Pázsit I. Stochastic theory of the fission chamber current generated by non-Poissonian neutrons. *Nuclear Science and Engineering.* 2016;**184**(4):537–550.

[168] Kitamura Y, Pázsit I and Misawa T. Determination of prompt neutron decay constant by time-domain fluctuation analyses of detector current signals. *Annals of Nuclear Energy.* 2018;**120**:691–706.

[169] Kitamura Y and Misawa T. Delayed neutron effect in time-domain fluctuation analyses of neutron detector current signals. *Annals of Nuclear Energy.* 2019;**123**:119–134.

[170] Szieberth M, Boros MI, Klujber G, *et al.* Feasibility demonstration of continuous signal-based neutron noise measurements by experiments and simulations (to be submitted). *Nuclear Instruments and Methods in Physics Research Section A: Accelerators, Spectrometers, Detectors and Associated Equipment.* 2025.

[171] Szieberth M, Nagy L, Klujber G, *et al.* Experimental demonstration of neutron fluctuation analysis based on the continuous signal of fission chambers: Neutron multiplicity and reactor noise measurements. In: *Proceedings of Mathematics & Computation (M&C) 2021.* American Nuclear Society; 2021. pp. 1752–1761. Available from: https//www.ans.org/pubs/proceedings/article-50146/.

[172] Papoulis A. *Probability, Random Variables and Stochastic Processes.* 3rd ed. New York: McGraw-Hill; 1991.

[173] Szieberth M, Klujber G, Boros MI, *et al.* Continuous signal-based neutron noise and multiplicity measurements – A nuclear innovation prize winner development (to be submitted). *EPJ N – Nuclear Science and Technology.* 2025.

[174] Hashemian HM. *Sensor Performance and Reliability.* North Carolina: ISA – Instrumentation, Systems, and Automation Society; 2005.

[175]   Hashemian HM. *Maintenance of Process Instrumentation in Nuclear Power Plants*. Berlin: Springer-Verlag; 2006.

[176]   Kerlin T, Miller L, Hashemian H, *et al*. *Temperature Sensor Response Characterization*. Electric Power Research Institute; 1980. NP-1486.

[177]   Thie J. Elementary Methods of Reactor Noise Analysis. *Nuclear Science and Engineering*. 1963;**15**(2):109–114.

[178]   Thie J. Nuclear reactor kinetics. *Nuclear Science and Engineering*. 1965;**23**(3):306.

[179]   Hashemian H and Petersen K. Experience with on-line measurement of response time of pressure transmitters using noise analysis. In: *Proceedings of the Sixth Symposium on Nuclear Reactor Surveillance and Diagnostics*, SMORN VI. Vol. 2. Gatlinburg, TN; 1991. pp. 68.01–68.12.

[180]   Hashemian H and Jiang J. Using the noise analysis technique to detect response time problems in the sensing lines of nuclear plant pressure transmitters. *Progress in Nuclear Energy*. 2010;**52**(4):367–373.

[181]   Hashemian H. *The Noise Analysis Technique for Testing Pressure Sensor Response Time*; 2010. SciTopics.

[182]   International Atomic Energy Agency. *Management of Ageing of I&C Equipment in Nuclear Power Plants*. Vienna, Austria: IAEA; 2000.

[183]   International Atomic Energy Agency. On-line monitoring for improving performance of nuclear power plants Part 1: Instrumentation channel monitoring and Part 2: Process and component condition monitoring and diagnostics. Vienna; 2008.

[184]   International Electrotechnical Commission. Nuclear power plants – Instrumentation and control systems important to safety: Management of ageing. Geneva; 2007. IEC 62342.

[185]   Pázsit I and Glöckler O. On the neutron noise diagnostics of pressurized water reactor control rod vibrations. I. Periodic vibrations. *Nuclear Science and Engineering*. 1983;**85**(2):167–177.

[186]   IAEA2011. Core knowledge on instrumentation and control systems in nuclear power plants. No. NP-T-3.12 in Nuclear Energy Series. Vienna: IAEA; 2011.

[187]   Wooten B. *Instrument Calibration and Monitoring Program Volume 1: Basis for the Method*. Palo Alto, CA: Electric Power Research Institute (EPRI); 1993. EPRI TR-103436-V1.

[188]   Davis E, Funk D, Hooten D and Rusaw R. *On-Line Monitoring of Instrument Channel Performance*. Palo Alto, CA: Electric Power Research Institute (EPRI); 2000. TR-104965-R1, NRC SER, EPRI 1000604, ADAMS Accession Number ML003734509.

[189]   Thomasson T, Shumaker B, Hashemian H, *et al*. First principles model of a simulation flow loop in support of on-line monitoring implementation in next generation nuclear power plants. In: *Proceedings of the American Nuclear Society 9th International Topical Meeting on Nuclear Plant Instrumentation, Control & Human-Machine Interface Technologies (NPIC&HMIT)*. Charlotte, NC; 2015.

[190] Shumaker B and Hashemian H. Resolving the regulatory issues with implementation of online monitoring technologies to extend calibration intervals of process instruments in nuclear power plants. In: *Proceedings of the American Nuclear Society 11th International Topical Meeting on Nuclear Plant Instrumentation, Control and Human–Machine Interface Technologies (NPIC&HMIT)*. Orlando, FL; 2019.

[191] Heo GY. Condition monitoring using empirical models: Technical review and prospects for nuclear applications. *Nuclear Engineering and Technology*. 2007;**40**:49–68.

[192] Fantoni P. Experiences and applications of PEANO for on-line monitoring in power plants. *Progress in Nuclear Energy*. 2005;**46**:3–4.

[193] Hashemian H, Shumaker B and Morton G. Online monitoring technology to extend calibration intervals of nuclear plant pressure transmitters. AMS; 2021. AMS-TR-0720R2-A. NRC ADAMS Accession Number ML21235A493.

[194] Hashemian A, Shumaker B, Gavin T, *et al.* Developments in online monitoring technologies for autonomous microreactor operations. In: *Proceedings of the American Nuclear Society 12th Nuclear Plant Instrumentation, Control and Human–Machine Interface Technologies (NPIC&HMIT)*; 2021.

[195] Hashemian H, Mitra C, Shumaker B, *et al.* Online monitoring in small modular reactors (SMRs). In: *Presented at the American Nuclear Society 2013 Annual Meeting*. Atlanta, GA; 2013.

[196] Shumaker B, Hashemian A, Hashemian H. Development of online monitoring technologies for autonomous microreactor operations. In: *Presented at the Technical Meeting on Instrumentation and Control, and Computer Security for Small Modular Reactors and Microreactors, International Atomic Energy Agency (IAEA)*. Vienna, Austria; 2022.

[197] Goldberg SM and Rosner R. *Nuclear Reactors: Generation to Generation*. American Academy of Arts and Sciences; 2011.

[198] OECD Nuclear Energy Agency. *Technology Roadmap for Generation IV Nuclear Energy Systems*; 2014.

[199] Power Magazine. NRC Approves Construction of First Electricity-Producing Gen IV Reactor in the U.S.; 2024. https://www.powermag.com.

[200] POWERnews. Green Light for Project Pele, Defense Department's Mobile Nuclear Microreactor Demonstration; 2022.

[201] Upadhyaya BR, Mehta C, Hines JW, *et al.* Monitoring pump parameters in small modular reactors using electric motor signatures. *ASME Journal of Nuclear Engineering and Radiation Science*. 2017;**3**(1): 011007-1–011007-7.

[202] Upadhyaya BR, Lish MR and Hines JW. Development of instrumentation and control systems for the integral inherently safe light water reactor. University of Tennessee; 2016. Prepared for the DOE Prime Contract No. DE-AC07-05ID14517 with Georgia Institute of Technology.

[203]   Upadhyaya BR, Lish MR, Hines JW, *et al.* Instrumentation and control strategies for an integral pressurized water reactor. *Nuclear Engineering and Technology.* 2015;**47**(2):148–156.

[204]   Upadhyaya BR, Kitamura M and Kerlin TW. Multivariate signal analysis algorithms for process monitoring and parameter estimation in nuclear reactors. *Annals of Nuclear Energy.* 1980;**7**:1–11.

[205]   Poornapushpakala S, Gomathy C, Sylvia JI, *et al.* Design, development and performance testing of fast response electronics for eddy current flowmeter in monitoring sodium flow. *Flow Measurement and Instrumentation.* 2014;**38**:98–107.

[206]   Sharma P, Kumar SS, Nashine BK, Development, computer simulation and performance testing in sodium of an eddy current flowmeter. *Annals of Nuclear Energy.* 2010;**37**(3):332–338.

[207]   Kerlin TW and Upadhyaya BR. *Dynamics and Control of Nuclear Reactors.* Cambridge, MA: Elsevier Academic Press; 2019.

[208]   Haubenreich PN and Engel JR. Experience with the molten salt reactor experiment. *Nuclear Applications and Technology.* 1970;**8**:118–136.

[209]   Singh V, Wheeler AM, Lish MR, *et al.* Nonlinear dynamic model of molten-salt reactor experiment – validation and operational analysis. *Annals of Nuclear Energy.* 2018;**113**:177–193.

[210]   Singh V, Lish MR, Chvala O, *et al.* Dynamic modeling and performance analysis of a two-fluid molten salt reactor system. *Nuclear Technology.* 2018;**202**(1):15–38.

[211]   Kairos Power. Kairos Power; 2024. https://kairospower.com.

[212]   NRC. Hermes 2 Kairos Power Reactor; 2024. https://nrc.gov/reactors/non-power/new-facility-licensing/hermes2-kairos.html.

[213]   Perillo SRP, Upadhyaya BR and Li F. Control and instrumentation strategies for multi-modular integral nuclear reactor systems. *IEEE Transactions on Nuclear Science.* 2011;**58**(5):2442–2451.

[214]   Treece R and Tilak S. Edge computing fundamentals. Automation.com, a monthly publication of the International Society of Automation. 2025 January/February; 19–24.

[215]   RDI Technologies. RDI Technologies; 2024. https://rditechnologies.com.

[216]   Baloh FJ, Kenney ES. Localisation of in-core disturbance with ex-core detectors. *Nuclear Applications.* 1969;**6**:232–237.

Index

.

www.ingramcontent.com/pod-product-compliance
Lightning Source LLC
Chambersburg PA
CBHW050510190326
41458CB00005B/1492